Lecture Notes in Computer Science 9143

Commenced Publication in 1973
Founding and Former Series Editors:
Gerhard Goos, Juris Hartmanis, and Jan van Leeuwen

More information about this series at http://www.springer.com/series/7411

Symeon Papavassiliou · Stefan Ruehrup (Eds.)

Ad-hoc, Mobile, and Wireless Networks

14th International Conference, ADHOC-NOW 2015
Athens, Greece, June 29 – July 1, 2015
Proceedings

 Springer

Editors
Symeon Papavassiliou
National Technical University of Athens
Athens
Greece

Stefan Ruehrup
FTW - Telecommunications Research
    Center Vienna
Vienna
Austria

ISSN 0302-9743   ISSN 1611-3349 (electronic)
Lecture Notes in Computer Science
ISBN 978-3-319-19661-9   ISBN 978-3-319-19662-6   (eBook)
DOI 10.1007/978-3-319-19662-6

Library of Congress Control Number: 2015939919

LNCS Sublibrary: SL5 – Computer Communication Networks and Telecommunications

Springer Cham Heidelberg New York Dordrecht London

Printed on acid-free paper

Springer International Publishing AG Switzerland is part of Springer Science+Business Media
(www.springer.com)

Preface

The International Conference on Ad-Hoc Networks and Wireless (ADHOCNOW) is one of the mot well-known series of events dedicated to research on wireless sensor networks and mobile computing. Since its creation in 2002, the conference has been held 13 times in six different countries. Its 14th edition in 2015 was held in Athens, Greece, from June 29 to July 1.

The 14th ADHOC-NOW attracted 52 submissions of which 25 papers were accepted for presentation after rigorous reviews by external reviewers, Technical Program Committee members, and discussions among technical program chairs. All submissions received at least three reviews, with an average of 4.2 reviews per submission. The review process was guided by the chairs of the conference tracks. In addition, three invited papers were contributed by invited speakers.

ADHOC-NOW traditionally covers a wide spectrum of topics across all networking layers for sensor networks, localization in various networking environments, and with applications in several domains. The 14th ADHOC-NOW featured four tracks, structuring the contributions into the following topics:

- Ad-hoc and Wireless Networks and Localization (main track)
- Efficient, Reliable, and Secure Smart Energy Networks (special track)
- Emerging Communications, Networking and Computing technologies for VANETs 2.0 (special track)
- Distributed Computing with Mobile Agents (special track)

We would like to thank all the people involved in the production of these proceedings. First of all, we are grateful to the members of the Technical Program Committee and the external reviewers for their help in providing detailed reviews of the submissions. We also thank Springer's team once again for their great assistance from the start of the submission process until the proceedings production. Last, but not least we thank all the members of the Organizing Committee, who helped in preparing and organizing the event and putting together an excellent program.

April 2015

Symeon Papavassiliou
Stefan Ruehrup

Organization

Conference Chair

Symeon Papavassiliou   National Technical University of Athens (NTUA), Greece

Technical Program Committee Co-chairs

Antonella Molinaro   University Mediterranea of Reggio Calabria, Italy
Jiannong Cao   The Hong Kong Polytechnic University, Hong Kong, SAR China
Vasileios Karyotis   National Technical University Athens, Greece

Special Track Committee Co-chairs

Yan Zhang   Simula Research Laboratory, Norway
Aris Pagourtzis   National Technical University of Athens, Greece

Program Chairs for the Track on "Efficient, Reliable, and Secure Smart Energy Networks"

Nikos Passas   University of Athens, Greece
Marco di Renzo   CNRS, France
Stefano Tennina   WEST Aquila, Italy
Dionysis Xenakis   University of Athens, Greece

Program Chairs for the Track on "Emerging Communications, Networking and Computing Technologies for VANETs 2.0"

Antonella Molinaro   University Mediterranea of Reggio Calabria, Italy
Claudia Campolo   University Mediterranea of Reggio Calabria, Italy
Riccardo Scopigno   Istituto Superiore Mario Boella, Turin, Italy

Program Chairs for the Track on "Distributed Computing with Mobile Agents"

Euripides Markou   University of Thessaly, Greece
Peter Widmayer   ETH Zurich, Switzerland

Publicity Chairs

Amiya Nayak   University of Ottawa, Canada
Xinbing Wang   Shanghai Jiao Tong University, P. R. China
Sandra Sendra   Universidad Politecnica de Valencia, Spain

Submission and Registrations Chair

Vasileios Karyotis   National Technical University of Athens, Greece

Proceedings Chair

Stefan Ruehrup   FTW — Telecommunications Research Center Vienna,
                           Austria

Local Arrangements Co-chairs

Eirini-Eleni Tsiropoulou   National Technical University of Athens, Greece
Eleni Stai   National Technical University of Athens, Greece

Web Chairs

Vasileios Karyotis   National Technical University of Athens, Greece
Milos Stojmenović   Singidunum University, Belgrade, Serbia

Technical Program Committee

Evangelos Bampas   LaBRI, University of Bordeaux, France
Carlos Calafate   Universitat Politecnica de Valencia, Spain
Juan-Carlos Cano   Universitat Politecnica de Valencia, Spain
Danny Krizanc   Wesleyan University, USA
Pierre Leone   University of Geneva, Switzerland
Nathalie Mitton   Inria, France
Katerina Potika   Santa Clara University, USA
Eirini Eleni Tsiropoulou   National Technical University of Athens, Greece
Volker Turau   Hamburg University of Technology, Germany
Dionysis Xenakis   University of Athens, Greece
Christos Xenakis   University of Piraeus, Greece
Michel Barbeau   Carleton University, Canada
Georgios Karopoulos   University of Athens, Greece
Vasileios Karyotis   National Technical University of Athens, Greece
Evangelos Kranakis   Carleton University, Canada
Pietro Manzoni   Universitat Politecnica de Valencia, Spain
Eleni Stai   National Technical University of Athens, Greece
Stefano Tennina   WEST Aquila S.r.l., Italy
Marica Amadeo   University Mediterranea of Reggio Calabria, Italy

Vasos Vassiliou   University of Cyprus, Cyprus
Alexey Vinel   Halmstad University, Sweden
Cheng Wang   Tongji University, China
Konrad Wrona   NATO, The Netherlands
Stella Kafetzoglou   National Technical University of Athens, Greece
Xufei Mao   Tsinghua University, China
Weigang Wu   Sun Yat-sen University, China
Qin Xin   University of the Faroe Islands, Faroe Islands

External Reviewers

Cristóvão Cruz   Daibo Liu
Nikos Deligiannis   Matoula Petrolia
Dejan Drajic   Vladan Rankov
Sebastian Ebers   Longfei Shangguan
Milan Erdelj   Thouraya Toukabri Gunes
Nenad Gligoric   Caglar Terzi
Marco Gramaglia   Jiliang Wang
Stevan Jokic   Yubo Yan
Nicholas Kaminski   Shanfeng Zhang
Christina Karousatou   Xiaolong Zheng
Emin Yiğit Köksal   Eike von Tils
Zhong Li   Lan Zhang

Contents

Distributed Computing with Mobile Agents

Efficient, Reliable, and Secure Smart Energy Networks

**Emerging Communications, Networking and Computing Technologies
for VANETs 2.0**

Routing, Connectivity, and Resource Allocation

A Dynamic Topology Control Algorithm for Wireless Sensor Networks

Gerry Siegemund, Volker Turau[✉], and Christoph Weyer

Institute of Telematics, Hamburg University of Technology,
Hamburg, Germany
{gerry.siegemund,turau,c.weyer}@tuhh.de

Abstract. Topology control algorithms (TCAs) are used in wireless sensor networks to reduce interference by carefully choosing communication links. Since the quality of the wireless channel is subject to fluctuations over time TCAs must repeatedly recompute the topology. TCAs ensure quick adjustment to new or deteriorating links while preventing precipitant changes due to transient faults. This paper contributes a novel dynamic TCA that provides a compromise between agility and stability, and constructs connected topologies for low latency routing. Furthermore, it enforces memory restrictions and is of high practical relevance for real sensor network hardware.

1 Introduction

The goal of topology control algorithms (TCAs) for wireless sensor networks (WSNs) is to dynamically form a graph G_c representing the communication network to be used by applications while maintaining global graph properties such as k-connectedness, low node degree, or small diameter [13]. G_c is a subgraph of the graph G representing the physical topology, i.e., G contains all possible communication links independent of their quality. From a communication point of view topology control aims to achieve reduced interference, less energy consumption, and increased network capacity. The essence is to make a deliberate choice on each node's neighbors, i.e., the physical topology is thinned out (e.g., Fig. 1).

Many TCAs disregard fluctuations over time, i.e., they keep a computed topology forever. It is well known that wireless links frequently disintegrate while others are established [19]. In particular in WSNs with resource-constrained low-power radios, this makes multi-hop routing as required in data collection applications, a challenge [11]. The origin of these fluctuations are typically environmental changes, e.g., changing weather conditions or moving obstacles. Raising the transmission power increases the number of direct communication partners, introducing benefits but also disadvantages. A high transmission power leads to an increased energy consumption. Secondly, a bigger number of neighbors places a heavy burden on the MAC protocol (e.g., for CSMA/CA protocols based on RTS/CTS). To continuously find usable links, link qualities have to be assessed. Woo et al. suggest to estimate link qualities dynamically through a *link quality estimator* (LQE) using metrics such as PRR or RSS. Due to the

© Springer International Publishing Switzerland 2015
S. Papavassiliou and S. Ruehrup (Eds.): ADHOC-NOW 2015, LNCS 9143, pp. 3–18, 2015.
DOI:10.1007/978-3-319-19662-6_1

limited memory in WSNs link status statistics must be maintained in neighbor-hood tables with constant space regardless of the cell density [19].

We argue that a TCA for a WSN must continuously react to changing link qualities. In doing so it must provide agility and stability. Once the quality of a link drops significantly the number of retransmissions will rise, leading to more contention with negative consequences for the network. Thus, TCAs should quickly discard low quality links. While agility is desirable, from an application's point of view stable topologies are preferred since otherwise the provided service may not be guaranteed, e.g., routing towards a target node fails. Once a topology satisfies a *goodness criterion*, stability demands to keep the topology unchanged as long as possible, i.e., not to react on transient faults (e.g., burst errors) prematurely. Agility demands to promptly update a topology as soon as the goodness criterion is violated permanently.

Stability is crucial if neighborhood information is used to compute support-structures, e.g., spanning trees. Consider the execution of a self-stabilizing, dis-tributed spanning tree algorithm for a network with an unstable link e (see Fig. 2). Here a node makes decisions based exclusively on its *local view* (own and neighbor states). Repeated inclusion and exclusion of link e causes the parent of node 2 to oscillate. Such instabilities may lead to loops, with severe impacts on tree routing. Excluding edge e permanently from G_c would stabilize the tree.

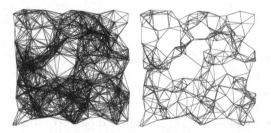
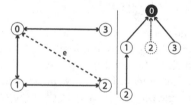

Fig. 1. Physical topology (225 random placed nodes) thinned out by the proposed TCA

Fig. 2. Influence of instability on tree routing

This paper proposes a dynamic TCA that maximizes several criteria (see below) and achieves a good compromise between stability and agility. Existing algorithms often make use of geographic information. We make no such assump-tions nor do we vary the transmission power of nodes. A preliminary version of the proposed TCA, without technical detail or evaluation, was published in [16]. The proposed TCA was successfully used in the implementation of a publish subscribe system for a WSN [15].

2 Topological Criteria

The distributed system is modeled as an undirected graph $G = (V, E)$, where edges represent possible communication links. A distributed implementation of a LQE periodically computes a quality value $Q(e)$ for each link e. To cope with the time dependent nature of the communication links the set of edges at time t is denoted by $E(t)$. Link (u, v) is in $E(t)$ iff nodes u and v can communicate directly at time t. Thus, $G(t) = (V, E(t))$ denotes the communication graph at time t. A TCA dynamically computes a subgraph $G_c(t) = (V, E_c(t))$ of $G(t)$ (i.e., $E_c(t) \subseteq E(t)$) satisfying the following criteria.

1. **Link quality:** $Q(e) \geq Q_{min}$ for each $e \in E_c(t)$ and any time t. Link qualities below Q_{min} trigger with high probability a large number of retransmissions leading to more contention and thus, reduce the overall throughput.

2. **Symmetry:** Links in $E_c(t)$ should allow for bidirectional communication. Unidirectional links are considered useless because ACKs cannot be sent directly.

3. **Connectivity:** If $G(t)$ is connected then $G_c(t)$ should also be connected.

4. **Bounded degree:** Each node in $G_c(t)$ should have at most C_N neighbors regardless of the size of V or $E(t)$. An implementation of criterion 1 and 2 requires to store data about each potential neighbor (e.g., the link quality). Thus, the available resources limit the number of nodes that can be maintained.

5. **k-Spanner:** The distance between two nodes in $G_c(t)$ should be at most k-times the corresponding distance in $G(t)$. A more practical criterion is to demand that the average length of paths between any pair of nodes should be small.

Static TCAs do not consider temporal aspects of link behavior, i.e., they do not react to link changes or failures. A simple remedy is to repeatedly apply a static TCA. This can lead to strong discontinuities of the topology, including oscillation and long adherence to low quality links. We therefore demand:

6. **Stability:** If $e \in E_c(t_0)$ and $Q(e) \geq Q_{min}$ at time t_0 then e should remain in $E_c(t)$ for $t \in [t_0, t_0 + T_N]$ (T_N a given constant). Furthermore, e should remain in $E_c(t)$ for $t \geq t_0$ as long as $Q(e) \geq Q_{min}$ and the other criteria are fulfilled. A neighbor of a node should not be instantaneously replaced when a neighboring node with a higher link quality emerges.

7. **Agility:** If $E_c(t)$ does not satisfy the above listed criteria and there exists $e \in E(t) \backslash E_c(t)$ that can improve one criterion then e should be included into $E_c(t)$. Similarly, if there exist $e \in E_c(t_0)$ with $Q(e) < Q_{tol}$ (a given constant for tolerable quality) then e should be discarded from E_c during $[t_0, t_0 + T_N]$.

The last two criteria can not be optimized simultaneously.

3   Related Work

TCAs for WSNs are surveyed in [9,13]. One approach is to adapt the transmission power. Reducing the transmission range reduces interference. Many algorithms minimize the transmission range until the resulting topology satisfies criteria 1–3. Algorithm ATPC permanently collects link quality histories and builds a model for each neighbor [6]. This model represents the correlation between transmission power levels and link qualities providing the basis for a dynamic TCA. Liu et al. optimize energy usage by reducing the number of neighbors [7]. The k-neighbors algorithm also belongs to this category [1]. All nodes initially use the maximal transmission power. It is decreased as long as each node has at least k neighbors (criterion 4). A node is considered a neighbor if a message can be successfully exchanged. The challenge is to find the minimal value for k satisfying criterion 3. Blough et al. report that for networks with 50–500 nodes $k = 9$ is sufficient. Their simulations suggest that k can be set to 6 independently of the network's size, if a small percentage of disconnected nodes is tolerated. The k-neighbors algorithm guarantees criterion 2. RETC [5] builds a rooted tree as the overlay topology. The number of children in the tree is not bounded (contradicting criterion 4), furthermore, RETC violates criterion 5 and 6.

The XTC protocol bases its decisions on static link qualities such as euclidean distance or energy consumption [18]. Each node orders its neighbors based on this metric. The immediate neighbor set is deduced by this order. A link between two nodes is used if there is no third node within reach that has equal or higher quality to both nodes. The resulting topologies satisfy criterion 3 whenever the maxpower communication graph does. XTC does not satisfy criterion 4. XTC was extend in [3] by adding all links that exceed a certain link quality to the original XTC links to the disprofit of criterion 3. The resulting algorithm XTC_{RLS} proved in simulations to be superior to XTC and the k-neighbors algorithm with respect to throughput and energy consumption. XTC and foremost XTC_{RLS} have a high memory consumption.

Many algorithms collect potential neighbors into a list with constant size, thus, comply with criterion 4. If the list is full then either no new neighbor can be added or neighbors have to be removed. Woo uses several replacement schemes [19]. In an experiment with 220 nodes, neighbor list sizes of 40 were necessary to satisfy criterion 3. Since links are not being evaluated before insertion, the biggest shortcoming is that neighbors reachable via poor links may be added (violating criterion 1). Even though poor links will be evicted at some point, the fluctuation is high, resulting in an unstable topology (criterion 6).

To comply with criterion 1, TCAs rely on link quality estimators. Common to all LQEs is that they periodically broadcast messages with information to assess the link quality (e.g., a sequence number to detect lost packets). If a LQE uses PRR to evaluate the communication quality between two nodes, several messages are necessary to get a representative value. Usually this PRR is not shared among neighbors. To the contrary, the *Expected Transmission Count Metric* (ETX) [2] does exactly this to determine link asymmetries (criterion 2).

TinyOS contains LEEP as a TCA [4]. LEEP uses a fixed neighborhood table size and ETX as the LQE. Each node sends out its current knowledge about all its neighbors. The eviction policy for full tables is to replace the node with the least ETX value. If all neighbors have roughly the same ETX value, small fluctuations in link qualities can lead to frequent topology updates violating criterion 6. Furthermore, even high density networks may be unconnected because LEEP chooses its neighbors exclusively based on the ETX value. Table 1 gives a short

Table 1. Comparison of TCAs

	Our TCA	RETC [5]	EELTC [17]	STC [14]	CONREAP [8]	LEEP [4]	XTC [18]
Distributed	YES	YES	NO	YES	NO	YES	YES
Dynamic	YES	YES	NO	NO	NO	YES	NO
Stable	YES	NO	YES	YES	YES	NO	YES
Agile	YES	YES	NO	NO	NO	YES	NO
Memory depends on Δ	NO	YES	-	YES	-	NO	YES

comparison of different TCAs. The majority of them considers static networks. In contrast, our adaptive approach regards dynamic networks, i.e., it incorporates varying link qualities and the possibility of adding and removing nodes.

4   The Topology Control Algorithm

The proposed TCA forms a stable network topology $G_c(t)$ on top of a physical network despite fluctuating link qualities and re-/disappearing nodes. It is a self-organizing algorithm, i.e., it continuously adapts the logical topology to the physical conditions. Each node selects a limited number C_N of nodes as neighbors (criterion 4). The selection aims to realize the above listed criteria.

Each node maintains a subset of the nodes it can receive messages from in three disjoint lists A, S, and N. Nodes in $v.N$ form the current neighborhood of node v. The other two lists are used to provide agility and stability. Acceptance into one of these lists depends on the communication quality of the corresponding link as determined by an LQE. To assess links, every node periodically broadcasts a message (see Sect. 4.1). These messages allow the LQE of all receiving nodes to estimate the link's quality represented as a numerical value Q. A link is considered useful if $Q \geq Q_{min}$ (criterion 1). Note that the LQE requires the data of a few periods to estimate the quality (e.g., to calculate the PRR). For this purpose each node collects (unidirectional) information about links in a list A – the *assessment list* – of length C_A. After being assessed with a quality of at least Q_{min}, a node can be promoted to list S – the *standby list*. S is organized as a priority queue based on link quality.

Nodes in this list are standby nodes to improve the *local topology* defined by the nodes in list N – the *neighborhood list*. The local topology LT_v of a node v is the subgraph of $G_c(t)$ induced by the nodes in the 2-hop neighborhood of v excluding edges between nodes with distance 2 to v. A node v can construct LT_v from the information contained in its list N.

Agility and stability (i.e., criteria 6 and 7) are provided by carefully promoting nodes to and deleting them from N (see Fig. 3). This process is controlled by two periodic timers. Upon expiration of the *assessment timer* T_A a node checks if the quality of the nodes in A is sufficient to get promoted to S (see Sect. 4.2).

4.1 Processing of Periodic Messages

The periodically broadcasted messages by a node v include, besides the data required by the LQE, the identifiers in v's list N, i.e., message size is $O(C_N)$. A node v stores for each $w \in v.N$ the list of w's current neighborhood as reported in w's last message. With the help of this information a node computes its local topology. The type of information stored in S per node is the same as in N (e.g., the neighbors of each entry). This leads to a total storage demand of $O(C_N^2)$.

Upon expiration of the *neighborhood timer* T_N a node checks whether nodes in S can be used to improve the fulfillment of criteria 1, 2, 3, and 5 for nodes in N (see Sect. 4.2). The main data structures for a node are the three lists A, S, and N of length C_A, C_S, and C_N, respectively. For each link to be assessed, A contains the sender's identifier and information required to estimate the link quality. N contains the identifiers of at most C_N nodes along with some additional information such as the links' quality measures. Nodes in N are regarded as the node's current neighborhood in G_c. List and timer lengths have to be chosen according to the node density and fluctuation rate.

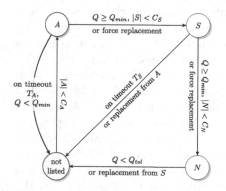

Fig. 3. Transitions between A, S, and N.

Upon the reception of a periodic message from a node w a node v checks if w is already contained in A, S, or N. In this case the LQE updates the value of the link quality. If w's identifier is already contained in $v.S$ or $v.N$ then the neighborhood data of w is updated for the respective list. The received information allows v to deduce whether the link to w is bidirectional, by checking if its own identifier is contained in the received list N. If w is not contained in any of the lists A, N, or S and $|A| < C_A$ (i.e., space available in A) then w is inserted into A. If A is full then the link to w is not considered at this instant, with the goal to improve stability, i.e., not to prematurely replace good links.

4.2 Periodic Processing of Lists

A node remains only for a fixed time interval T_A in A. T_A depends on the number of messages the LQE requires to assess a link and on the number of nodes within the node's transmission range. After time T_A the LQE has assessed the link's quality Q and the entry is removed from A to allow other links to be evaluated (see Algorithm 1). If $Q < Q_{min}$ then it is discarded, otherwise it is considered as a candidate for being a neighbor of v, i.e., it may be included into S. A node w removed from A is inserted into S if $Q \geq Q_{min}$ and $|S| < C_S$. If S is already full then w replaces the entry u in S with the lowest quality value provided the quality of u is lower than that of the link to w, otherwise it is discarded.

With period T_N each node updates lists N and S as follows (see Algorithm 2). A node is discarded from N if the quality of the link falls below Q_{tol}. We define $Q_{tol} < Q_{min}$ to spare a newly promoted node from being evicted from N due to a minor degradation of its link quality (criterion 6). If $|N| < C_N$ then the best $C_N - |N|$ nodes of S will be promoted to N. A promotion is only performed if the link's quality is above Q_{min} (link qualities may decrease over time). A node $p \in S$ may also oust a node in N if this increases the rank of the logical topology (see Sect. 5). For this purpose the nodes in S with a quality value above Q_{min} are considered in decreasing order. If there exists $q \in N$, such that the local topology defined by $N \backslash \{q\} \cup \{p\}$ has higher rank than the local topology defined by N, node p replaces q in N and is removed from S. The replaced node is not immediately reinserted into S, yet it may later reappear. The rationale for this behavior is as follows: Node q was replaced in N because other nodes formed a higher ranked topology. Therefore, it should not be reinserted too soon to give other nodes a chance to further improve the topology's rank and to avoid oscillation, i.e., to foster stability. This concept realizes a temporary blacklist. For similar reasons nodes only remain for a fixed period T_S in S. If they are not promoted during that time, then they are discarded to make room for other nodes from A which may improve the rank. The lengths of T_N and T_S control the agility of our TCA.

upon *expiration of timer T_A for $p \in A$*
do
    A.remove(p);
    if $p.Q \geq Q_{min}$ **then**
        if $|S| < C_S$ **then**
            S.add(p);
        else
            $q := \min_Q S$;
            if $q.Q < p.Q$ **then**
                S.remove(q); S.add(p);

Algorithm 1: Promoting nodes from A to S

with *Period T_N* **do**
    for *all $p \in S$ in descending order* **do**
        if $p.Q \geq Q_{min}$ **then**
            if $|N| < C_N$ **then**
                N.add(p); S.remove(p);
        else
            $q :=$
            replacementCandidate(N, p);
            if $q \neq null$ **then**
                N.remove(q); N.add(p);
                S.remove(p);

Algorithm 2: Promoting nodes from S to N

The overall memory consumption of our TCA consist of two bytes for the neighbor id, the 2-hop neighbor ids ($2C_N$ and $2C_S$), the quality value (4 bytes), the sequence number of the last received message (1 byte), and two bytes for Algorithm 3, times $C_N + C_S$. Thus, in total $C_N(9 + 2C_N) + C_S(9 + 2C_S) + 7C_A$. In the experiments described in Sect. 6 we used $C_N = 7, C_S = 3$ and $C_A = 4$, thus, 234 bytes to store the complete information about the 2-hop neighborhood.

5   The Rank of a Local Topology

It remains to explain how Algorithm 2 decides to replace a node in N, i.e., the rank of a local topology LT_v as implemented by function *replacementCandidate(p, N)*. The function evaluates four metrics capturing criteria 1, 2, 3, and 5.

5.1   Minimizing Paths Length

To assess criterion 5 for a local topology LT_v the metric *weight* is introduced. Nodes in LT_v are called *level 1* nodes and all other nodes, except v are called *level 2* nodes of LT_v (see Fig. 4(a)). Thus, in LT_v every level 1 node w is adjacent to v and those level 2 nodes that w reported in its last message to v. LT_v contains at most $1 + C_N + C_N^2$ nodes. C_v denotes the set of level 1 and 2 nodes of v, henceforth called the *cover* of v. To construct a global topology with a small average path length we use the following heuristics. Each node v attempts to maximize the number of nodes in C_v. Among the local topologies with the same number of nodes the one which minimizes the distances from v to the nodes in its cover is preferred, i.e., $Len(LT_v) = \sum_{w \in C_v} dist(v, w)$ is minimized. Here $dist(v, w)$ is 1 if w is a level 1 node and 2 otherwise. To satisfy a weaker form of criterion 5, each node aims to minimize Len as shown in the following.

To test whether $p \in S$ can increase the rank of LT_v, p is inserted into N and the enlarged local topology LT_v' is analyzed. Let u be a level 1 node in LT_v'. The impact of removing edge (v, u) from LT_v' is assessed by an integer $\omega(u)$. A level 2 neighbor of u is called *dependent* if it is not adjacent to another level 1 node in LT_v'. Denote by $dep(u)$ the number of dependent neighbors of u in LT_v'. Removing the link to a level 1 neighbor u increases the distance from v to all $dep(u)$ dependent neighbors of u. In Fig. 4(a) $dep(u_3) = 1$ and $dep(u_2) = 2$. Now $\omega(u)$ is defined as follows:

1. $\omega(u) = dep(u) + 1$ if $N(u) \cap N(v) \neq \emptyset$: u is adjacent to another level 1 node of LT_v'
2. $\omega(u) = 2(dep(u) + 1)$ if $N(u) \cap N(v) = \emptyset$ and not all neighbors of u that are level 2 nodes are dependent.
3. $\omega(u) = C(dep(u) + 1)$ if $N(u) \cap N(v) = \emptyset$ and all level 2 nodes that are neighbors of u are dependent, here C is a constant larger than C_N.

The weight of a level 1 node u of LT_v' expresses the degradation of the topology when edge (v, u) is removed. In the first two cases the number of nodes in the local topology remains constant, but $Len(LT_v')$ is increased by $\omega(u)$. In the third case u and its dependent neighbors are no longer connected to v. Since $C > C_N$ this case always gets a higher weight than the other cases. Consider the extended local topology of node v depicted in Fig. 4(a): $\omega(u_1) = 3$ and $\omega(u_4) = 2$ (first case), $\omega(u_2) = 6$, $\omega(u_3) = 4$ (second case), and $\omega(p) = 4C$ (third case). Nodes within the shaded regions will experience the same degradation of path length to v if the link from v to the corresponding level 1 node is removed.

To decide whether $p \in S$ improves the rank, the weight of each node in the extended local topology LT_v' is computed. If $\omega(p)$ is larger than the smallest weight of the nodes in N then p improves the rank and p replaces the node in N with the lowest weight (see Fig. 4(a)).

5.2   Connected Components to Identify Bridges

Minimizing metric Len for each node does not guarantee connectivity, i.e., criterion 3. Consider Fig. 4(b) with $C_N = 2$. List $v.N$ contains u_1 and u_2. Since the weights of u_1, u_2, and p are equal to 3, v will not replace u_1 or u_2 by p. However, edge (v, p) maybe a bridge, i.e., (v, p) is required to form a connected topology. Without further information it is impossible for v to know that p must be included in N. This requires some kind of global knowledge.

We use a distributed algorithm that labels the connected components of the topology G_c in parallel to the main routine. The algorithm labels every node of a connected component with the smallest id of all nodes within this component (see Algorithm 3). On order to provide connectivity a node selects preferentially links to nodes from a different connected component. The algorithm is based on [10]. A node v maintains two variables: m holds the smallest id of a node reachable from v and d stores the distance to that node. Upon the reception of a periodic message a node first executes the code to update its lists. Afterwards it updates m and d with Algorithm 3.

There is a situation that requires special treatment: When the node with the smallest id leaves a connected component all other nodes refer to an id not contained in the component. This id is only eliminated when a node with a smaller id joins the component. In case this id also marks another component then the two components cannot be distinguished. Variable d solves this issue, its value grows unlimitedly in the component. The remedy is to reset d to 0 and m to the node's id if d grows beyond the diameter of the network. This requires each node to know an upper bound K of the diameter.

In the situation depicted in Fig. 4(b) this algorithm enables node v to establish that p belongs to a different connected component. v will therefore replace v_1 or v_2 by p to join the two components. Note that once a bridge is part of the local topology it has a high weight. This is because the third definition of a weight applies. Thus, it is very unlikely that a bridge will be broken again.

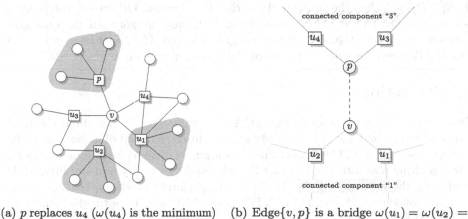

(a) p replaces u_4 ($\omega(u_4)$ is the minimum)   (b) Edge$\{v, p\}$ is a bridge $\omega(u_1) = \omega(u_2) = \omega(p)$

Fig. 4. Two local topologies illustrating dependent neighbors and a bridge. Level 1 and 2 nodes are depicted as squares and circles, respectively.

if $(v.d = 0 \wedge v.m \neq v.id) \vee (v.d \neq 0 \wedge v.m = v.id) \vee v.m > v.id \vee v.d \geq K \vee$
  $(v.d > 0 \wedge (\nexists u \in v.N : u.m = v.m \wedge u.d + 1 = v.d))$ **then**
  $v.m := v.id; v.d := 0; //*$
  [f]Reset
if $\exists u \in v.N : u.m < v.m \wedge u.d + 1 < K$ **then**
  $v.m := \min\{u.m \mid u \in v.N\}; v.d := \arg\min_u\{u.m \mid u \in v.N\}.d + 1; //*$
  [f]Smaller identifier
if $\forall u \in v.N : u.m \geq v.m \wedge \exists u \in v.N : u.m = v.m \wedge u.d + 1 < v.d$ **then**
  $v.d := \min\{u.d \mid u.m = v.m\} + 1; //*$
  [f]Smaller distance

Algorithm 3: Labeling Connected Components

5.3   Rank of a Local Topology

With these preliminary considerations we can formulate the concept of improving the rank of the local topology of a node v by a node p. The replacement candidate is defined by a priority based scheme with the following four categories.

Category 1	Link (v, p) is bidirectional and N contains a unidirectional link.
(Symmetry)	$R_1 = \{w \in N \mid (v, w)$ is unidirectional$\}$ (criterion 2).
Category 2	$p.m \neq v.m$ and N contains a node w with $w.m = v.m$.
(Conn. Compo.)	$R_2 = \{w \in N \mid w.m = v.m\}$ (criterion 3).
Category 3	N contains a node w with $\omega(p) > \omega(w)$.
(Weight)	$R_3 = \{w \in N \mid \omega(p) > \omega(w)\}$ (criterion 5).
Category 4	N contains a node w with $p.Q > w.Q$.
(Quality)	$R_4 = \{w \in N \mid p.Q > w.Q\}$ (criterion 1).

Function $replacementCandidate(p, N)$ implements the selection of a replacement candidate for a prospective neighbor p. Let i be the smallest integer such that $R_i \neq \emptyset$. If $|R_i| = 1$ then the single node in R_i is the replacement. Otherwise, the remaining categories are used to determine the replacement candidate, i.e., to narrow R_i down to a single node. Let $j > i$ be minimal, such that, $R_i \cap R_j \neq \emptyset$. If $|R_i \cap R_j| = 1$ then the single node in $R_i \cap R_j$ is chosen. Otherwise, the process is repeated, i.e., $R_i \cap R_j \cap R_k$ is considered. If the last category still does not give a unique candidate the most recent entry is selected. If $R_i = \emptyset$ for $i = 1, \ldots, 4$ the function returns *null* and none of the current neighbors is replaced.

6   Evaluation

First the dynamic behavior of our TCA was evaluated. We measured the time required to produce a stable topology after addition/removal of nodes. Secondly, a comparison with XTC was performed focusing on memory requirements. Next, our algorithm was compared to LEEP with respect to agility and stability, while analyzing the influence of the TCAs on the performance of applications using G_c. We implemented our TCA, XTC, and LEEP and performed simulations using OMNeT++ with the MiXiM framework. To simulate the wireless channel, slow-fading[1] was applied. Regular and random node placements with varying

---

[1] MiXiM parameters: Logarithmic-normal shadowing (*mean* = 0.5 dB, *standard deviation* = 0.25 dB), integrated path-loss model (*alpha* = 3.5), and random bit errors (*lower bound* = 10^{-8}).

node numbers and densities were simulated. The simulations used HoPS [12] as the LQE, hence, criterion 1 is handled by it. HoPS evaluates the channel by observing package loss through sequence numbers.

To reduce traffic, state messages are piggybacked with application messages. To decrease the probability of collisions a node randomly perturbs the time between two broadcasts. The required storage per node is proportional to C_N^2. As argued above, we expect C_N to be in the range of 5–10 (see [1]). Thus, message size and required storage are fitting for current hardware.

An important requirement for a TCA is that each node is reachable from every other node in the created topology (criterion 3). Therefore, we define the metric *connectedness* Γ as the ratio of the number of pairs of nodes connected by a path and the total number of pairs of nodes. Note that $\Gamma = 1$ iff all nodes can communicate pairwise via multi-hop routing.

6.1 Dynamic Behavior

First we simulated the overall performance of our algorithm, mainly in consideration of criterion 3, 4, 6, and 7. We evaluated the quality of the created topology (i.e., Γ) and the stability of the algorithm for a square grid topology with 196 nodes in a high density setting. The transmission radius was chosen such that on average each node could communicate with 33 other nodes. The TCA was used with $C_N = 7$, $C_A = 4$, and $C_S = 3$, while $T_A = 10$s. A round denotes the time it takes for all nodes to broadcast a maintenance message. The round length depends on node density and must be chosen sufficiently large to avoid collisions and retransmissions (here length of a round was 1s). Figure 5(a) shows the median of the number of link changes per round as a measure for stability. The results of 100 simulation runs are averaged. In all runs the topology became stable after 250–300 rounds. After that time about 10 links changed per round, this corresponds to about 1.4 % of all links. In particular, after the stabilization on average in about 50 % of all rounds no link change occurred.

We also measured the time the algorithm needs to create a connected topology. Figure 5(a) shows, that after 300 rounds $\Gamma \geq 0.98$. The simulations exhibited very small variations of the stabilization time. Furthermore it can be seen that once connected the topology stays that way. Transient faults cause link changes but the algorithm is robust enough to cope with them. This shows the stability and agility (criteria 6 and 7) characteristics of our approach.

Next we added/removed nodes to/form the network. We evaluated the *outgoing convergence time* ϕ_{out}, i.e., the time it takes between removing nodes from the deployment and their eviction from all lists. The *incoming convergence time* ϕ_{in} denotes the time it takes until new nodes are added bidirectionally into N of at least one node. ϕ_{out} and ϕ_{in} depend on T_N and T_A; the latter is strongly influenced by the used LQE and the node density. HoPS was configured, such that, after ten rounds the quality of a link could reach the value Q_{min}.

To evaluate ϕ_{in} we considered a random topology with 190 nodes. A random topology was used because the arbitrary removal of nodes has a bigger impact on the stability. Grid topologies exhibit a high degree of connectedness. They

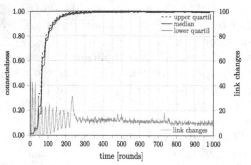 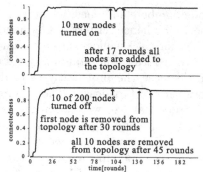

(a) Performance of our TCA in a dense setup. Top down the variation of 100 simulations between third quartile, median, and first quartile are depicted. In gray: Median of link changes.

(b) ϕ_{in} and ϕ_{out} in case of adding (top) (removing (bot)) 10 nodes in a stable phase.

Fig. 5. Analysis of our TCA in respect to connectedness, agility, and stability.

are k-connected with high k values. The minimum, average, and maximum node degree in G was 2, 23, and 55, respectively, to ensure that the random removal of nodes can disconnect the network. The simulations were executed with $C_N = 6$, $C_S = 4$, and $C_A = 4$. In a first experiment ten nodes were added after 100 rounds ($\Gamma = 1$). The placement of these 10 nodes was random. Figure 5(b) shows that $\phi_{in} = 17$ rounds, after that time Γ had reached value of 1 again.

To evaluate ϕ_{out} 10 randomly selected nodes were removed from the set of 200 nodes of the previous deployment after round 100. Note that in this case HoPS requires at most 45 rounds for the quality value of a removed node, measured by all neighbors, to fall below Q_{tol} (this corresponds to a removal of a link after 30 missed messages). Thus, ϕ_{out} is about twice as long as ϕ_{in}. The values of ϕ_{in} and ϕ_{out} can be adjusted by the LQE to better fit application requirements.

6.2   Comparison with XTC

XTC is a static TCA making a direct comparison with our TCA with respect to criteria 6 and 7 is impossible. Furthermore, concerning criterion 1, XTC does not make use of a minimum link quality which will cause problems in the context of the usability of this approach in WSNs. XTC does comply with criterion 3.

Figure 6 addresses the quality of XTC with respect to criterion 4. For multiple communication ranges square grids of 16–100 nodes were simulated. The purpose of these different ranges is to consider physical topologies with varying densities. Communication range 2 covers simulations in which nodes on average have 8.5 neighbors in G and 17.5 in the case of a range of 3. Figure 6 presents results with ranges 2 and 3. The average as well as the maximum number of neighbors chosen by XTC is depicted. To compare XTC with our TCA we computed the minimum value of C_N that leads to a connected topology G_c for each setting. For this purpose we ran ten simulations for each communication range and grid.

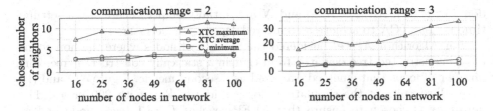

Fig. 6. XTC compared to our algorithm

We continuously decreased the value for C_N as long as all created topologies were stably connected; we excepted this value as minimum. Note that since $C_N \geq C_S \geq C_A$, the required memory is bounded.

The simulations revealed that the average number of neighbors chosen by XTC is about as big as the minimum value for C_N (Fig. 6). On the other hand the maximum value of neighbors chosen by XTC grows proportional to the average node degree in the physical topology. This demonstrates a weak point of XTC for a practical usage, its memory footprint is too large. Since the average number of neighbors and the maximum number of neighbors differ severely, it is obvious that some nodes only choose very few neighboring nodes. In fact, on a regular basis nodes only choose a single neighbor. The removal of one node will therefore often disconnect the network.

6.3 Topology G_c as a Basis for Other Algorithms

We also compared our TCA to TinyOS's neighbor-based topology control algorithm LEEP [4], because of its wide usage. Firstly, we evaluated the connectedness (criterion 3). In sparse networks, where the choice of neighbors is trivial, LEEP performed comparably to our approach. High density networks were much more challenging. Simulations with up to 200 nodes in very dense grids with high link variation showed weaknesses of LEEP. Since LEEP by default uses neighbor tables of size 10 a random selection of neighbors generates a connected topology with very high probability (we omit plots due to space restrictions). Nevertheless, our approach outperformed LEEP in special setups, e.g., very high density or clustered topologies only connected by bridges.

LEEP quickly removes nodes from its neighbor table when the quality measure drops below the bidirectional ETX value of all other neighbors. The LEEP specification does not state a minimum ETX value to remove unserviceable nodes. Hence, if there are no other nodes to replace an unserviceable node, it will remain in LEEP's neighbor table. Disconnected or frequently changing topologies lead to problems for applications. It is very likely that link changes degrade the quality of applications. Therefore, we tested how LEEP and our approach influence applications. Especially self-stabilizing algorithms require stable neighborhood relations to perform well. To evaluate this we chose two self-stabilizing algorithms: $\mathcal{A}_1$ constructs a spanning tree and $\mathcal{A}_2$ builds a maximum independent set (MIS). All lists A, S, and N are initially empty. Therefore, a settling time for our TCA to achieve a connected topology is necessary.

For Algorithm $\mathcal{A}_1$ we considered the number of nodes where the hop count in the constructed tree deviated from that in a statically constructed tree. The number of loops that emerged was used as a second metric. Figure 7(a) and (b) show the results of the simulations for a grid topology with 200 nodes. The density was chosen to ensure that LEEP reaches $\Gamma = 1$ (each node had at most 20 communication partners). Figure 7(a) shows that during the stabilizing phase algorithm $\mathcal{A}_1$ is disrupted constantly and errors are inevitable. Loops showed up throughout the complete simulation. The quality of the output of $\mathcal{A}_1$ is considerably improved when executed on top of our TCA (Fig. 7(b)).

Due to the stability of our TCA the performance of $\mathcal{A}_2$ increases significantly, as depicted in Fig. 7(c,d). Due to LEEP's constant link changing the independent set never stabilizes. We compared two different communication ranges to show the difference between our approach and LEEP more clearly. On the left 30 nodes can on average be reached by a single node, 14 on the right. The used metric shows the percentage of nodes in a correct state. For low densities our TCA allows $\mathcal{A}_2$ to produce a MIS that was disturbed only twice after the settling time. Running $\mathcal{A}_2$ on top of LEEP did at no time lead to an error free result.

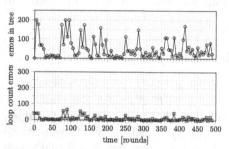 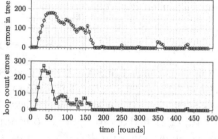

(a) Executing algorithm $\mathcal{A}_1$ on top of LEEP does not lead to a stable tree.

(b) The proposed topology algorithm allows Algorithm $\mathcal{A}_1$ to produce a tree that is almost stable after about 170 rounds.

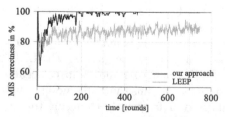 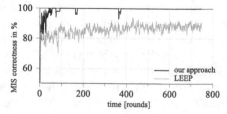

(c) Correctness of the MIS produced by algorithm $\mathcal{A}_2$ on top of LEEP and our TCA (high density).

(d) Correctness of the MIS produced by algorithm $\mathcal{A}_2$ on top of LEEP and our TCA (low density).

Fig. 7. Comparison LEEP as basis for $\mathcal{A}_1$ and $\mathcal{A}_2$ versus our approach

7 Conclusion

This paper presented a novel dynamic topology control algorithm which aims at optimizing global properties of the topology such as connectedness while keeping the number of neighbors of each node below a given value. The algorithm has a small footprint and is of high practical relevance for real sensor network hardware. Simulations have shown that the algorithm provides a good compromise between agility and stability. Evaluation was conducted through simulation. The proposed TCA is used as basis for the self-stabilizing publish subscribe system described in [15].

Acknowledgments. This research was funded by the Deutsche Forschungsgemeinschaft (DFG), contract number TU 221/6-1.

References

1. Blough, D.M., Leoncini, M., Resta, G., Santi, P.: The k-neighbors approach to interference bounded and symmetric TC in ad hoc networks. IEEE Trans. Mob. Com. **5**(9), 1267–1282 (2006)
2. Couto, D.D., Aguayo, D., Bicket, J.C., Morris, R.: A high-throughput path metric for multi-hop wireless routing. In: MOBICOM, pp. 134–146 (2003)
3. Fuchs, F., Völker, M., Wagner, D.: Simulation-based analysis of topology control algorithms for wireless ad hoc networks. In: Even, G., Rawitz, D. (eds.) MedAlg 2012. LNCS, vol. 7659, pp. 188–202. Springer, Heidelberg (2012)
4. Gnawali, O.: The link estimation exchange protocol (LEEP), TinyOS Extension Proposal (2007)
5. Lee, C.Y., Shiu, L.C., Lin, F.T., Yang, C.S.: Distributed topology control algorithm on broadcasting in WSN. J. Netw. Comput. App. **36**(4), 1186–1195 (2013)
6. Lin, S., Zhang, J., Zhou, G., Gu, L., Stankovic, J.A., He, T.: ATPC: adaptive transmission power control for wireless sensor networks. In: Proceedings of the 4th International Conference on Embedded Networked Sensor Systems, SenSys 2006, pp. 223–236 (2006)
7. Liu, B.H., Gao, Y., Chou, C.T., Jha, S.: An energy efficient select optimal neighbor protocol for wireless ad hoc networks. In: 29th IEEE International Conference on Local Computer Network, pp. 626–633 (2004)
8. Liu, Y., Zhang, Q., Ni, L.M.: Opportunity-based topology control in wireless sensor networks. IEEE Trans. Parallel Distrib. Syst. **21**(3), 405–416 (2010)
9. Manolopoulos, Y., Katsaros, D., Papadimitriou, A.: Topology control algorithms for wireless sensor networks: a critical survey. In: Proceedings of the 11th International Conference on Computer System and Technology, New York, pp. 1–10 (2010)
10. Pandit, V., Dhamdhere, D.: Self-stabilizing Maxima Finding on General Graphs. IBM T.J. Watson Research Center, New York (2002)
11. Puccinelli, D., Gnawali, O., Yoon, S.H., Santini, S., Colesanti, U., Giordano, S., Guibas, L.: The impact of network topology on collection performance. In: Marrón, P.J., Whitehouse, K. (eds.) EWSN 2011. LNCS, vol. 6567, pp. 17–32. Springer, Heidelberg (2011)

12. Renner, C., Ernst, S., Weyer, C., Turau, V.: Prediction accuracy of link-quality estimators. In: Marrón, P.J., Whitehouse, K. (eds.) EWSN 2011. LNCS, vol. 6567, pp. 1–16. Springer, Heidelberg (2011)
13. Santi, P.: Topology control in ad hoc WSN. ACM Comput. Surv. **37**, 164–194 (2005)
14. Sethu, H., Gerety, T.: A new distributed topology control algorithm for wireless environments with non-uniform path loss and multipath propagation. Ad Hoc Netw. **8**(3), 280–294 (2010)
15. Siegemund, G., Turau, V., Maâmra, K.: A self-stabilizing publish/subscribe middleware for wireless sensor networks. In: Proceedings of the International Conference on Networked Systems (2015)
16. Siegemund, G., Turau, V., Weyer, C., Lobs, S., Nolte, J.: Brief announcement: agile and stable neighborhood protocol for WSNs. In: Proceedings of the 15th International Symposium on Stabilization, Safety, and Security of Distributed Systems, pp. 376–378 (2013)
17. Tashtarian, F., Haghighat, A., Honary, M., Shokrzadeh, H.: A new energy-efficient clustering algorithm for wireless sensor networks. In: International Conference on Software, Telecommunications and Computer Networks, pp. 1–6. IEEE (2007)
18. Wattenhofer, R., Zollinger, A.: XTC: a practical topology control algorithm for ad-hoc networks. In: Proceedings of the 18th International Parallel and Distributed Processing Symposium, p. 216 (2004)
19. Woo, A., Tong, T., Culler, D.: Taming the underlying challenges of reliable multi-hop routing in sensor networks. In: Proceedings of the First International Conference on Embedded Networked Sensor Systems, pp. 14–27 (2003)

Geographic GReedy Routing with ACO Recovery Strategy GRACO

Mouna Rekik[1,2]([✉]), Nathalie Mitton[1], and Zied Chtourou[2]

[1] Inria Lille – Nord Europe, Villeneuve-d'Ascq, France
{mouna.rekik,nathalie.mitton}@inria.fr
[2] CMERP, Sfax, Tunisia
ziedchtourou@cmerp.net

Abstract. Geographic routing is an attractive routing strategy in wireless sensor networks. It works well in dense networks, but it may suffer from the void problem. For this purpose, a recovery step is required to guarantee packet delivery. Face routing has widely been used as a recovery strategy since proved to guarantee delivery. However, it relies on a planar graph not always achievable in realistic wireless networks and may generate long paths. In this paper, we propose **GRACO**, a new geographic routing algorithm that combines a greedy forwarding and a recovery strategy based on swarm intelligence. During recovery, *ant* packets search for alternative paths and drop pheromone trails to guide next packets within the network. GRACO avoids holes and produces near optimal paths. Simulation results demonstrate that GRACO leads to a significant improvement of routing performance and scalability when compared to the literature algorithms.

Keywords: Wireless sensor networks · Geographic routing · Guaranteed delivery · Swarm intelligence · Ant colony optimization

1 Introduction

Wireless sensor networks (WSN) consist of a large number of densely deployed sensors that have communication, computing, and sensing capacities [1]. Although sensors have power and memory constraints, they are multi-functional with sensing, wireless communication, computation capabilities and low-cost devices. For these reasons, WSNs are widely used in many fields as military surveillance, disaster prediction, and environment monitor [1]. WSNs require efficient routing protocols that adapts to the unpredictable and highly dynamic environment. The network topology may change dynamically due to node mobility, node failure and various physical properties related to the propagation channel (e.g., obstructions, noise, and power limitations) [2].

Geographic routing [3] is an attractive routing technique for large scale wireless sensor networks due to its low overhead, high scalability and memory-less features. Unlike topology-based routing, it uses only local information about the geographic location of nodes to determine, at each step, the next node to forward the packet. Greedy geographic routing schemes route data closer to the

© Springer International Publishing Switzerland 2015
S. Papavassiliou and S. Ruehrup (Eds.): ADHOC-NOW 2015, LNCS 9143, pp. 19–32, 2015.
DOI:10.1007/978-3-319-19662-6_2

destination at each step of the routing. This can lead to stuck nodes in case of the current node fails to find a closer node to the destination. Hence, geographic routing algorithms usually combine a greedy forwarding strategy with a recovery mechanism to solve void problem.

In this paper, we present GRACO a new geographic routing protocol that combines a modified greedy forwarding (GR) phase and an Ant-Colony-Optimization (ACO)-based recovery strategy. GRACO greedy forwarding is assisted by pheromone trails from previous recovery phases, which will prevent to return to the same stuck node. The ACO based recovery phase will use ants to discover alternative paths if greedy is not possible. The *ant* packets are sent around the void searching for route to a closer node than the stuck node to destination, then the protocol can switch back to greedy mode and until the destination or a new void. GRACO is a geographic routing algorithm that dynamically avoids holes and maintains near optimal paths around them. Results show that GRACO can reduce path lengths up to 60 % compare to state of the art.

The remainder of this paper is as follows. Section 2 discusses the most relevant related work in geographic routing, recovery techniques and ACO based routings. Section 4 presents GRACO. Section 5 analyzes simulation results where CRACO and GFG routing are compared. Finally, Sect. 6 concludes this work.

2   Related Work

This section browses the different concepts of the literature used in GRACO.

2.1   Geographic Routing

Geographic routing [3] is a routing concept that exploits geographic information instead of topological connectivity. Generally, only one-hop geographic information is needed to make routing decision. Geographic routing is thus memory-less, does not require the establishment or maintenance of complete routes and nodes do not have to store routing table or write additional information in the packet. There is no need to transmit routing messages to update route states either. Its localized and stateless features make it simple and scalable. A geographic forwarding strategy defines the next hop to which forward the packet using only geographic information. Greedy, MFR [4], NFP [5] and Compass [6] are the most famous geographic forwarding strategies, they differ in the way to exploit geographic information to forward a packet toward destination: the geographic distance, the largest projection, the shortest hop or the angle respectively. Greedy [7] forwards the message to the neighbor that minimizes the Euclidean distance to the destination in each step [3]. Geographic Greedy forwarding is loop free. Indeed, at each step, the message has to move toward the destination and thus can not loop by going through a node it has already visited [3]. However, Greedy forwarding may not always be possible if all neighboring nodes are further away from the destination than the sender itself. This problem is called communication void, local maximum phenomenon or local minimum phenomenon. It is caused by deployment holes where the forwarding process is blocked at a node called stuck node. The occurrence of hole can be caused by many factors, such as sparse deployment, physical obstacles, node failures, communication jamming, power exhaustion, and animus interference [8]. An alternative mode called recovery mode is then needed to guarantee delivery otherwise the packet has to be discarded and the delivery fails.

2.2   Recovery Techniques

Many studies have been focusing on the communication void problem and different solutions that guarantee delivery have been introduced. The most famous and used one is face routing [9], the first geographic routing algorithm to guarantee message delivery without flooding [3]. Face routing [3,10] is applied on a plane sub-graph of the network graph, a sub-graph where no edges intersect each other. A plane graph divides the plane into faces. The line segment between the source node and the destination node intersects some faces then, the packets will be forwarded along the boundaries of these faces. Given the advantages of face routing, it has been combined with a greedy routing approach to guarantee delivery. Indeed, when the greedy fails to forward a packet, face routing is used as a recovery mechanism. Afterward, several routing algorithms using the combination greedy face routing were proposed [3]. Although face guarantees delivery, it is energy-consuming since it may generate long detours and makes the packets follow a succession of short edges [11].

2.3   Ant Colony Optimization

Ant colony optimization (ACO) is a bio-inspired approach from ants foraging behavior. Indeed, real ants are able to find the shortest path between their nest and a food source without any visible, central and active coordination mechanisms. They drop pheromones, a natural chemical substance, on the path. The path optimization is achieved by exploiting the pheromone quantity deposited. Then, ants select a path based on the pheromone concentration deposited on the set of paths found. The higher the concentration of pheromone on a path, the greater the probability to select it. This indirect communication mechanism is called *stigmergy*. In addition to that, real ants show an impressive behavior when finding obstacles on their way. Actually, they are able not only to avoid obstacles, but also to find a shortest path around them.

ACO based approaches are very effectively applied to NP-hard problems and results in good optimization. Networking field is one of many domains that investigated ACO-approaches to design multi-objective and multi-constraint routing protocol and solve issues like mobility, path optimization, resource utilization and energy awareness. ACO was mainly proposed by Dorigo in his thesis [12,13], and it was widely used to solve network data routing problems. Many routing protocols based on ACO meta-heuristic were proposed in the literature [14–17]. ARA (Ant Colony Routing Algorithm) [18] the first ACO-based routing algorithm aims to reduce routing overhead and most of the existing ACO based routing techniques in WSN and mobile ad-hoc networks are derived from this algorithm [19]. ACO based routing uses two types of agents, forward ant packets and backward ant packets. The first type of ants is used to discover paths toward destination and the second one aims to drop pheromone trails on the established path between the source node and the destination node. Although ACO-based routing algorithms solve many problems such as multiconstraints and multiobjective routing, in addition to path optimization, these algorithms usually produce high overhead, since, in order to converge to an optimized solution, they need a colony-like behavior, i.e. a huge number of ant-like packets.

In this paper, we introduce a new geographic routing algorithm that combines a modified greedy forwarding and a recovery technique based on ACO.

3   Notations and System Models

We assume that all nodes are aware of their location through an hardware device such as GPS or any other location mean.

We model the wireless sensor network as a directed graph G = (V,E), composed of a finite set V of sensors, called also nodes, and a finite set E of links. There exists a wireless link between nodes u and v ($uv \in E$) if U and v are within transmission range of each other, i.e. $|Uv| <= R$, where $|Uv|$ represents the Euclidean distance between U and v. The physical set of nodes which are in the transmission range of node U is noted $N(U)$ and called the neighborhood of node U. $N(U) = \{v \in V$ such as $|Uv| \leq R\}$. We note $|N(U)|$ the number of neighbors of U. We also define $N_D(U)$ the subset of $N(U)$ in which each node is nearer from node D than U itself, i.e. such that: $N_D(U) = \{v \in N(U)$ such as $|vD| \leq |UD|\}$. The directed link from a node U to a node v is noted $\overrightarrow{Uv}$.

In the following, we call "current node" the node that has a packet to route and we use $NextNode$ to refer to the next hop in the path of a packet. We note $C(A, r)$ the circle C of radius R and center at A.

4   GRACO

4.1   GRACO Principles

GReedy with ACO based recovery routing protocol (GRACO) is a geographic routing algorithm that combines two modes: a greedy mode and an ACO-based recovery mode. A data packet is first routed using a modified greedy routing that accounts for the pheromone trails. If a pheromone trail exists for a given destination, it is used to select the next node. Otherwise, the next hop is selected using plain greedy forwarding strategy. However, if the packet reaches a stuck node, the node launches an ACO-based recovery and sends some exploratory ants (Fant) to find a path. The stuck node waits for a Bant to come back, if so, a path is established and data packets can be sent. The data packet will be routed using the same strategy as the Fant until arriving to the unstuck node, and then switch to greedy forwarding again until its destination or another stuck node, in which case, the same mechanisms applies again. Algorithm 1 depicts the GRACO behavior. Both steps are now detailed in the following sections.

Algorithm 1. GRACO(N,D) - Run at each node U upon reception of a message $M(N, D)$

1: **if** U=D **then**
2:     EXIT # U *is the final destination of* M*. The routing has succeeded.*
3: **else**
4:     **if** U$\neq$ N **then**
5:         Discard M - EXIT # U *is not needed in the routing of* M
6:     **else**
7:         # U *is the recipient of the packet, in charge of forwarding it to D.*
8:         **if** $existPh(U, D) =$ **true then**
9:             $NextNode \leftarrow getNextNodePH(U, D)$ # *There already exists a pheromone trail to* D
10:        **else**
11:            **if** $N_D(U) \neq \emptyset$ **then**
12:                $NextNode \leftarrow$ N st $||UD| - |ND|| = max_{v \in N_D(U)}||UD| - |vD||$
13:            **else**
14:                # *Greedy mode fails,* U *launches the recovery mode.*
15:                $NextNode \leftarrow$ Recovery(U,D)
16:            **end if**
17:        **end if**
18:        Return $NextNode$
19:    **end if**
20: **end if**

4.2 Ph-assisted Greedy Forwarding

The greedy mode of GRACO consists of a variant of the plain greedy forwarding (GF). GF is enhanced with the use of pheromone trails from older recoveries. Consider a node S that wants to send a data packet to D. Before applying GF, S checks whether it has recorded pheromone trails to D in order to avoid returning to the same stuck node. If so, that means the greedy failed to progress a previous data packet during a previous attempt for the same destination D and a recovery mode was been launched. For that reason, S will use these pheromone trails to send the packet instead of the GF. Otherwise, If there is no pheromone trail for D, S proceeds by using the greedy method. In the example presented in Fig. 1, S sends a data packet to D, the packet is routed using greedy forwarding until arriving to N_4 where it finds pheromones to D, then it uses a pheromone based forwarding which helps the packet to avoid the stuck node N_5. Thus, GRACO's greedy mode is a Ph-assisted greedy forwarding.

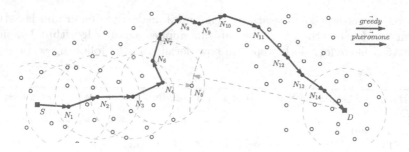

Fig. 1. Ph-assisted greedy forwarding

4.3   ACO Based Recovery

Similarly to other ACO based algorithms, the ACO recovery uses two types of ants to solve a problem: Fants (Forward-ants) to discover the environment, and Bants (Backward-ants) to "mark" the solutions found. In addition, GRACO relies on a concept of zones which plays an important role that we introduce in the following.

Zones. Using the concept of zones [20], a node divides its neighborhood into 4 zones based on its position and the position of D. Consider a destination node D, each node U partitions its neighbors into two main zones: the positive progress zone, later called $zone_1$, and a negative progress zone. As shown in Fig. 2, $zone_1$ is represented by the intersection of the two circles $C_1(U, R)$ and $C_2(D, |DU|)$. It gathers nodes in $N_D(U)$. Then, the negative progress zone is then partitioned into 3 sub-zones: $zone_2$, $zone_3$ and $zone_4$.

Let α be the positive progress angle and β the negative progress angle, α is the angle $(\overrightarrow{UB}, \overrightarrow{UA})$ where A and B are the points of intersection of two circles $C_1(U, R)$ and $C_2(D, |DU|)$, and $\beta = 2\Pi - \alpha$ (Fig. 3). For a node $v \in N(U)$, θ is the angle $(\overrightarrow{UD}, \overrightarrow{Uv})$. Node v belongs to:

- $zone_2$ if $\frac{\alpha}{2} < \theta \leq \frac{\alpha}{2} + \frac{\beta}{3}$
- $zone_4$ if $\frac{\alpha}{2} + \frac{\beta}{3} < \theta \leq \frac{\alpha}{2} + 2.\frac{\beta}{3}$
- $zone_3$ if $\frac{\alpha}{2} + 2.\frac{\beta}{3} < \theta \leq \frac{\alpha}{2} + \beta$

Pheromone Initialization. Before starting the route establishment phase, the amount of pheromone deposited on the links must be initialized. Each node U assigns a pheromone trail to each of its outgoing links, i.e., an initial pheromone value ϕ_{oj} is assigned for each neighbor v in $N(U)$ as $j \in \{1, 2, 3, 4\}$ and $zone_j$ being the zone in which node v lies, knowing that $\phi_{o1} \geq \phi_{o2}(= \phi_{o3}) \geq \phi_{o4}$. The motivation is that, in most cases, shortest paths pass through the neighbors whose directions are closer to the direction of the destination. Thus, the initial pheromone values being bigger as the zone is closer to the destination, direction

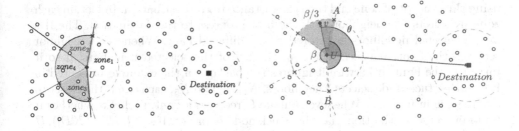

Fig. 2. Different zones of U for the destination node D

Fig. 3. Illustration of α, β and θ

will favor ants to choose these zones. As a result, this phase leads to a faster convergence to a shortest path most of the time. The ACO-based recovery strategy assumes that each node maintains a PhTable, a table of pheromone values assigned to its outgoing links for different destinations. Whenever a node receives a packet for a specific destination D, it searches in its PhTable for pheromone for D. If such pheromone trail exists, it will be used to choose the next hop. Otherwise, a pheromone initialization process is launched.

The pheromone initialization process attributes pheromone trails to all the outgoing links of a node. However, the not updated pheromone trails will be evaporated as the time passes. If a link is unused, the pheromone level on it should be completely evaporated by the end, thus, the corresponding PHTable entree is deleted. In this way, the pheromone evaporation process minimises the amount of data stored in the nodes.

Route Establishment. The route establishment phase is accomplished using two types of ants: the Fant and the Bant. The main role of the Fant is to explore the neighborhood in order to find an alternative path. Furthermore, it drops, on its way, a pheromone track to the stuck node to be used later by the Bant. The Bant establishes the route to the destination found by the Fant by dropping pheromone trails when it goes back to the stuck node. The route establishment starts at the stuck node K by diffusing (broadcasting) an Fant to the neighbors. The Fant will be considered and forwarded only by 3 neighbors that are selected

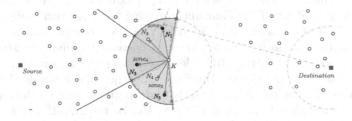

Fig. 4. Fant is considered by the closest node in each zone.

using the concept of zone and then the distance to the destination. In fact, in each zone, K selects the neighbor with the best progress to the destination. The IDs of the selected neighbors are stored in the diffused Fant. Whenever a neighbor receives the Fant, it check if it is in the list of the selected neighbors, if so it forwards the Fant, otherwise it ignores the packet. In the example presented in Fig. 4, the stuck node selects 3 neighbors, N_1, N_3 and N_5, each one is the closer to destination in its zone. Whenever a node U receives a Fant, it checks whether it is closer to the destination than the Stuck node K or not. If so ($|UD| < |KD|$), U sends back a Bant, the recovery can end. Otherwise, U forwards the ant. Similar to other ACO routing algorithms [19], a node forwards a Fant to the next node using a stochastic decision which is based on the values of pheromone trails to select the next hop. Suppose that a Fant is currently residing in node U. U has k neighbors v_1, v_2, ..., v_k. We will note Φ_i the amount of pheromone assigned to v_i (or the link $\overrightarrow{Uv_i}$). The neighbors of U will be partitioned into 4 zones. Consider Φ_{zone_i} is the maximum pheromone amount assigned to the neighbors of $zone_i$, $\Phi_{zone_i} = \max \{\Phi_j \setminus v_j \in zone_i\}$. In order to forward a Fant, first of all, U selects a zone with probability P_{zone_i}:

$$P_{zone_i} = \frac{\Phi_{zone_i}}{\sum_{j=1}^{4} \Phi_{zone_j}} \tag{1}$$

U chooses a node v_i in this selected zone with probability P_i:

$$P_i = \frac{\Phi_i}{\sum_{j=1}^{|N(D)|} \Phi_j} \tag{2}$$

Using the concept of zones, the ants are not completely blind, as in the usual ACO based algorithms, they will be directed, but not forced, to the right direction. Besides the PHTable mentioned before, each node maintains also a back routing table, the BRTable, to store information that will be used later by Bants to go back to the stuck node. Whenever a Fant arrives to a node from one of its neighbors, an entry is added to the BRTable. Figure 5 shows an example of a Fant F trying to bypass a void. F is launched at the stuck node, and forwarded to N_1, N_2, N_3 until reaching unstuck node N_4. To summarize, Algorithm 2 shows the forwarding strategy of the Fant until arriving to an unstuck node.

When a Fant reaches an unstuck node N, a Bant is sent back to the stuck node. An unstuck node is a node closer to destination than the stuck node. After adding an entry to the BRTable, N extracts the information from the Fant, creates a Bant and deletes the Fant. Subsequently, N sends the Bant to the stuck node K. The role of the Bant is to drop pheromone on the path found in order to establish a track from K to N. Unlike the traditional ACO based algorithm, the Bant will not return to the stuck node using the same path of the corresponding Fant, but it will use the pheromone dropped by other Fants

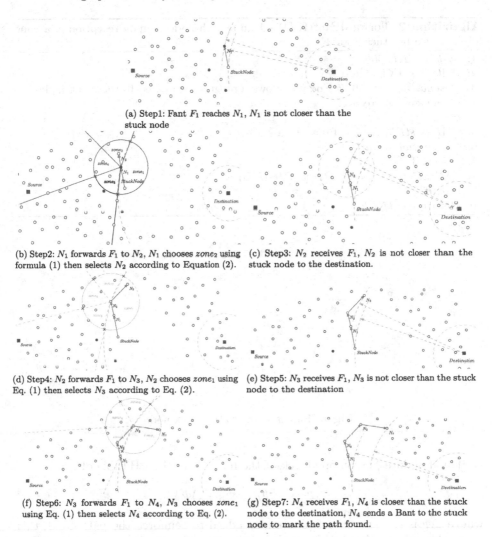

(a) Step1: Fant F_1 reaches N_1, N_1 is not closer than the stuck node

(b) Step2: N_1 forwards F_1 to N_2, N_1 chooses $zone_2$ using formula (1) then selects N_2 according to Equation (2).

(c) Step3: N_2 receives F_1, N_2 is not closer than the stuck node to the destination.

(d) Step4: N_2 forwards F_1 to N_3, N_2 chooses $zone_1$ using Eq. (1) then selects N_3 according to Eq. (2).

(e) Step5: N_3 receives F_1, N_3 is not closer than the stuck node to the destination

(f) Step6: N_3 forwards F_1 to N_4, N_3 chooses $zone_1$ using Eq. (1) then selects N_4 according to Eq. (2).

(g) Step7: N_4 receives F_1, N_4 is closer than the stuck node to the destination, N_4 sends a Bant to the stuck node to mark the path found.

Fig. 5. An example of a Fant searching for a path around a void

that used at least one node of its path. In fact, the Bant will not only search for a path in the BRTable, but also it will have the ability to choose the best path found (in our case the shortest). In the example presented in Fig. 6, the Fants find two valid paths to an unstuck node, the first one presented in green is a 4-hop-length path and the second one in pink is a 16-hop-length path. A Bant, in this example, chooses to use the green path to go back to K since it is the shortest path to K. On its way back to the stuck node, the Bant updates pheromone trails in the PHTables. Consider a Bant arrives to node A from node

Algorithm 2. ForwardFant(K,D) - Run at each node U upon reception of a Fant for D issued by stuck node K

1: *update_BRTable(U)*
2: **if** $|UD| < |KD|$ **then**
3:     *send_Bant(K, D)* # Node U allows a progress compared to the stuck node and can stop the recovery.
4: **else**
5:     **if** $existPh(U, D) =$ **false then**
6:         *initialisePh(U, D)*
7:     **end if**
8:     $NextNode \leftarrow getNextNodePH(U, D)$
9: **end if**
10: Return $NextNode$

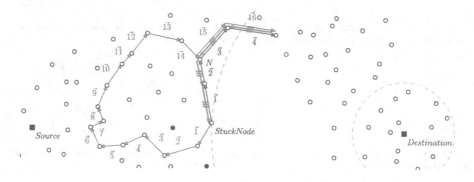

Fig. 6. Bant returns to the stuck node

B, it will update pheromone track of the link $\overrightarrow{AB}$ in the PHTable of A:

$$\Phi_{\overrightarrow{AB}} = \Phi_B = \Phi_B + \Delta\Phi$$

where $\Delta\Phi$ is the amount of pheromone added to reinforce the path to D, this value depends on the quality of the path. Algorithm 3 sums up the Bant forwarding.

5   Simulations and Results

In order to evaluate the performance of the GRACO algorithm, we compare it to the GFG algorithm under the WSNet simulator [21]. We generate random topologies in a region of 300×300 m region with 250 to 900 nodes and $R = 25$ m. To evaluate the impact of voids recovery to performance, we run simulations on different void diameters. We evaluate the algorithm according to end-to-end delay, data delivery cost, delivery rate and hop count. The simulation results of GRACO are given with 95 % confidence intervals.

Algorithm 3. ForwardBant(K,D) - Run at node U upon reception of a Fant for K
1: **Returns** $NextNode$ # the next hop
2: $update\_PHtable(U)$
3: **if** $U = K$ **then**
4:    kill Bant # the Bant reaches the stuck node.
5: **else**
6:    $NextNode \leftarrow getNextNodeBRtable(U, D)$
7: **end if**

5.1 End-to-End Delay

The end-to-end delay is the time interval between a given source sends a packet and the destination receives it. In case of a set of data packets, it represents the delay between the source sends the first data packet and the destination receives the last one. Sources and destinations are randomly chosen in a way they are in different sides of a void. Figure 7 shows the end-to-end delay varying with the number of data packets sent. We can see that the end-to-end delay of GFG algorithm is always higher than GRACO. This is mainly due to the fact that GFG tries to create a path around the void every time a packet tries to cross the void, even if the packet has the same source and destination as the previous one. However, GRACO launches the recovery only the first time when the packet is trying to bypass the void, and then the next packets will find pheromone traces for the paths found in the previous recovery phase. When the number of data packets sent increases, the difference between the delay produced by GFG and the one produced by GRACO become very large. Thus, GRACO delivers the data packets faster than GFG.

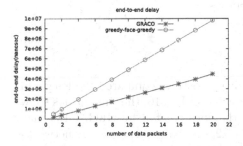
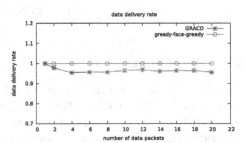

Fig. 7. End-to-End delay varying with the number of data packets sent

Fig. 8. Delivery rate varying with the number of data packets sent

5.2 Data Delivery Cost

Data delivery cost is the number of packets sent in the network to deliver a data packet from a source to a destination, including recovery cost. Figure 9a presents the data delivery cost varying with the number of data packets sent. For the particular case of a single packet, the data delivery cost used by GRACO is higher than GFG. This is explained by the number of packets sent during the recovery phase in GRACO is bigger than GFG in one recovery mode. However, when the number of data packets sent between the same source and the same destination increases, GRACO becomes cheaper than GFG since GRACO launches the recovery mode only when the first data packet is sent, the next ones will use the routes already found. As for GFG, it launches the recovery mode every time a data packet is sent. Thus, the cost on GRACO for multiple packets between the same source and destination is the same as for only one packet, in contrary with GFG, the cost used to deliver multiple data packet will be the cost of sending one packet multiplied by the number of data packets sent. In Fig. 9b, it can be seen that the cost per packet decreases when the number of data packets increases. On the other hand, the cost of GFG to deliver a single packet remains constant.

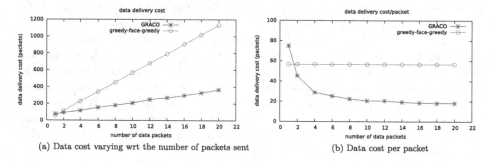

(a) Data cost varying wrt the number of packets sent

(b) Data cost per packet

Fig. 9. Data cost

5.3   Hop Count

Besides the end-to-end delay and data delivery cost, we compared the length of paths found by GRACO and GFG. As presented in Fig. 10, the simulations show that GRACO creates shorter paths than the ones created by GFG. GFG may generate extremely long path in some cases [22] specially in low density networks as shown in Fig. 10. In the other hand, the ACO and the zone concept used in GRACO help the algorithm to create shorter paths than GFG, these paths are usually optimal or near optimal paths.

5.4   Packet Delivery Rate

Packet delivery rate is the ratio of data packets successfully received by their destinations to all data packets sent by the sources. From Fig. 8, we can see that GFG delivers all packets sent successfully. Although GRACO couldn't maintain its delivery rate at 100 %, it has a high delivery rate (more than 95 %). The main

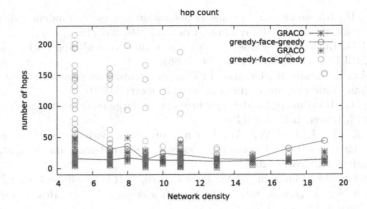

Fig. 10. hop count

possible reasons for packet loss is that, in the case of multiple voids, GRACO could fail in cascading recoveries leading to sporadic data delivery failures.

6   Conclusion

In this paper, we presented GRACO, a new geographic routing protocol able to handle the void problem with very low added complexity. It combines a pheromone assisted greedy forwarding phase and an ACO based recovery phase to avoid holes and recover from local minima. The proposed routing protocol is fully localized, distributed, scalable, and does not require any graph structure or routes information to be maintained. Besides the efficient hole avoidance, simulations results show that GRACO delivers data packets faster and with better end-to-end delay, data delivery cost and hop count compared to GFG routing protocol. GRACO provides also a reliable delivery performance. GRACO still has some potential for future improvement. The pheromone update rule has to be improved in order to account for multiple recoveries in a single path. We intend to consider GRACO as a routing solution for Advanced Metering Infrastructure (AMI) communications in the context of smart grid. For this purpose, we plan to analyze communication requirements in this type of networks, and to implement an extension of GRACO that takes into account the QoS required.

References

1. Akyildiz, I.F., Su, W., Sankarasubramaniam, Y., Cayirci, E.: A survey on sensor networks. IEEE Com. Mag. **40**, 102–114 (2002)
2. Dvir, A., Carlsson, N.: Power-aware recovery for geographic routing. In: Wireless Communications and Networking Conference (WCNC) (2009)
3. Mitton, N., Razafindralambo, T., Simplot-Ryl, D.: Position-based routing in wireless ad hoc and sensor networks. In: Nikoletseas, S., Rolim, J.D.P. (eds.) Theoretical Aspects of Distributed Computing in Sensor Networks, pp. 447–477. Springer, Heidelberg (2010)

4. Takagi, H., Kleinrock, L.: Optimal transmission ranges for randomly distributed packet radio terminals. IEEE Trans. Com. **32**, 246–257 (1984)
5. Hou, T.C., Li, V.O.K.: Transmission range control in multihop packet radio networks. IEEE Trans. Com. **34**, 38–44 (1986)
6. Kranakis, E., Singh, H., Urrutia, J.: Compass routing on geometric networks. In: Canadian Conference on Computational Geometry (1999)
7. Finn, G.G.: Routing and addressing problems in large metropolitan-scale internetworks. ISI Research Report (1987)
8. Jiang, Z., Ma, J., Lou, W., Wu, J.: An information model for geographic greedy forwarding in wireless ad-hoc sensor networks. In: Conference on Computer Communications (Infocom) (2008)
9. Sanavullah Shri Alarmelu, M.Y., Poonkuzhali, V.R.: An efficient void handling technique for geographic routing in manet: a survey. J. Adv. Res. Comput. Sci. Softw. Eng. **2**(12), 164–170 (2012)
10. Guan, X.: Face Routing in Wireless Ad-hoc Networks. Ph.D. thesis. University of Toronto (2009)
11. Gouvy, N., Mitton, N., Zheng, J.: Greedy Routing Recovery Using Controlled Mobility in Wireless Sensor Networks. In: International Conference on Ad hoc Networks and Wireless (ADHOC-NOW) (2013)
12. Dorigo, M., Di Caro, G.: The ant colony optimization meta-heuristic. In: Corne, D., Dorigo, M., Glover, F. (eds.) New Ideas in Optimisation. McGraw-Hill, London (1999)
13. Dorigo, M., Stützle, T.: The ant colony optimization metaheuristic: algorithms, applications, and advances. In: Glover, F., Kochenberger, G.A. (eds.) Handbook of Metaheuristics, pp. 250–285. Springer, US (2003)
14. Zungeru, A.M., Ang, L.M., Seng, K.P.: Classical and swarm intelligence based routing protocols for wireless sensor networks: a survey and comparison. J. Netw. Comput. Appl. **35**, 1508–1536 (2012)
15. Uchhula, V., Bhatt, B.: Article: comparison of different ant colony based routing algorithms. IJCA Spec. Issue MANETs **2**, 97–101 (2010)
16. Sutariya, D., Kamboj, P.: A survey of ant colony based routing algorithms for manet. Eur. Scientific J. **3**, 82–91 (2014)
17. Huang, R., Guanghui, X.: Swarm intelligence-inspired adaptive routing construction in wsn. In: Wireless Communications, Networking and Mobile Computing (WiCOM) (2010)
18. Gunes, M., Sorges, U., Bouazizi, I.: Ara-the ant-colony based routing algorithm for manets. In: International Conference on Parallel Processing Workshops (2002)
19. Shirkande, S.D., Vatti, R.A.: Aco based routing algorithms for ad-hoc network (wsn, manets): a survey. In: Communication Systems and Network Technologies (CSNT) (2013)
20. Kamali, S., Opatrny, J.: Posant: a position based ant colony routing algorithm for mobile ad-hoc networks. In: International Conference on Wireless and Mobile Communications (ICWMC), (2007)
21. Fraboulet, A., Chelius, G., Fleury, E.: Worldsens: development and prototyping tools for application specific wireless sensors networks. In: Symposium on Information Processing in Sensor Networks (IPSN), April 2007
22. Seada, K.: Modeling and analyzing the correctness of geographic face routing under realistic conditions. Elsevier Ad Hoc Netw. J. Spec. Issue Recent Adv. Wirel. Sens. Netw. **5**, 855–871 (2007)

Scheduling Connections via Path and Edge Multicoloring

Evangelos Bampas[1]([✉]), Christina Karousatou[2], Aris Pagourtzis[3], and Katerina Potika[4]

[1] LaBRI, UMR 5800, Univ. Bordeaux, 33400 Talence, France
evangelos.bampas@labri.fr
[2] LIF, Aix-Marseille University and CNRS, Marseille, France
christina.karousatou@lif.univ-mrs.fr
[3] School of Electrical and Computer Engineering,
National Technical University of Athens, Athens, Greece
pagour@cs.ntua.gr
[4] Santa Clara University,
500 El Camino Real, Santa Clara, CA 95053, USA
katerina.potika@gmail.com

Abstract. We consider path multicoloring problems, in which one is given a collection of paths defined on a graph and is asked to color some or all of them so as to optimize certain objective functions. Typical objectives are: (a) the minimization of the average, over all edges, of the maximum-multiplicity color when the number of colors is given (MINAVGMULT-PMC), (b) the minimization of the number of colors when the maximum multiplicity for each edge is given (MIN-PMC), or (c) the maximization of the number of colored paths when both the number of colors and a maximum multiplicity constraint for each edge are given (MAX-PMC). Such problems also capture edge multicoloring variants (such as MINAVGMULT-EMC, MIN-EMC, and MAXEMC) as special cases and find numerous applications in resource allocation, most notably in optical and wireless networks, and in communication task scheduling.

Our contribution is two-fold: On the one hand, we give an exact polynomial-time algorithm for MIN-PMC on spider networks with even admissible color multiplicities on each edge. On the other hand, we present an approximation algorithm for MINAVGMULT-PMC in star networks, with a ratio strictly better than 2; our algorithm uses an appropriate path orientation. We also show that any algorithm which

---

Research supported in part by the ALGONOW project, co-financed by the European Union (European Social Fund ESF) and Greek national funds through the Operational Program Education and Lifelong Learning of the National Strategic Reference Framework (NSRF) Research Funding Program: THALES, Investing in knowledge society through the European Social Fund.
E. Bampas—Partial support by the ANR project DISPLEXITY (ANR-11-BS02-014).
C. Karousatou—This work was done while Christina Karousatou was with the School of Electrical and Computer Engineering, National Technical University of Athens, Greece.

© Springer International Publishing Switzerland 2015
S. Papavassiliou and S. Ruehrup (Eds.): ADHOC-NOW 2015, LNCS 9143, pp. 33–47, 2015.
DOI:10.1007/978-3-319-19662-6_3

is based on path orientation cannot achieve an approximation ratio better than $\frac{7}{6}$. Our results apply to the corresponding edge multicoloring problems as well.

1   Introduction

In the class of path coloring problems, one is given a set of simple paths on a graph and is asked to assign a color to each of them so as to optimize a certain objective function. Path coloring problems have been studied extensively in the context of routing and wavelength assignment in optical networks, as well as in several other applications varying from compiler optimization to vehicle scheduling. In standard path coloring problems, edge-intersecting paths must receive distinct colors. In a meaningful generalization, path multicoloring problems were defined and studied in [1,9,18,21]. Note that, by the term 'path multicoloring', it is meant that edge-intersecting paths may receive identical colors. This is in contrast to standard path coloring.

Various optimization objectives have been studied in the context of path multicoloring. A usual one is to minimize the number of colors used, under the assumption that, for each edge, an upper bound is given on the maximum color multiplicity allowed on that edge, i.e., on the maximum number of paths using this edge that can receive the same color. We will refer to this upper bound as the *admissible color multiplicity* for that edge. This problem is called MINIMUM PATH MULTICOLORING (MIN-PMC) and it is formally defined as follows:

*Problem 1 (*MINIMUM PATH MULTICOLORING, MIN-PMC*).*

Instance: $\langle G, \mathcal{P}, \mu \rangle$, where $G = (V, E)$ is an undirected graph, $\mathcal{P}$ is a set of undirected simple paths on G, and $\mu : E \to \mathbb{N}$ is a function that maps each edge to the admissible color multiplicity on that edge.

Feasible solution: an assignment of one color from $\{1, \ldots, w\}$ to each path in $\mathcal{P}$, where w is some positive integer, so that on every edge e, each color is used by at most $\mu(e)$ paths.

Goal: minimize the number of colors w.

A well-studied variant [1,9,21,25] assumes a bound on the number of available colors and asks for a coloring such that the sum of edge multiplicities (equivalently, average edge multiplicity) over all edges of the graph is minimized. Here, the *multiplicity* of an edge refers to the maximum number of paths of the same color that share that edge, with respect to a certain coloring of paths. We will call this latter problem MINIMUM AVERAGE MULTIPLICITY PATH MULTICOLORING (MINAVGMULT-PMC). It is formally defined as follows:

*Problem 2 (*MINIMUM   AVERAGE   MULTIPLICITY   PATH   MULTICOLORING, MINAVGMULT-PMC*).*

Instance: $\langle G, \mathcal{P}, w \rangle$, where $G = (V, E)$ is an undirected graph, $\mathcal{P}$ is a set of undirected simple paths on G, and $w \in \mathbb{N}$ is the number of available colors.

Feasible solution: a coloring of $\mathcal{P}$ with w colors.

Goal: minimize $\sum_{e \in E} \mu(e)$, where $\mu(e)$ is the multiplicity of the maximum-multiplicity color on e.

Note that, in optical network terminology, MINAVGMULT-PMC corresponds to minimizing the number of parallel fiber links needed to accommodate a given set of requests with a given set of wavelengths, assuming that no wavelength collisions are allowed within the same fiber. A wavelength collision occurs when two communication requests are routed on the same fiber and modulated on the same wavelength. Other known variants include maximizing the number of satisfied requests, given constraints on both the number of colors and the admissible color multiplicities of edges (MAXIMUM PATH MULTICOLORING (MAX-PMC)), and minimizing the maximum edge multiplicity. The formal definition of MAX-PMC follows:

Problem 3 (MAXIMUM PATH MULTICOLORING, MAX-PMC).

Instance: $\langle G, \mathcal{P}, \mu, w \rangle$, where $G = (V, E)$ is an undirected graph, $\mathcal{P}$ is a set of undirected simple paths on G, $\mu : E \to \mathbb{N}$ is a function that maps each edge to the maximum admissible color multiplicity on that edge, and $w \in \mathbb{N}$ is the number of available colors.

Feasible solution: a coloring of a subset $\mathcal{Q} \subseteq \mathcal{P}$ with w colors, so that on every edge e, each color is used by at most $\mu(e)$ paths.

Goal: maximize $|\mathcal{Q}|$.

Apart from the optical networks setting, the above problems find applications in scheduling communication requests or tasks in generic networks. In such scenarios, colors represent time slots and edge multiplicity represents congestion on the corresponding network link, assuming that requests receiving the same color are served simultaneously. Therefore, MIN-PMC is the problem of minimizing the number of time slots, given the available bandwidth per link, and MINAVGMULT-PMC corresponds to minimizing the average congestion given the number of available time slots.

Yet another interesting application comes through the known equivalence of path (multi)coloring in stars to edge (multi)coloring in general graphs. It is noteworthy that this equivalence is approximation-preserving. Therefore, one defines the corresponding MIN-EMC, MINAVGMULT-EMC and MAX-EMC problems by replacing the path intersection relation with edge incidence to the same vertex; (vertex) multiplicity now refers to the number of edges that are incident to a vertex and receive the same color. Note that MIN-EMC has been considered by Coffman et al. [15] under the name 'f-coloring' as a problem of scheduling file transfers among servers that can support a certain number of parallel transfers each, in a minimum number of time slots. Drawing on their approach, we consider MINAVGMULT-EMC as a way to address the following problem: given a graph where edges represent file transfer requests, and assuming that the number of requests that can be served simultaneously by a particular node depends on the number of available ports, find a scheduling of the requests within a given number time-slots so as to minimize the total number of ports needed to satisfy all requests. Thus, solving MINAVGMULT-EMC can help a network administrator in allocating efficiently the necessary number of ports, or a network designer who has access (in advance) to demand statistics among the nodes in designing a more efficient network.

Edge (multi)-coloring problems have a self-evident connection to channel assignment problems in multi-channel wireless networks [16]. For example, different colors may represent distinct communication channels and it is possible to have parallel transmissions from the same transceiver, provided that distinct channels are used. Color multiplicity in such considerations can model the number of time frames needed to complete a given set of communication tasks. Therefore, solving MINAVGMULT-EMC amounts to minimizing the average time per node, while MIN-EMC corresponds to minimizing the number of channels needed to accomplish all tasks, given the time availability of each particular component of the network. Finally, MAX-EMC describes the situation where both the number of channels and the desired or available time per node are provided and the goal is to maximize the number of served communication requests. Under an orthogonal perspective, in a single-channel model color multiplicity may represent the available bandwidth, whereas colors may stand for time frames. Under this approach, solving MIN-EMC can help reduce the maximum time needed, MINAVGMULT-EMC minimizes the average bandwidth used, and MAX-EMC aims at maximizing the number of satisfied requests, given the bandwidth per node and the time availability. For other, more specific applications of edge multicoloring to wireless channel assignment see [4,14] and references therein.

It makes sense to consider both undirected and directed versions of the above problems. In this paper, we provide results mainly for the undirected versions. Nonetheless, we will make use of some known results for directed versions of the problems. Throughout the text, when we refer to a particular problem, we will refer by default to its undirected version and we will state it explicitly whenever we refer to the directed version.

Related Work. Path coloring and path multicoloring problems have been extensively studied during the last twenty years (see e.g. [2,5,7–9,19–21,24] and references therein). All variants are NP-hard in general graphs and even in simple topologies, e.g. stars and rings, whereas they can be solved optimally in chains. In addition, most variants are hard to approximate in general graphs within a constant factor. However, this is possible in stars and rings, and in some other simple topologies.

The MIN-PMC problem was defined independently in [9] (under the name MIN-COLORS-PMC) and in [6], where it was proved to admit a 4-approximation in trees. A (3/2)-approximation for stars was proposed in [19] and an asymptotic

(9/8)-approximation can be obtained directly from the equivalence of MIN-PMC in stars and the MIN-EMC problem in general graphs (aka f-coloring) and the result of Coffman et al. for f-coloring [15]. In [22], algorithms for MIN-PMC in spiders and caterpillars were proposed[1], achieving approximation ratios of 2 and 3 respectively. In contrast, it was shown in [22] that the directed version of MIN-PMC can be solved exactly in spiders.

A more recent result of Bian and Gu [5] states that MIN-PMC can be solved exactly in spiders with *uniform* and even admissible color multiplicities (here, uniform stands for 'the same on each edge of the graph'). In addition, they show that, for stars, the uniformity requirement can be removed and, therefore, MIN-EMC can also be solved optimally in graphs where the admissible color multiplicity is even. However, Bian and Gu also prove that, in the case of odd admissible color multiplicities, the situation is not similar: the problem becomes NP-hard.

Regarding MINAVGMULT-PMC, the problem was defined in [21], and independently in [25]. In [21], an exact algorithm for chain graphs was presented, as well as 2-approximation algorithms for rings and stars. These algorithms were later extended in [20] to cover the generalized version of MINAVGMULT-PMC with non-uniform multiplicity costs. To the best of our knowledge, the MINAVGMULT-EMC problem has not been studied as such so far, hence the best known result is the 2-approximation due to the approximation preserving equivalence with MINAVGMULT-PMC in stars, as explained previously.

Results for MAX-PMC appear in [9], where a 2.54-approximation for trees is proposed, in [23], where the authors give an exact algorithm for chains and 2-approximation algorithm for rings and stars, and in [5], where they provide a 1.58-approximation algorithm for the problem in spiders and an exact algorithm for spiders with uniform and even admissible color multiplicity.

Finally, path multicoloring problems were also studied in a non-cooperative setting in [2], where the social cost depends on the maximum color multiplicity. A more general framework for studying path multicoloring under selfish considerations was proposed in [3].

Our Results. In this work, we take some steps towards coping with the path and edge multicoloring problems described above.

Our first contribution is that we manage to show that MIN-PMC can be solved optimally in spiders with non-uniform even admissible color multiplicity on each edge. This represents an improvement with respect to the result of Bian and Gu [5], as it removes the uniformity requirement. As a byproduct, we obtain an exact algorithm for MAX-PMC in such graphs, i.e., spiders with non-uniform even color multiplicity. This latter result holds for any number of given colors.

We then turn our focus to MINAVGMULT-EMC(equivalent to MINAVGMULT-PMC in stars) and start by rewriting an algorithm of [21] in edge multicoloring

---

[1] A *spider* graph is a tree with at most one node of degree strictly greater than 2. A *caterpillar* is a tree in which all nodes of degree 2 or more lie on the same simple path.

terms. This algorithm, which achieves an approximation factor of 2, has a locally near-optimal behaviour: the maximum multiplicity at each node differs from the optimal by at most one. We take advantage of this property and first show that MINAVGMULT-EMC can be solved exactly if the number of available colors equals 2. Next, we further fine-tune the algorithm, by adding a random orientation step together with a derandomization process, in order to come up with an algorithm having an approximation ratio of $2 - \frac{1}{2^w}$, where w is the number of available colors. Since the ratio of 2 of the algorithm in [21] is tight, our results represent the first, to the best of our knowledge, algorithm with an approximation ratio strictly better than 2, for both MINAVGMULT-EMC in general graphs and MINAVGMULT-PMC in stars.

We conclude by proving a $\frac{7}{6}$ lower bound for all orientation-based algorithms for MINAVGMULT-EMC, i.e., for algorithms that attempt to solve the (unoriented) edge (or path) multicoloring problem by computing an orientation of the given multigraph (or set of paths) and then applying the algorithm from [21].

Remark 1 (Notation). Throughout the paper, we will use the notation L_e for the *load* of edge e with respect to a given path set, i.e., the number of paths which contain edge e. Also, d_v will denote the degree of a node v in the context of an undirected graph, whereas for directed graphs we will use d_v^{in} (resp. d_v^{out}) for the in-degree (resp. out-degree) of node v.

Some of the proofs are omitted due to lack of space and they are deferred to the full version of the paper.

2   Min-PMC and Max-PMC on Spiders

In this section, we present exact algorithms for MIN-PMC and MAX-PMC on spiders with non-uniform even admissible color multiplicity.

2.1   Minimizing the Number of Colors

With respect to the MIN-PMC problem in any graph, one observes immediately the following lower bound:

Fact 1. *No* MIN-PMC *instance can be colored with fewer than* $w_{lb} = \max_e \left\lceil \frac{L_e}{\mu(e)} \right\rceil$ *colors.*

It is known [8,24] that MIN-PMC is NP-hard when G is a star and $\mu(e) = 1$ for all edges. Bian and Gu [5] prove that for every odd k, MIN-PMC is NP-hard when G is a star and $\mu(e) = k$ for all edges. They also give a polynomial-time algorithm for the case where G is a spider and every leg of the spider has uniform and even admissible color multiplicity. We extend their algorithm in order to remove the uniformity requirement and we show that MIN-PMC is polynomial-time solvable on spiders with even admissible color multiplicity.

We reduce the problem to the *directed* version of MIN-PMC, which is known to be polynomial-time solvable on spiders [22]. Formally, the directed version of the problem is defined as follows:

*Problem 4 (*DIRECTED MINIMUM PATH MULTICOLORING, DIR-MIN-PMC*).*

Instance: $\langle G, \vec{\mathcal{P}}, \mu \rangle$, where $G = (V, E)$ is an undirected graph, $\vec{\mathcal{P}}$ is a set of directed simple paths on G, and $\mu : E \to \mathbb{N}$ is a function that maps each edge to the admissible color multiplicity on that edge for each direction.

Feasible solution: a coloring of $\vec{\mathcal{P}}$ so that on every edge e, each color is used by at most $\mu(e)$ paths in each direction.

Goal: minimize the number of colors.

In order to perform the reduction, we need to decide on a direction for each undirected path in the original MIN-PMC instance $\langle G, \mathcal{P}, \mu \rangle$. We accomplish this by adding unit-length dummy paths on edges with odd load, and then considering the multigraph H which has the same node set as G and contains one edge (u, v) for each path with endpoints u and v. It can be shown that the addition of dummy paths ensures that all nodes of H have even degree. Therefore, the Euler partition algorithm of [10] will partition the edges of G into closed paths. We then perform an arbitrary orientation of each closed path in the Euler partition. This assigns a direction to each edge of H, which in turn corresponds to a direction for each path in $\mathcal{P}$ and for each dummy path. Figures 1, 2 and 3 illustrate the reduction.

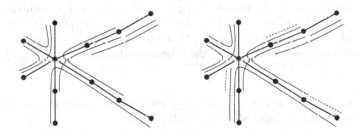

Fig. 1. Original MIN-PMC instance (left) and instance with added dummy paths (right).

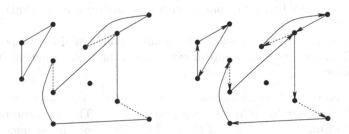

Fig. 2. The corresponding graph H (left) and an orientation of the closed paths in its Euler partition (right).

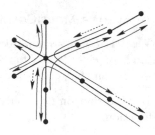

Fig. 3. The DIR-MIN-PMC instance.

Algorithm 1. An exact algorithm for MIN-PMC in spiders with even admissible color multiplicity

---

Input: an instance $\langle G, \mathcal{P}, \mu \rangle$ of MIN-PMC, where $G = (V, E)$ is a spider and $\mu(e)$ is even for all e

Output: an optimal solution using w_{lb} colors

1: For any edge with odd load, add one unit-length dummy path on that edge. Let $\mathcal{P}'$ be the resulting set of paths.
2: Construct the multigraph $H = (V, E')$, where E' contains one edge (u, v) for each path in $\mathcal{P}'$ with endpoints u and v.
3: Use the algorithm EP in [10] to find an Euler partition of H. Orient each closed path of the Euler partition in an arbitrary manner, thus obtaining a directed multigraph $\vec{H}$. Assign a direction to each path in $\mathcal{P}'$ according to the orientation of the corresponding edge in $\vec{H}$, thus obtaining a set $\vec{\mathcal{P}'}$ of directed paths on G.
4: Find an optimal coloring c of the DIR-MIN-PMC instance $\langle G, \vec{\mathcal{P}'}, \mu' \rangle$, where $\mu'(e) = \frac{\mu(e)}{2}$ for all e, using the algorithm in [22, Theorem 6].
5: Return the coloring c restricted to the original paths in $\mathcal{P}$.

Theorem 1. *Algorithm 1 is an exact polynomial-time algorithm for* MIN-PMC *on spiders with even admissible color multiplicity.*

2.2   Maximizing the Number of Satisfied Requests

A corollary of Theorem 1 is that the MAX-PMC problem is also optimally solvable in polynomial time on spiders with (non-uniform) even admissible color multiplicity.

In [5], Bian and Gu propose a deterministic polynomial-time algorithm for MAX-PMC on spiders with uniform even admissible color multiplicity, which works as follows:

1. First, it computes a maximum-size subset $\mathcal{Q}$ of $\mathcal{P}$ such that every edge e of the spider is used by at most $w\mu(e)$ paths in $\mathcal{Q}$. This is accomplished by solving optimally an instance of the Call Control problem on spiders, via an algorithm developed in [5].
2. Then, $\mathcal{Q}$ is colored optimally with w colors using the algorithm for MIN-PMC on spiders with uniform even admissible color multiplicity.

By plugging in Algorithm 1 instead in step 2, we can solve MAX-PMC on spiders with non-uniform admissible color multiplicity. Thus, we have the following:

Theorem 2. *There exists an exact polynomial-time algorithm for* MAX-PMC *on spiders with even admissible color multiplicity.*

3 Reducing Average Color Multiplicity on Stars

In this section, we present our results on the approximability of MINAVGMULT-PMC on stars. We will only be interested in MINAVGMULT-PMC instances in which all paths in $\mathcal{P}$ are of length 2. In view of the following proposition, we can essentially ignore paths of length 1 if we ensure that our algorithm produces solutions whose cost is at most a constant times $\mathrm{OPT}_{LB}(\mathcal{I}) = \sum_{e \in E} \left\lceil \frac{L_e}{w} \right\rceil$ (a lower bound on the cost of any optimal solution of $\mathcal{I}$):

Proposition 1. *Let $\mathcal{A}$ be a polynomial-time algorithm for* MINAVGMULT-PMC *instances with paths of length 2 which, for any such instance $\mathcal{I}$, produces a solution with cost* $\mathrm{SOL}_{\mathcal{A}}(\mathcal{I}) \leq \alpha \cdot \mathrm{OPT}_{\mathcal{LB}}(\mathcal{I})$*, where $\alpha \geq 1$ is a constant. Then, there exists a polynomial-time algorithm $\mathcal{B}$ with the same approximation guarantee for general instances $\mathcal{I}'$ of* MINAVGMULT-PMC.

It has been pointed out before in the literature [8,11] that there exists an easily computed bijection between pairs $(G, \mathcal{P})$, where G is a star and $\mathcal{P}$ is a set of paths of length 2 on G, and multigraphs G', such that the conflict graph of $\mathcal{P}$ is isomorphic to the line graph of G'. Therefore, MINAVGMULT-PMC can be equivalently cast as the following problem:

*Problem 5 (*MINIMUM AVERAGE MULTIPLICITY EDGE MULTICOLORING, MINAVGMULT-EMC*).*

Instance: $\langle G, w \rangle$, where $G = (V, E)$ is an undirected multigraph and $w \in \mathbb{N}$ is the number of available colors.

Feasible solution: a coloring of E with w colors.

Goal: minimize $\sum_{v \in V} \mu(v)$, where $\mu(v)$ is the multiplicity of the maximum-multiplicity color over all edges incident to v.

Fact 2. *The multiplicity of the maximum multiplicity color incident to node v is at least $\left\lceil \frac{d_v}{w} \right\rceil$, thus the minimum cost for any* MINAVGMULT-EMC *instance is at least $\sum_{v \in V} \left\lceil \frac{d_v}{w} \right\rceil$.*

For any fixed $w \geq 3$, MINAVGMULT-EMC is NP-hard via a straightforward reduction from the decision version of the classical edge coloring problem on w-regular graphs, which is known to be NP-complete [13,17]. Nomikos et al. [21] propose a 2-approximation algorithm which we restate as Algorithm 2 in MINAVGMULT-EMC terms (the algorithm was originally stated in MINAVGMULT-PMC terms). The analysis in [21] is tight, as there exists a

---

Algorithm 2. A 2-approximation algorithm for MINAVGMULT-EMC [21]

Input: an instance $\langle G, w \rangle$ of MINAVGMULT-EMC, $G = (V, E)$
Output: a 2-approximate solution

1: Assign an arbitrary direction to each edge of G.
2: For each $v \in V$, group its out_v outgoing edges into $\left\lceil \frac{\mathrm{out}_v}{w} \right\rceil$ groups of at most w edges each, and let V_{out} denote the set of all groups of outgoing edges. Similarly, for each $v \in V$, group its in_v incoming edges into $\left\lceil \frac{\mathrm{in}_v}{w} \right\rceil$ groups of at most w edges each, and let V_{in} denote the set of all groups of incoming edges.
3: Construct the bipartite multigraph $H = (V_{\mathrm{out}} \cup V_{\mathrm{in}}, A)$, where for each edge in E, A contains one edge joining its outgoing group to its incoming group. The maximum degree of H is bounded by w.
4: Use an algorithm for edge coloring of bipartite multigraphs (e.g. [10]) to compute an edge coloring of H with w colors.
5: Assign to each edge of G the color of the corresponding edge in H.

---

family of instances on which Algorithm 2 computes a solution with cost exactly twice the optimum: $\{\langle C_k, w \rangle : k \geq 2 \text{ and } w \geq 2\}$, where C_k is the ring graph with k nodes. Indeed, if the directions assigned in step 1 are such that each node has in-degree 1 and out-degree 1, then the resulting solution will have cost $2k$, whereas the optimum solution has cost k.

If we have only two available colors, the problem can be solved exactly in polynomial time: The Euler partition algorithm in [10] computes a partition of the edges of a multigraph into open and closed paths, with the property that each vertex of odd degree is the end of exactly one open path, and each vertex of even degree is the end of no open paths. Note, then, that if we color the edges of each path of the Euler partition alternately with the two available colors, the resulting coloring will have the property that the edges incident to each even-degree node will be equipartitioned into two same-colored subsets, whereas the edges incident to each odd-degree node will be partitioned into two same-colored subsets whose sizes differ by exactly one. This implies that the cost incurred by each node v will be exactly $\left\lceil \frac{d_v}{2} \right\rceil$, thus the solution is optimal in view of Fact 2. We have thus proved the following:

Theorem 3. *There exists an exact polynomial-time algorithm for* MINAVG MULT-EMC *with two available colors.*

3.1 An Approximation Algorithm with Ratio Strictly Better than 2

We first show that if we assign random directions to the edges of a multigraph with n nodes, then the resulting orientation will have, in expectation, at least $\frac{1}{2w} \cdot n$ nodes such that, if we subsequently execute steps 2–5 of Algorithm 2 on this orientation, each of them will contribute the minimum possible cost to the cost of the solution: $\left\lceil \frac{d_v}{w} \right\rceil$. Then, we show how to derandomize this procedure in order to obtain a deterministic algorithm for MINAVGMULT-EMC with approximation ratio at most $\left(2 - \frac{1}{2w} \right)$.

As a preliminary observation, consider the simplest conceivable randomized algorithm for MINAVGMULT-EMC, i.e. the one that assigns a random color with uniform probability to each edge. Unfortunately, this algorithm performs quite poorly in the following family of instances: The multigraph contains two nodes with w parallel edges joining them, where w is the number of available colors. As is well known [12], the maximum multiplicity color will appear $\Theta\left(\frac{\log w}{\log\log w}\right)$ times with high probability, whereas the optimum cost is 2. This motivates us to search for a randomized algorithm with a better performance guarantee.

Definition 1 (Locally Optimal Nodes). *Let $\langle G, w \rangle$ be an instance of the* MINAVGMULT-EMC *problem, let $\vec{G}$ be an orientation of G, and let v be a node of G. We say that v is locally optimal if the following condition holds:*

$$\left(d_v^{\text{in}} \bmod w = 0\right) \vee \left(d_v^{\text{out}} \bmod w = 0\right) \vee \left(\left(d_v^{\text{in}} \bmod w\right) + \left(d_v^{\text{out}} \bmod w\right) > w\right)$$

The following lemma is implicit in the analysis in [21].

Lemma 1 ([21].) *In any solution computed by Algorithm 2, each node v incurs a cost of exactly $\lceil \frac{d_v}{w} \rceil$ if it is locally optimal with respect to the directions assigned after step 1, or $\lceil \frac{d_v}{w} \rceil + 1$ if it is not locally optimal.*

In other words, Algorithm 2 incurs an additional cost, with respect to the lower bound of Fact 2, of exactly one for each non-locally-optimal node. The existence of an algorithm that computed an orientation of G in which at least some fraction of the nodes were guaranteed to be locally optimal would yield an approximation algorithm for MINAVGMULT-EMC with ratio strictly better than 2. We now consider the algorithm that assigns a random direction to each edge of G and then executes steps 2–5 of Algorithm 2.

Definition 2. *$S(d, w)$ denotes the set of integers i such that, if exactly i out of d incident edges to a d-degree node are incoming and $(d - i)$ incident edges are outgoing, then the node is locally optimal assuming we have w colors.*

Lemma 2. *If d is a multiple of w, then $S(d, w) = \{i \cdot w : 0 \le i \le \frac{d}{w}\}$.*

Lemma 3. *If $d = r \cdot w + x$, where $0 < x < w$, then $S(d, w)$ is exactly the set: $\bigcup_{i=0}^{r} \{i \cdot w + j : j \in \{0, x, x+1, \ldots, w-1\}\} \cap \{0, 1, \ldots, d\}$.*

Lemma 4. *Let $\langle G, w \rangle$ be an instance of MINAVGMULT-EMC and let $\vec{G}$ be a random orientation of G in which each edge receives each of the two possible directions with probability $\frac{1}{2}$. The expected fraction of locally optimal nodes in $\vec{G}$ is at least $\frac{1}{2w}$.*

Theorem 4. *There exists a $\left(2 - \frac{1}{2w}\right)$-approximation algorithm for MINAVG MULT-EMC.*

Proof. By Lemma 4, if we assign random directions to the edges of G, we get at least $\frac{1}{2^w} \cdot n$ locally optimal nodes in expectation. This algorithm can be derandomized by a straightforward application of the method of conditional expectations. Indeed, if we assume that the orientation of a subset of the edges has already been fixed, we can compute in polynomial time the probability that a fixed node v of degree d will be locally optimal if we assign the rest of the directions randomly, as follows: Assume that a of its incident edges have already been oriented as incoming to v, and b of its incident edges have already been oriented as outgoing from v. Then, the probability that the node will be locally optimal is $\frac{1}{2^{d-a-b}} \cdot \sum_{s:s+a \in S(d,w)} \binom{d-a-b}{s}$.

Therefore, the algorithm which examines edges in an arbitrary order, and to each edge assigns the direction which maximizes the expected number of locally optimal nodes under the current partial orientation, runs in deterministic polynomial time and produces an orientation with at least $\frac{1}{2^w} \cdot n$ locally optimal nodes. Taking also into account Lemma 1, this implies that, if we execute steps 2–5 of Algorithm 2 on this orientation, we will obtain a solution with cost SOL that can be expressed as follows: (let V denote the set of nodes of G and $\mathcal{O}$ denote the set of locally optimal nodes)

$$
\begin{aligned}
\text{SOL} &= \sum_{v \in \mathcal{O}} \left\lceil \frac{d_v}{w} \right\rceil + \sum_{v \in V \setminus \mathcal{O}} \left(\left\lceil \frac{d_v}{w} \right\rceil + 1 \right) \\
&= \sum_{v \in V} \left\lceil \frac{d_v}{w} \right\rceil + |V \setminus \mathcal{O}| \\
&\leq \sum_{v \in V} \left\lceil \frac{d_v}{w} \right\rceil + \left(1 - \frac{1}{2^w} \right) \cdot |V|.
\end{aligned}
\tag{1}
$$

If OPT is the cost of an optimal solution, then Eq. 1, Fact 2, and the observation that OPT $\geq |V|$ yield: SOL $\leq \left(2 - \frac{1}{2^w} \right) \cdot$ OPT. □

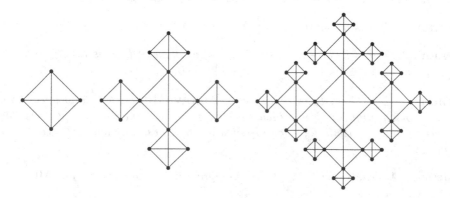

Fig. 4. The graphs G_1, G_2, and G_3, as defined in Sect. 3.2.

3.2 A Lower Bound for Orientation-Based Algorithms

Let $\{G_i\}_{i\geq 1}$ be the infinite family of graphs where $G_1 = K_4$, and G_{i+1} is constructed by attaching to each degree-3 node of G_i a new clique K_4 (by identifying the degree-3 node of G_i to which the K_4 is attached with one of the nodes of the attached clique). Figure 4 illustrates the first few graphs in the family. For every $i \geq 1$, consider the instance $\mathcal{I}_i = \langle G_i, 3\rangle$ of MINAVGMULT-EMC.

A simple calculation reveals that the number of nodes n_i in G_i is $n_i = 2 \cdot (3^i - 1)$. In particular, for $i \geq 2$, the number of nodes of degree 3 in G_i is $n_i' = 4 \cdot 3^{i-1}$, whereas the rest of the nodes have degree 6.

For every $i \geq 2$, there exists a coloring of the edges of G_i, such that each node v contributes exactly $\lceil \frac{d_v}{w} \rceil$ to the cost of the solution (Fig. 5). By Fact 2, this coloring is necessarily optimal. Therefore, the cost of the optimal solution for $\mathcal{I}_i$ is: $\text{OPT}_i = n_i' + 2(n_i - n_i') = \frac{8}{3} \cdot 3^i - 4$.

Fig. 5. An optimal coloring of K_4 with three colors. We obtain an optimal coloring for $\mathcal{I}_i$ by repeating this coloring for every K_4 clique contained in G_i.

On the other hand, for $i \geq 2$, for every orientation of G_i, out of every three nodes of degree 3 which belong to the same K_4 subgraph of G_i, at most two will be locally optimal. Indeed, these nodes form a triangle in G_i, and thus any orientation of the edges between them will result in at least one of the nodes having at least one outgoing and at least one incoming edge. Therefore, every MINAVGMULT-EMC algorithm which computes an orientation of the given multigraph and then executes steps 2–5 of Algorithm 2, will compute a solution for $\mathcal{I}_i$ with cost: $\text{SOL}_i \geq 2 \cdot \frac{n_i'}{3} + \frac{2n_i'}{3} + 2 \cdot (n_i - n_i') = \frac{28}{9} \cdot 3^i - 4$.

By the calculations for OPT_i and SOL_i and letting $i \longrightarrow +\infty$, we obtain:

Theorem 5. *No* MINAVGMULT-EMC *algorithm which computes an orientation of the given multigraph and then executes steps 2–5 of Algorithm 2 can have an approximation ratio which is better than $\frac{7}{6}$.*

4 Concluding Remarks

We gave exact algorithms for MIN-PMC and MAX-PMC on spiders with (non-uniform) even admissible color multiplicity, thus extending the algorithms from [5] which work only for spiders with uniform even admissible color multiplicity.

Furthermore, we gave an approximation algorithm for MinAvgMult-PMC on stars with approximation ratio at most $\left(2 - \frac{1}{2^w}\right)$. This improves a previous algorithm from [21], which guarantees an approximation ratio of 2. Note that the bound of 2 on the approximation ratio of the algorithm from [21] is tight. Having said that, one could easily modify that algorithm to guarantee an approximation ratio of $\left(2 - \frac{1}{|E|}\right)$, where $|E|$ is the number of edges of the star network. However, from a practical standpoint, in most applications the number of colors w represents a scarce network resource which needs to be utilized efficiently (e.g., in particular in the context of optical networks, w represents the number of available wavelengths per fiber, which is very limited due to technological constraints), whereas the size of the network can grow unbounded. Therefore, the approximation ratio of our algorithm represents a significant improvement over the previous algorithms.

Finally, we have provided a lower bound of $\frac{7}{6}$ on the approximation ratio of any algorithm which computes an orientation of the edges of the given multigraph and then executes steps 2–5 of Algorithm 2. In view of the large gap between this lower bound and the upper bound of Theorem 4, it makes sense to try to improve the performance of the edge orientation algorithm. For example, one could try to tweak the probabilities with which edges receive one of the two directions, in order to increase the expected number of locally optimal nodes.

References

1. Andrews, M., Zhang, L.: Minimizing maximum fiber requirement in optical networks. J. Comput. Syst. Sci. **72**(1), 118–131 (2006). http://dx.doi.org/10.1016/j.jcss.2005.08.001
2. Bampas, E., Pagourtzis, A., Pierrakos, G., Potika, K.: On a noncooperative model for wavelength assignment in multifiber optical networks. IEEE/ACM Trans. Netw. **20**(4), 1125–1137 (2012)
3. Bampas, E., Pagourtzis, A., Pierrakos, G., Syrgkanis, V.: Selfish resource allocation in optical networks. In: Spirakis, P.G., Serna, M. (eds.) CIAC 2013. LNCS, vol. 7878, pp. 25–36. Springer, Heidelberg (2013). http://dx.doi.org/10.1007/978-3-642-38233-8
4. Barrett, C.L., Kumar, V.S.A., Marathe, M.V., Thite, S., Istrate, G.: Strong edge coloring for channel assignment in wireless radio networks. In: 4th IEEE Conference on Pervasive Computing and Communications Workshops (PerCom 2006 Workshops), 13–17 March 2006, Pisa, Italy, pp. 106–110. IEEE Computer Society (2006). http://doi.ieeecomputersociety.org/10.1109/PERCOMW.2006.129
5. Bian, Z., Gu, Q.P.: Wavelength assignment in multifiber star networks. Networks **56**(1), 30–38 (2010)
6. Chekuri, C., Mydlarz, M., Shepherd, F.B.: Multicommodity demand flow in a tree. In: Baeten, J.C.M., Lenstra, J.K., Parrow, J., Woeginger, G.J. (eds.) ICALP 2003. LNCS, vol. 2719, pp. 410–425. Springer, Heidelberg (2003)
7. Erlebach, T., Jansen, K.: Maximizing the number of connections in optical tree networks. In: Chwa, K.-Y., Ibarra, O.H. (eds.) ISAAC 1998. LNCS, vol. 1533, p. 179. Springer, Heidelberg (1998)

8. Erlebach, T., Jansen, K.: The complexity of path coloring and call scheduling. Theor. Comput. Sci. **255**(1–2), 33–50 (2001)
9. Erlebach, T., Pagourtzis, A., Potika, K., Stefanakos, S.K.: Resource allocation problems in multifiber WDM tree networks. In: Bodlaender, H.L. (ed.) WG 2003. LNCS, vol. 2880, pp. 218–229. Springer, Heidelberg (2003)
10. Gabow, H.N.: Using Euler partitions to edge color bipartite multigraphs. Int. J. Comput. Inf. Sci. **5**(4), 345–355 (1976)
11. Golumbic, M.C., Jamison, R.E.: The edge intersection graphs of paths in a tree. J. Comb. Theory Ser. B **38**(1), 8–22 (1985)
12. Gonnet, G.H.: Expected length of the longest probe sequence in hash code searching. J. ACM **28**(2), 289–304 (1981)
13. Holyer, I.: The NP-completeness of edge-colouring. SIAM J. Comput. **10**(4), 718–720 (1981)
14. Hsu, C., Liu, P., Wang, D., Wu, J.: Generalized edge coloring for channel assignment in wireless networks. In: 2006 International Conference on Parallel Processing (ICPP 2006), 14–18 August 2006, Columbus, Ohio, USA, pp. 82–92. IEEE Computer Society (2006). http://dx.doi.org/10.1109/ICPP.2006.45
15. Coffman Jr., E.G., Garey, M.R., Johnson, D.S., LaPaugh, A.S.: Scheduling file transfers. SIAM J. Comput. **14**(3), 744–780 (1985). http://dx.doi.org/10.1137/0214054
16. Kyasanur, P., Vaidya, N.H.: Capacity of multichannel wireless networks under the protocol model. IEEE/ACM Trans. Netw. **17**(2), 515–527 (2009). http://doi.acm.org/10.1145/1552193.1552205
17. Leven, D., Galil, Z.: NP completeness of finding the chromatic index of regular graphs. J. Algorithms **4**(1), 35–44 (1983)
18. Li, G., Simha, R.: On the wavelength assignment problem in multifiber WDM star and ring networks. In: INFOCOM, pp. 1771–1780 (2000). http://www.ieee-infocom.org/2000/papers/222.ps
19. Nomikos, C., Pagourtzis, A., Potika, K., Zachos, S.: Fiber cost reduction and wavelength minimization in multifiber WDM networks. In: Mitrou, N.M., Kontovasilis, K., Rouskas, G.N., Iliadis, I., Merakos, L. (eds.) NETWORKING 2004. LNCS, vol. 3042, pp. 150–161. Springer, Heidelberg (2004)
20. Nomikos, C., Pagourtzis, A., Potika, K., Zachos, S.: Routing and wavelength assignment in multifiber WDM networks with non-uniform fiber cost. Comput. Netw. **50**(1), 1–14 (2006). http://dx.doi.org/10.1016/j.comnet.2004.11.028
21. Nomikos, C., Pagourtzis, A., Zachos, S.: Routing and path multicoloring. Inf. Process. Lett. **80**(5), 249–256 (2001)
22. Pagourtzis, A., Potika, K., Zachos, S.: Path multicoloring with fewer colors in spiders and caterpillars. Computing **80**(3), 255–274 (2007)
23. Potika, K.: Maximizing the number of connections in multifiber WDM chain, ring and star networks. In: Boutaba, R., Almeroth, K.C., Puigjaner, R., Shen, S., Black, J.P. (eds.) NETWORKING 2005. LNCS, vol. 3462, pp. 1465–1470. Springer, Heidelberg (2005)
24. Raghavan, P., Upfal, E.: Efficient routing in all-optical networks. In: Leighton, F.T., Goodrich, M.T. (eds.) STOC, pp. 134–143. ACM (1994)
25. Winkler, P., Zhang, L.: Wavelength assignment and generalized interval graph coloring. In: Proceedings of the Fourteenth Annual ACM-SIAM Symposium on Discrete Algorithms, January 12–14, 2003, Baltimore, Maryland, USA, pp. 830–831. ACM/SIAM (2003). http://dl.acm.org/citation.cfm?id=644108.644246

A Schedule Template Construction Technique
for Duty Cycled Sensor Networks

Van Ho and Ioanis Nikolaidis[✉]

Department of Computing Science, University of Alberta, Edmonton
AB T6G 2E8, Canada
{vhho,nikolaidis}@ualberta.ca

Abstract. We exploit the relatively predictable nature of wireless sensor networks that exhibit fixed topology and fixed application demands for off-line construction of TDMA schedules. In particular, we are able to account for the duty cycling (DC) behavior of the nodes, and hence for the time-varying properties of the underlying communication graph. The novelty lies in pursuing an alternative to a genuinely algorithmic, but notoriously computationally hard, scheduling approach. Specifically, we leverage the fact that the system is virtually a deterministic one and use a pre-simulation technique, detecting when the system has reached steady state, past which point the behavior is essentially periodic. We extract from the pre-simulation a periodic schedule template which can be subsequently used, with minor adjustments, as the TDMA schedule of all nodes in the network. We study the properties of the technique and analyze its performance in example duty-cycled networks.

Keywords: Duty cycling · TDMA scheduling · Time varying graphs

1 Introduction

In this paper we attempt an unorthodox view of constructing TDMA transmission schedules for arbitrary topology networks whose nodes duty cycle (DC) in a periodic fashion. Namely, rather than exploring a purely algorithmic approach we use a pre-simulation technique whereby, assuming all traffic sources are greedy, we expect a periodic behavior to develop at the steady state. We extricate this periodic behavior and, with minor adjustments, construct a template for the transmissions schedule. In short, the thesis of this paper is that in a network with periodic topology changes (as is the case with duty cycling), fed by periodic traffic arrivals, and with deterministic behavior, the system's pattern of transmissions in the steady state ought to be a periodic one. Seeing how the network transmissions occur during one such periodic "cycle" in the steady state is a template which can be repeated continuously in the form of a transmission schedule.

For the purposes of establishing a performance benchmark against which to compare the constructed schedule, we consider a lexicographic maximization formulation of the network throughput problem, i.e., in essence max-min fairness objective. The lexmax is implemented via a classic water-filling scheme. The complete process is as follows: (a) first we use a simple water-filling method to determine the per-flow

© Springer International Publishing Switzerland 2015
S. Papavassiliou and S. Ruehrup (Eds.): ADHOC-NOW 2015, LNCS 9143, pp. 48–61, 2015.
DOI: 10.1007/978-3-319-19662-6_4

rates, (b) the rates determined from the first step guide the generation of a slot-by-slot simulation, and (c) the steady state is detected and the (periodic) template of transmissions is extricated and forms the basis of the TDMA schedule. As long as the duty cycling behavior and the topology of the network is known, the described process can take place off-line and the resulting constructed schedule can be "downloaded" to the nodes for execution.

The process involves a number of non-trivial steps that can conspire to produce a less-than-perfect result. For example, step (b) is a simulation performed with global knowledge of the network topology and the state of the queues of all nodes, yet in order to produce a maximal number of concurrent transmissions (needed to exploit the spatial reuse in a large network) it resorts to using fast approximations to the Maximum Weighted Independent Set (MWIS). The approximation can result in "loss" of throughput compared to what would have been the per-slot optimal. Additionally there exists a problem of discretization of the rate allocation derived from the water-filling algorithm which is typically a rational number but due to the nature of slotted schedule allocation it has to be rounded to a ratio of an integer number of transmissions over the (generally short) schedule length period. We add the fact that not all rates are schedulable by virtue of the underlying interference graph [1] and that even periodically scheduling slots to express different rates at a single node is a hard problem in its own right [2]. To compound the complication, the underlying communication graph is a Time Varying Graph (TVG) [3] and the reader can appreciate why a pre-simulation is, after all, not as outlandish a strategy as it would appear at first. Our work is inspired by theoretical work in queueing systems, demonstrating the periodic behavior of the steady state of networks of queues (roughly corresponding to networks of nodes in our case) as described by Willie [4]. Albeit, our approach is a constructive one and it entails approximation steps because we need to maximize the concurrent transmissions subject to the constraints that links interfere with each other (contrary to the assumption of independent links in classical networks of queues), and that nodes operate in a half-duplex fashion.

Before we introduce the technique, we hasten to add that the chosen benchmark of lexmax allocation is only one of many possible benchmarks. We use it for reasons of its relation to max-min fairness but one could determine the rates based e.g., on sum-rate maximization, or other optimization objective. As long as step (a) produces feasible rate allocations under some performance objective, the subsequent steps (b) and (c) are indifferent to (a)'s strategy for calculating those rates.

The rest of the paper is organized as follows: Sect. 2 describes the network model and related notation, providing also small scale examples, and defining the benchmark performance of lexmax optimization. Section 3 describes the pre-simulation and steady state excision process. Section 4 provides results for regular topologies. Finally Sect. 5 concludes the paper by pointing to shortcomings of the method and possible extensions and improvements.

2 Network Model and Problem Formulation

2.1 Notation and Definitions

A duty-cycled wireless sensor network (DC WSN) is modeled as a set of nodes $\mathcal{N}$, a set of directed links $\mathcal{L}$, and a set of flows $\mathcal{F}$. Each node is characterized by its triplet $\langle \phi_n, \alpha_n, T_n \rangle$. In each triplet in unit of timeslots, T_n is its awake-sleep (ON-OFF) period, ϕ_n is the phase, i.e., the instant within T_n at which the node switches to ON, and α_n is the duration of its awake (ON) period. Note that depending on whether nodes are in the ON or OFF state, the communication link between two nodes may be time-dependent over time because a link "exists" at an instant in time if and only if both its endpoints are in the ON state at that particular time instant. For simplicity of presentation, in the following we assume nodes have the same period T. Time is assumed to be slotted and every packet has length of one slot. Also, for computational convenience, we divide period T into the least number $\mathcal{K}$ of phases $\mathcal{T}^{(k)}$, in unit of slots with $k \in \mathcal{K}$, such that during each of these phases the states (ON/OFF) of nodes are unchanged. Note that our computations only concern phases $\mathcal{T}^{(k)}$, in which there exists at least one active link, i.e., at least one link with both endpoints ON.

Each multi-hop flow $f \in \mathcal{F}$ is defined as $f = (f_s, f_d)$, where $f_s = o(f)$ is the source node and $f_d = d(f)$ is the destination node. We assume that all source rates are greedy in our work. Each directed link $l \in \mathcal{L}$ is specified by a pair of nodes as $l = (l_s, l_d)$, where $l_s = o(l)$ is the origin and $l_d = d(l)$ is the destination of link l. We also assume that each flow traverses a predetermined single-routing path, which is a sequence of directed links/arcs. For the reachability from the source node to the destination node, we also assume that there exist overlaps of the ON intervals between any two nodes defining an arc in a flow's routing path. We note that such overlaps may be brief and hence not all arcs along a routing path of a flow can sustain the same traffic rate. The less the overlap, the more constrained the possible achievable rate by the flow.

Given $\mathcal{F}, \mathcal{L}$ and $\mathcal{N}$ defined earlier, we define r_f to be the rate of flow f and denote $r_f^{(k)}(l)$ the rate of flow f, also called the sub-flow rate of flow f, over an active link l, i.e., both ends are active, during $\mathcal{T}^{(k)}$. Note that with the assumption of single-path routing, each sub-flow is corresponding to one link. For simplicity of notation, a link with both ends being active in a slot is called an ON link in the slot. Otherwise, a link is called an OFF-link in a given slot if at least one of its ends is in the OFF state in the slot. Note that an ON link exists only during the overlap between the ON periods of its two endpoints.

Sub-flows traversing an ON (or OFF) link are then called, respectively, ON (or OFF) sub-flows. One ON sub-flow contends with another ON-sub-flow if the receiver end of the former is within the transmission range of the transmitter of the other (and vice versa). Hence, two flows are contending if one flow has an ON sub-flow contending with another ON sub-flow of the other flow. The relationship of contending sub-flows is characterized by a sub-flow contention graph where vertices are corresponding to sub-flows and an edge between two sub-flows indicates the two are contending. We define a contention domain, in which each sub-flow contends with at least another sub-flow. Hence, there may be multiple contention domains in a sub-flow contention graph. The contention domains correspond to maximal cliques $q \in \mathcal{Q}$ (where $\mathcal{Q}$ is the set of maximal cliques in the contention graph). An implication of the time-varying nature of the graph is that the set of maximal cliques is also time-varying, because the contention graph changes to consist at any point in time of links that not only interfere because of the topology, but are also ON at that point in time.

2.2 Problem Formulation

The following formulation provides a performance benchmark for the rates to be achieved by each flow, given predetermined routing paths. The formulation for DC WSNs we provide is an extension of [5] where it was originally introduced for wireless mesh networks.

$$lexmax_{r_f \in \mathcal{R}} \langle r_f \rangle \tag{1}$$

$$\sum_{k \in \mathcal{K}; o(l)=n} r_f^{(k)}(l) - \sum_{k \in \mathcal{K}; d(l)=n} r_f^{(k)}(l) = \begin{cases} r_f, n = f_s \\ -r_f, n = f_d \\ 0, else \end{cases} \forall n \in \mathcal{N}, f \in \mathcal{F} \tag{2}$$

$$\sum_{f \in \mathcal{F}; l \in \mathcal{L}; r_f^{(k)}(l) \in q} r_f^{(k)}(l) \le r.T^{(k)} \quad \forall k \in \mathcal{K}, \forall q \in \mathcal{Q} \tag{3}$$

$$r_f \ge 0, r_f^{(k)}(l) \ge 0 \quad \forall f \in \mathcal{F}, l \in \mathcal{L}, k \in \mathcal{K} \tag{4}$$

Suppose there are m flows in $\mathcal{F}$, i.e., $f_1, \ldots, f_m \in \mathcal{F}$. We define r_f, (in bold), as a vector of flow rates, i.e., r_f $(r_{f_1}, \ldots, r_{f_m})$. Hence, we have $r_f \in \mathcal{R}$, which is a set of all feasible flow rates, i.e., satisfying constraints (2) to (4), and $\mathcal{R} \subseteq \mathbb{R}^m$, which is the set of all real m-vectors. Let $\langle r_f \rangle = (\langle r_f \rangle_1, \ldots, \langle r_f \rangle_m) r_f$ denote a version of vector $r_f = (r_{f_1}, \ldots, r_{f_m}) \in \mathbb{R}^m$ ordered in the non-decreasing order, i.e., for some permutation φ on the set $\{1, \ldots, m\}$ it holds that $\langle r_f \rangle_j = r_{f\varphi(j)}$ $r_{fj} = r_{f\varphi(j)}$ for $j = 1, \ldots, m$ and $\langle r_f \rangle_1 \le \ldots \le \langle r_f \rangle_m$.

The objective (1) denotes the lexicographic maximization of the sorted outcome vector r_f over $\mathcal{R}$, as defined in [6]. Constraint (2) specifies the flow conservation constraints which require that at any node, n, the rate of each flow summed over all phases is conserved. Constraint (3) specifies, for each phase, the link capacity restrictions that apply to contending sub-flows which are active in that phase. The rate $r = 1/T$ (transmissions/slot) is the maximum rate in a slot and $T^{(k)}$ is the length of phase k in number of slots.

Note that constraint (3) considers each maximal clique ($q \in \mathcal{Q}$) of the contention graph, as defined in each phase k. Those cliques are, in principle, different in each phase due to the time varying nature of the underlying graph. To illustrate the concept, we consider a simple duty-cycled network with four nodes $n \in \mathcal{N} = \{1, 2, 3, 4\}$, six directed wireless links in the link set $\mathcal{L} = \{l_1, l_2, l_3, l_4, l_5, l_6\}$ shown in Fig. 2a, and two opposite flows $f_1, f_2 \in \mathcal{F}$ traversing, respectively, the paths $\{l_1, l_2, l_3\}$ and $\{l_4, l_5, l_6\}$

indicated in Fig. 2b. The DC-configurations, i.e., $\langle \phi_n, \alpha_n, T_n \rangle$, of the nodes is as shown in Fig. 1, in which period T is divided into three phases $T^{(k)}$, in units of slots, with $k \in \mathcal{K} = \{1,2,3\}$, such that in each of these phases the ON/OFF state of the nodes is unchanged. In Fig. 2b, during each phase k, flow f_1 is represented by three sub-flow components $r_{f_1}^{(k)}(l_1)$, $r_{f_1}^{(k)}(l_2)$, and $r_{f_1}^{(k)}(l_3)$, and flow f_2 is represented by three sub-flow components $r_{f_2}^{(k)}(l_4)$, $r_{f_2}^{(k)}(l_5)$ and $r_{f_2}^{(k)}(l_6)$. Hence, the maximal cliques of the sub-flow contention graph (CG) which is made up from all sub-flows competing in each phase $T^{(1)}$, $T^{(2)}$, and $T^{(3)}$ are shown in Fig. 2c1, c.2, and c.3, respectively. Note that due to duty cycling, in each phase $T^{(k)}$, though the communication links of a path may not always exist simultaneously, over time, packets from the source nodes of the two flows can reach their corresponding destination nodes.

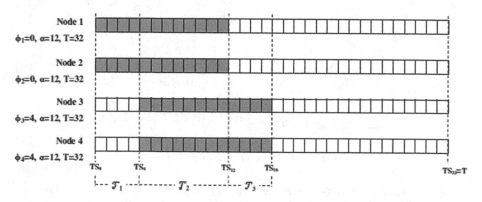

Fig. 1. A four node example of ON/OFF duty cycling with period T (ON slots are gray) (Colour figure online).

The LP problem is solved using a water-filling algorithm to numerically compute the per-flow rates. It is a variation of the standard water-filling algorithm [7]. The result produced is the lexicographically maximized (sorted) outcome vector r_f.

3   Pre-simulation and Periodic Pattern Excision

The network is simulated using greedy traffic sources, i.e., capable to provide as much traffic as could possibly be accommodated by transmission opportunities. The transmissions that can be scheduled to occur simultaneously in the network are determined by invocations of a fast approximation [9] to the Maximum Weighted Independent Set (MWIS) problem. The simulation progresses in a slotted fashion, in each slot, sub-flows with backlog (i.e., non-zero queues) and whose corresponding link is active (both endpoints ON) are assigned a weight to be used by MWIS. The weight assignment at timeslot i tries to capture two aspects of the selection of which sub-flow (and hence link) to schedule over others: (a) sub-flows on link/arc l, of a flow f whose transmissions scheduled so far, denoted by #$_f(i, l)$, lag compared to what should have been

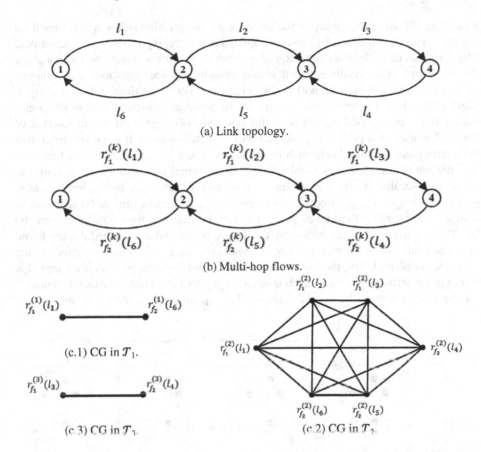

(a) Link topology.

(b) Multi-hop flows.

(c.1) CG in $\mathcal{T}_1$.

(c.3) CG in $\mathcal{T}_3$.

(c.2) CG in $\mathcal{T}_2$.

Fig. 2. The cliques of the contention graphs (CG) corresponding to each phase of Fig. 1.

the analytically derived rate r_f are given higher weight to "catch" up with the rate they are supposed to achieve, and (b) sub-flows whose corresponding link which will be ON for a period of $a(l)$ and the nearest future point at which the link will become OFF (either one of the endpoints entering the OFF state) is time $u(i, l)$ (seen as a function of current time i) are given higher weight the closest we are to the "deadline" of $u(i, l)$. Symbolically, the weight assigned to the arc for the purposes of MWIS execution is:

$$w = \max\{0, ir_f - \#_f(i, l)\} + (a(l) - (u(i, l) - i))$$

where the first component captures the sub-flow's lag and the second is an expression of the proximity to the link's next OFF deadline, increasing (hence becoming more "urgent") the closer the current time is to the deadline.

Without harm to generality and because the DC period is equal to T we anticipate the periodic steady state to develop over periods that are multiples of T (indeed as the experiments will show, in some cases, it is exactly T). We forego the discussion of how it is decided that the steady state has been reached, pointing to relevant literature on the topic, i.e., [8]. After executing in the steady state, the pre-simulation is interrupted (at points equal to a multiple of T) and a simple pattern matching procedure is performed. The decisions made by the simulation of whether a sub-flow transmits or not in a given slot are stored in a two dimensional vector whereby rows are sub-flows and columns are activity (transmission or not) taking place in a particular time slot. The array is illustrated in Fig. 3 where transmissions that occurred are represented as black circles (white if the corresponding sub-flow did not transmit). The most recent interval of length T is used as a prefix template that will be matched to see if it has occurred over the recent past (scanning backwards in time). Let us denote this as the T_{sample} template. The pattern of transmissions within T_{sample} is compared and assuming it is found to repeat periodically, it is evidence that not only during the T_{sample} prefix, but the entire pattern (of length $T_{schedule}$) between successive T_{sample} matches might be repeated. A complete comparison then takes place between the transmissions scheduled over the last $T_{schedule}$ intervals as delineated by the T_{sample} prefix and if the schedules are found in agreement, the most recent (later in simulation time) of them is excised as the schedule template. If not, the simulation continues and the same process is attempted at a later point in time. An exact match is not always possible, hence a mismatch "budget" for the comparison of the two last success $T_{schedule}$ intervals is accounted for.

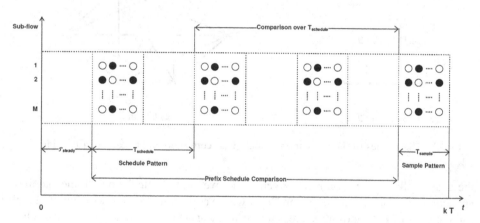

Fig. 3. Illustration of the pre-simulation based schedule excision process.

The careful reader will notice that the periodic pattern excision we perform is only with respect to the pattern of transmissions constructed by the pre-simulation. Clearly, this is a source of approximation because the total state of the system (even if deterministic) is guided by the queueing behavior, i.e., the backlogs at the nodes. Yet, the pattern excision we perform disregards matching the queue backlogs. In other words, the excision we perform is based on partial state matching of the system. As long as the queues are on average at close to zero occupancy (which itself is a sign that the match between analytical rate and simulated rate are approximately equal) we anticipate little impact from ignoring the queue state and the results seem to confirm this view.

A final technicality is that the excised schedule pattern (whose periodic repetition will be the schedule executed by the nodes) does not necessarily obey the flow conservation constraint. For example the excised schedule does not necessarily offer the same number of transmissions for inbound traffic of a flow to a node as it does for outbound traffic of the same flow from that node. This is because the (analytical) flow conservation property pertains to an infinite time horizon but fluctuations (imbalances) are always possible over small intervals of time as long as they are cancelled out in the long run. To address this point we use an ad hoc strategy of equating the inbound and outbound transmission for all sub-flows of a flow either by making it a criterion during the pattern matching process (option A) that has to be satisfied as much as possible or enforcing it after the excision has taken place (option B) by trimming away the transmissions that result in the imbalance (at the loss of some throughput).

4 Evaluation

In order to evaluate the ability of the proposed technique to produce schedules that capture the desired performance objectives, we restrict ourselves to examples of regular topologies and experiment with different per-node phases. There is no restriction to just regular topologies for the application of the proposed scheme, but the reader can follow easily the layout, phase relation, and flow paths on a regular topology.

We consider a 3×3 and a 4×4 topology as shown in Fig. 4. Circular arcs are used to pictorially depict which nodes have the same phases (e.g., in the 3×3 topology nodes 3, 5, and 7 have the same phase). The duty cycle period is 32 slots and the ON intervals are all 12 slots long. Given that the phases can be staggered differently, we explore the impact of staggering by various "phase gaps" (Fig. 4c.1 and c.2) and indicate the phase (counted in slots) at which the corresponding node switches to ON. Additionally, a number of flows are simulated in each configuration following certain patterns groups (PT-1 to PT-3 for 3×3 and PT-4 for 4×4 topology) shown in Fig. 4b.1 to b.4 noting that the routes in each pattern group were computed based on single-shortest-path time-varying routing. The path directions of the four pattern groups are distinctly different; in patterns PT-1 and PT-4, the directions are along the sides of the grid; in pattern PT-2, the directions cross the center of the grid; in pattern PT-3, the directions are parallel with the diagonals of the grid.

Another facet of the experiments is the phase relation between adjacent nodes. We consider three schemes:

Synchronized Phases (SP), in which all the phases are synchronized to start at the same point in time. This is a benchmark value which essentially eliminates the time-varying nature of the communication graph.

Fixed Ladder Phases (FLP), in which any two adjacent nodes in the grid (vertically or horizontally) have their phases staggered 2 slots apart. This scheme is meant to test the impact of flow patterns and their interference on the schedule construction but with a fairly benign impact by the phase differences, i.e., adjacent ON periods overlap significantly over relatively long periods of time.

Varied Ladder Phases (VLP), in which we vary the staggering of the phases from 2 to 10 slots (Fig. 4.c.1 and c.2 for topologies 3×3 and 4×4 respectively), creating increasingly "difficult" short periods over which links are active. Note that when the VLP scheme is used, the number of flows in each flow pattern is kept maximized and

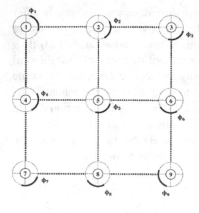

Flow Id	Routing Path
F₁	$1 \rightarrow 2 \rightarrow 3$
F₂	$4 \rightarrow 5 \rightarrow 6$
F₃	$7 \rightarrow 8 \rightarrow 9$
F₄	$1 \rightarrow 4 \rightarrow 7$
F₅	$2 \rightarrow 5 \rightarrow 8$
F₆	$3 \rightarrow 6 \rightarrow 9$

(b.1) Flow pattern PT-1

Flow Id	Routing Path
F₁	$1 \rightarrow 4 \rightarrow 5 \rightarrow 8 \rightarrow 9$
F₂	$4 \rightarrow 5 \rightarrow 6$
F₃	$7 \rightarrow 8 \rightarrow 5 \rightarrow 6 \rightarrow 3$
F₄	$8 \rightarrow 5 \rightarrow 2$

(b.2) Flow pattern PT-2

Flow Id	Routing Path
F₁	$4 \rightarrow 5 \rightarrow 2$
F₂	$7 \rightarrow 8 \rightarrow 5 \rightarrow 6 \rightarrow 3$
F₃	$8 \rightarrow 9 \rightarrow 6$
F₄	$4 \rightarrow 7 \rightarrow 8$
F₅	$1 \rightarrow 4 \rightarrow 5 \rightarrow 8 \rightarrow 9$
F₆	$2 \rightarrow 5 \rightarrow 6$

Phase Gaps	ϕ_1, ϕ_2, ϕ_3 ϕ_4, ϕ_5, ϕ_6 ϕ_7, ϕ_8, ϕ_9
2	0 , 2 , 4 2 , 4 , 6 4 , 6 , 8
4	0 , 4 , 8 4 , 8 , 12 8 , 12 , 16
6	0 , 6 , 12 6 , 12 , 18 12 , 18 , 24
8	0 , 8 , 16 8 , 16 , 24 16 , 24 , 0
10	0 , 10 , 20 10 , 20 , 30 20 , 30 , 8

(a.1) 3x3 Grid topology   (b.3) Flow pattern PT-3   (c.1) Phase scheme for G3x3

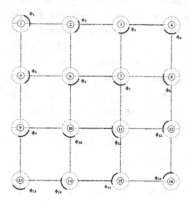

Flow Id	Routing Path
F₁	$1 \rightarrow 2 \rightarrow 3 \rightarrow 4$
F₂	$5 \rightarrow 6 \rightarrow 7 \rightarrow 8$
F₃	$9 \rightarrow 10 \rightarrow 11 \rightarrow 12$
F₄	$13 \rightarrow 14 \rightarrow 15 \rightarrow 16$
F₅	$1 \rightarrow 5 \rightarrow 9 \rightarrow 13$
F₆	$2 \rightarrow 6 \rightarrow 10 \rightarrow 14$
F₇	$3 \rightarrow 7 \rightarrow 11 \rightarrow 15$
F₈	$4 \rightarrow 8 \rightarrow 12 \rightarrow 16$

Phase Gaps	$\phi_1, \phi_2, \phi_3, \phi_4$ $\phi_5, \phi_6, \phi_7, \phi_8$ $\phi_9, \phi_{10}, \phi_{11}, \phi_{12}$ $\phi_{13}, \phi_{14}, \phi_{15}, \phi_{16}$
2	0 , 2 , 4 , 6 2 , 4 , 6 , 8 4 , 6 , 8 , 10 6 , 8 , 10 , 12
4	0 , 4 , 8 , 12 4 , 8 , 12 , 16 8 , 12 , 16 , 20 12 , 16 , 20 , 24
6	0 , 6 , 12 , 18 6 , 12 , 18 , 24 12 , 18 , 24 , 30 18 , 24 , 30 , 4
8	0 , 8 , 16 , 24 8 , 16 , 24 , 0 16 , 24 , 0 , 8 24 , 0 , 8 , 16
10	0 , 10 , 20 , 30 10 , 20 , 30 , 8 20 , 30 , 8 , 18 30 , 8 , 18 , 28

(a.2) 4x4 Grid topology   (b.4) Flow pattern PT-4   (c.2) Phase scheme for G4x4

Fig. 4. Topologies, flows and phases used in the simulations.

unchanged, i.e., in PT-1, PT-2, PT-3 and PT-4 the number of flows are 6, 4, 6 and 8, respectively.

In all cases, and as a matter of convention, instead of denoting the specific flows that comprise a certain mix, we create larger sets of flows by adding more flows in their numerical order. That is, sets of 1, 2, 3, etc. flows comprise correspondingly of the flows {F1}, {F1, F2}, {F1, F2, F3}, etc. Finally, note that because of the time varying nature of the underlying communication graph, the set of parameters used here that guides the duty cycling, generates a great deal of actual link topologies.

Representative results are summarized in Figs. 5 (for PT-1 and PT-2) and 6 (for PT-3 and PT-4). The **#F or Gap** is the number of flows for phase schemes SP and FLP, or the stagger gap for phase scheme VLP. **Pre-Sim** is the total throughput of the constructed excised periodic schedule. **Numerical** is the total throughput as derived from the water filling algorithm. **Fairness** is a fairness index between what should be the achieved rates (as per water-filling) and what is achieved by the excised schedule. **Tschedule/T** expresses how many multiples of the duty cycle period is the length of the excised periodic schedule. **Mismatch** captures the fraction as percentage of sub-flows who did not match exactly during the prefix comparison. **Err** captures the fraction of sub-flows that did not match exactly during the complete schedule length comparison. The note field indicates whether Option A or Option B for correcting the flow balance was necessary in the corresponding case and which of the two options produced the best results.

The virtually identical numbers in the Pre-Sim and Numerical columns demonstrate that the process of schedule excision is a reasonable approach to produce accurate schedules. Despite the fact that a key component of the pre-simulation is the invocation of an approximation to MWIS [9] the fact is that a schedule is constructed via multiple (one per slot) invocation of MWIS, hence what matters is less the worst-case behavior of the approximation and more the average performance.

Leaving MWIS aside, the process of matching the T_{sample} prefix cannot be expected to lead to perfect matches. For this reason, we allowed a maximum of mismatch (seen as difference in transmissions scheduled for a sub-flow between two compared patterns) of 10 % which turned out to be very pessimistic, i.e., we did not reach that degree of mismatch. Given that the prefix is only a fraction of the possible eventual schedule, the mismatch is generally larger (denoted by Err in our performance tables) when comparing complete $T_{schedule}$ periods but again not as large as we had originally anticipated.

Another aspect of the results is that the produced schedule of duration $T_{schedule}$ is a small multiple of T (the DC period). The less loaded the network with flows, the smaller this multiple tends to be, but there are exceptions. Fundamentally, the relation $T_{schedule}/T$ reveals the impact of the queue state. That is, the queues, even though they tend to behave periodically, exhibit dynamics that develop across multiple T periods, as traffic possibly enters (inflating the queue) in one period to be delivered (not necessarily completely) to the next hop in the next period.

Across the board, we note that the first set of patterns PT-1 and PT-2 (Fig. 5) rarely resulted in excised schedule templates that did not exhibit flow balance. The flow imbalance (and hence the need for corrective action) was more prevalent in PT-3 and PT-4. Moreover when flow balance correction had to be applied, it was always for schedules that were larger than a single T period, i.e., who had some queueing dynamics influencing successive T periods. We conjecture from what we have seen so far that the problem lies with flows that, as per the numerical findings, ought to have been given a tiny fraction of the data rate. Hence situations in which one such flow is scheduled (during the simulation) for a single transmission for one of its sub-flows over a long run of time slots, is a flow that will likely demonstrate a mismatch

Pattern	#F or Gap	Pre-Sim	Numerical	Fairness	Tschedule/T	Mismatch	Err	Scheme	Notes
PT-1	1	0.1875	0.1875	1	1	0	0	SP	
PT-1	2	0.1875	0.1875	1	1	0	0	SP	
PT-1	3	0.28125	0.28125	1	1	0	0	SP	
PT-1	4	0.28125	0.28125	1	4	0	0	SP	
PT-1	5	0.234375	0.234375	1	2	0	0	SP	
PT-1	6	0.28125	0.28125	1	2	0	0	SP	
PT-1	1	0.1875	0.1875	1	1	0	0	FLP	
PT-1	2	0.21875	0.21875	1	1	0	0	FLP	
PT-1	3	0.328125	0.328125	1	2	0	0	FLP	
PT-1	4	0.3359375	0.34375	0.99	4	0	0	FLP	Option B
PT-1	5	0.3125	0.3125	1	1	0	0	FLP	
PT-1	6	0.328125	0.328125	1	4	0	0	FLP	
PT-1	2	0.328125	0.328125	1	4	0	0	VLP	
PT-1	4	0.375	0.375	1	1	0	0	VLP	
PT-1	6	0.375	0.375	1	1	0	0	VLP	
PT-1	8	0.234375	0.234375	1	4	0	0	VLP	
PT-1	10	0.09375	0.09375	1	2	0	0	VLP	
PT-2	1	0.125	0.125	1	1	0	0	SP	
PT-2	2	0.140625	0.140625	1	4	0	0	SP	
PT-2	3	0.140625	0.140625	1	2	0	0	SP	
PT-2	4	0.125	0.125	1	1	0	0	SP	
PT-2	1	0.140625	0.140625	1	2	0	0	FLP	
PT-2	2	0.171875	0.171875	1	4	0	0	FLP	
PT-2	3	0.140625	0.140625	1	2	0	0	FLP	
PT-2	4	0.15625	0.15625	1	4	0	0	FLP	
PT-2	2	0.15625	0.15625	1	4	0	0	VLP	
PT-2	4	0.15625	0.15625	1	4	0	0	VLP	
PT-2	6	0.125	0.125	1	1	0	0	VLP	
PT-2	8	0.0625	0.0625	1	2	0	0	VLP	
PT-2	10	0.03125	0.03125	1	4	0	0	VLP	

Fig. 5. Simulation results with PT-1 and PT-2.

during the excision process, or will exhibit a flow imbalance in the excised schedule. Indirect evidence to this end is the fact that when the flow balance corrections (options A or B) were taken, after removing the flow imbalance, certain flows ended up being (unfairly) victimized, i.e. losing the little allocation they had. Also, the less the overlap (the larger the phase stagger between nodes) the links remain active only briefly, resulting in smaller rates assigned to flows traversing them, as they become more "bottlenecked" leading to smaller (and hence problematic to schedule as just mentioned) rate allocations.

Pattern	#F or Gap	Pre-Sim	Numerical	Fairness	$T_{schedule}/T$	Mismatch	Err	Scheme	Notes
PT-3	1	0.1875	0.1875	1	1	0	0	SP	
PT-3	2	0.140625	0.140625	1	4	0	0	SP	
PT-3	3	0.140625	0.140625	1	2	0	0	SP	
PT-3	4	0.1875	0.1875	1	2	0	0	SP	
PT-3	5	0.15625	0.15625	1	1	0	0	SP	
PT-3	6	0.140625	0.140625	1	4	0	0	SP	
PT-3	1	0.15625	0.15625	1	1	0	0	FLP	
PT-3	2	0.140625	0.140625	1	4	0	0	FLP	
PT-3	3	0.1875	0.1875	1	1	0	0	FLP	
PT-3	4	0.1953125	0.1953125	1	4	0	0	FLP	
PT-3	5	0.1875	0.1953125	1	5	0	0	FLP	Option A
PT-3	6	0.1875	0.1875	1	1	0	0	FLP	
PT-3	2	0.1875	0.1875	1	1	0	0	VLP	
PT-3	4	0.1953125	0.1953125	1	4	0	0	VLP	
PT-3	6	0.234375	0.25	0.98	2	0	0	VLP	Option A
PT-3	8	0.125	0.140625	0.96	2	0	0	VLP	Option A
PT-3	10	0.0625	0.0703125	0.96	4	0	0	VLP	Option A
PT-4	1	0.125	0.125	1	1	0	0	SP	
PT-4	2	0.1875	0.1875	1	1	0	0	SP	
PT-4	3	0.28125	0.28125	1	1	0	0	SP	
PT-4	4	0.375	0.375	1	1	0	0	SP	
PT-4	5	0.3515625	0.3515625	1	4	0	0	SP	
PT-4	6	0.284375	0.328125	0.97	10	0	0.03	SP	Option B
PT-4	7	0.328125	0.328125	1	2	0	0	SP	
PT-4	8	0.375	0.375	1	2	0	0	SP	
PT-4	1	0.140625	0.140625	1	2	0	0	FLP	
PT-4	2	0.21875	0.21875	1	2	0	0	FLP	
PT-4	3	0.28125	0.28125	1	1	0	0	FLP	
PT-4	4	0.375	0.375	1	1	0	0	FLP	
PT-4	5	0.3515625	0.4296875	0.98	4	0	1.14	FLP	Option B
PT-4	6	0.3515625	0.375	0.96	4	0	0	FLP	Option B
PT-4	7	0.328125	0.328125	1	2	0	0	FLP	
PT-4	8	0.375	0.375	1	4	0	0	FLP	
PT-4	2	0.375	0.375	1	4	0	0	VLP	
PT-4	4	0.375	0.375	1	2	0	0	VLP	
PT-4	6	0.5	0.5	1	1	0	0	VLP	
PT-4	8	0.3125	0.3125	1	4	0	0	VLP	
PT-4	10	0.125	0.125	1	2	0	0	VLP	

Fig. 6. Simulation results with PT-3 and PT-4.

5 Conclusions

It is widely accepted that wireless medium scheduling is a hard problem even if several simplifications are performed. In this study we opted for situations where the topology is fixed and known but the actual link dynamics are dependent on the periodic duty cycling of the nodes. The time-varying nature of the underlying communication graph compounds the complexity of determining a single periodic TDMA-like schedule. We attempted to solve the scheduling problem by performing an off-line pre-simulation of the system for which we have strong evidence that, due to the periodicity of all input factors, a similar periodicity should be exhibited by the steady state. Part of this periodic steady state behavior is the transmission decisions. We excise the transmissions scheduled during the simulation to create a template which can then be subsequently downloaded to the nodes.

We evaluated our approach based on a number of regular topologies and duty cycling behaviors. Even though we restricted the evaluation to homogeneous (same duty cycling period) nodes, the process outlined here can be extended to situations with different duty cycle characteristics albeit at the cost of computing the combined (based on the least common multiple of the different cycles) repeated pattern across all nodes. Furthermore, our technique does not need to comply to rate allocation derived under lexmax/maxmin criteria. Other objectives can be used instead. The rate allocations, regardless of how they are calculated, as long as they are feasible, simply act as a target for the pre-simulation process.

The most evident shortcoming of the excision process is that it ignores the state of the queues whose influence is only accounted for indirectly by excising scheduling patterns that are multiples of the duty cycle period. The reader can question whether the state of queues ought to have been a component of the state comparison/matching. We have, so far, avoided matching the queue state in addition to the transmissions because it is unclear whether matching the exact number of packets in the queue is necessary (or just that there are some) and for the added complexity that this would bring about (the buffers can be as many as the flows times the number of links in the network). Future work will elaborate the inter-dependency between queues and transmissions and how this is manifested in the schedule template excision. A fascinating extension is the inclusion of network coding (and scheduling when to code and when not) which we have found in the past [10] to be a means to improve the throughput when, due to duty cycling, throughput is "lost" because nodes are in OFF state.

References

1. Zuyuan, F., Bensaou, B.: Fair bandwidth sharing algorithms based on game theory frameworks for wireless ad-hoc networks. In: INFOCOM 2004, pp. 1284–1295 (2004)
2. Bar-Noy, A., Dreizin, V., Patt-Shamir, B.: Efficient algorithms for periodic scheduling. Comput. Netw. **45**(2), 155–173 (2004)
3. Casteigts, A., Flocchini, P., Quattrociocchi, W., Santoro, N.: Time-varying graphs and dynamic networks. In: Frey, H., Li, X., Ruehrup, S. (eds.) ADHOC-NOW 2011. LNCS, vol. 6811, pp. 346–359. Springer, Heidelberg (2011)

4. Willie, H.: Periodic steady state of loss systems with periodic inputs. Adv. Appl. Probab. **30** (1), 152–166 (1998)
5. Shabdanov, S., et al.: Joint routing, scheduling, and network coding for wireless multihop networks. In: WiOpt 2011, pp. 33–40 (2011)
6. Nace, D., Pioro, M.: Max-min fairness and its applications to routing and load-balancing in communication networks: a tutorial. IEEE Comm. Surv. tutor. **10**(4), 5–17 (2008)
7. Bertsekas, D., Gallager, R.: Data Networks, 2nd edn. Prentice Hall, Englewood Cliffs (1992)
8. Eickhoff, M., McNickle, D., Pawlikowski, K.: Detecting the duration of initial transient in steady state simulation of arbitrary performance measures. In: ValueTools 2007 (2007). Article #42
9. Sakai, S., Togasaki, M., Yamazaki, K.: A note on greedy algorithm for the maximum weighted independent set problem. Discret. Appl. Math. **126**(2–3), 313–322 (2003)
10. Ho, V., Nikolaidis, I.: Tradeoffs of combining network coding and duty cycling in WSNs. In: CNSR 2011, pp. 231–238 (2011)

On the Impact of Network Evolution on NUM Resource Allocation Problems in Wireless Multihop Networks

Eleni Stai[✉], Vasileios Karyotis, and Symeon Papavassiliou

School of Electrical and Computer Engineering,
National Technical University of Athens (NTUA),
Zografou, 15780 Athens, Greece
{estai,vassilis}@netmode.ntua.gr, papavass@mail.ntua.gr

Abstract. Network churn is one of the primary reasons for network topology evolution, creating a dynamic environment that all designed and developed mechanisms have to cope with. In this work, we focus on studying the impact of network evolution on the resource allocation mechanisms in wireless multihop networks, and thus set the scene for more pragmatic network optimization. More specifically, we study the impact of topology evolution in the form of node and edge churn on the Network Utility Maximization (NUM) mechanism, where the latter is employed as a form of intelligent resource allocation framework and implementation. Our study in this paper serves as a stepping stone for a more realistic consideration of cross-layer design, capable of coping with the dynamic nature of the evolutionary changes that take place in each network inevitably. This work aspires to stimulate the interest and further research on improving and developing network optimization and control mechanisms under more realistic operational conditions.

Keywords: Network Utility Maximization · Topology evolution · Network churn · Cross-layer design · Solution optimality and convergence

1 Introduction

Topology evolution is an inevitable process in communications networks, mainly due to user habits and environmental factors. Users enter and leave the network based on their work and social needs, which physically corresponds to nodes/edges added/deleted in the topology. In specific types of networks, e.g. wireless ad hoc, environmental factors might hamper the quality of wireless links, thus leading again to node/edge churn, namely addition or deletion of edges and potentially disconnecting the corresponding nodes they link, i.e. node churn, as well.

Churn-based network modification has been considered in the past, mainly from the perspective of topology evolution and the impact it might have on the network type itself [1]. Nevertheless, it has not been considered and taken into

© Springer International Publishing Switzerland 2015
S. Papavassiliou and S. Ruehrup (Eds.): ADHOC-NOW 2015, LNCS 9143, pp. 62–75, 2015.
DOI:10.1007/978-3-319-19662-6_5

account in the design and the actual operation of a network. For instance, the impact of churn on resource allocation mechanisms that have been developed for static networks, in principle, has not been analyzed in the past. The conditions under which the resource allocation mechanisms are affected or not by network churn should be studied, since under churn, the operational environment and thus the resource allocation problem is modified while the corresponding mechanisms compute/assign the resources.

The problem can become even more complicated when resource allocation relies on cross-layer techniques for ensuring more efficient operation and resource utilization. This is especially the case in wireless multihop networks, where designers are often forced to break the barriers across protocol layers and exploit cross-layer information for more efficient allocation of resources across the layers as well. Thus, especially for these types of networks, the impact of network evolution on cross-layer resource allocation mechanisms should be studied and taken into account in the design of future wireless multihop networks.

In this paper we will address exactly the aforementioned concerns for the corresponding wireless distributed networks and cross-layer approaches. More specifically, we will study the impact of network topology evolution in the convergence behavior of the cross-layer algorithms for resource allocation in wireless multihop networks. We adopt Network Utility Maximization (NUM) because it allows capturing the cross-layer features of wireless multihop networks in a holistic framework, which in principle allows obtaining optimal solutions in a feasible manner. However, until now such problems were developed for fixed topologies, namely networks where the nodes and edges remain fixed as well. In this paper, we will study the behavior of these schemes when network churn dictates the evolution of the network.

More precisely, in this work, we will study the convergence behavior of the NUM-based cross-layer network operation with respect to achieving and retaining the optimal source rates, while the set of relay nodes within the network topology evolves and/or the set of the interconnections among all nodes varies. The set of source-destination pairs is maintained throughout the examined time-period of the network's operation. Therefore, during the network operation, the available paths connecting the source-destination pairs vary in number and length (in number of hops). This path variation is a realistic fact in wireless multihop networks where nodes and links may be rendered periodically unavailable due to channel conditions, battery reserves, etc. We conclude that under the examined modes of network evolution, the convergence behavior of the NUM-based cross-layer scheme for the network operation is not significantly affected, allowing for efficient deployment of such NUM-based solutions for the network operation, even under source-destination connection paths' dynamically changing conditions, caused due to environmental factors.

The structure of the rest of the paper is as follows. Section 2 reviews relevant works available in the literature and distinguishes our contribution from them, while Sect. 3 explains the considered system model and assumptions. Section 4 presents and explains the studied NUM problem, while Sect. 5 describes the network churn modeling framework. Section 6 provides extensive results from our numerical evaluations, and finally, Sect. 7 summarizes the obtained outcomes and discusses future work.

2   Related Work

Evolving networks have been studied, especially lately, under the auspices of Network Science [1]. Scale-free networks are characteristic examples of topologies that emerged as evolving graphs on the basis of preferential attachment growth [1]. Network churn may also arise due to nodes that may be removed with some probability, or nodes that may join the network. Additionally, existing links may be destroyed and new links may be created [1,2]. Contrary to all these works that focus on the evolution of the topology of the network, our work will focus on the impact of this topological evolution on the protocols and operation of the network.

Especially we will be concerned with resource allocation mechanisms and more specifically with cross-layer techniques in wireless multihop networks. Multiple works have focused on the improvement of the operation of these networks by exploiting cross-layer approaches to determine the optimal operational points, e.g. [2–9], while in this paper we will focus on the impact of topology evolution on these types of mechanisms and their behavior.

Among all the available cross-layer approaches, we will be especially concerned with Network Utility Maximization (NUM) techniques, targeted for wireless multihop networks [3–9]. The concept behind these techniques lies in the formulation of an optimization problem targeting at the maximization of a utility function that expresses the satisfaction of the users of the network by the assigned resources over the layers. This strategy can be applied in all types of communications networks that follow a layered protocol stack approach, as can be found in [10] and references therein. But especially in evolving multihop networks, the environment is changing fast compared to the algorithms' convergence and therefore, it is required to investigate the impact of evolution over several NUM problem formulations and study the suitability of the latter for those networks as well.

Even though the literature regarding NUM is vast, not many works have addressed the impact of network evolution on NUM problem formulation and solution, and especially in wireless multihop networks. Some scattered approaches have addressed more specific aspects of the impact of network evolution on NUM problems, e.g. what the change of flows (i.e. source-destination pairs) implies for the NUM formulation, obtained solutions, etc. [11,12]. This work aspires to perform a study that can serve as a stepping stone for further analysis of the impact of network evolution on NUM and in general on cross-layer resource allocation problems in wireless networks.

3   System Model and Assumptions

We consider a wireless multihop network represented by a random geometric graph, with N nodes and E edges. Time is assumed slotted and at each time slot, t, decisions for routing, scheduling and congestion control are taken by the nodes. The scheduling/routing algorithm determines which set of non-interfering links is going to transmit and which flows will be served by each link, while the congestion control determines the new arrival packet rate at each source-node.

The packets generated exogenously are assumed to arrive at source-node i with rate λ_i^d, where $\lambda_i^d \neq 0$ only if i is a source of packets for destination d. The latter fact is denoted as $i \in S_r(d)$, where $S_r(d)$ stands for the set of sources for node d. λ_i^d is re-adapted by the congestion control algorithm at each time slot t and for this reason the arrival packet rates are time-dependent ($\lambda_i^d(t)$). We denote with $\mu_{ij}(t)$ the maximum possible communication traffic on the link (i,j) at time t, where $\mu_{ij}(t) = \mu_{max} > 0$ if the link (i,j) is scheduled at time slot t, while otherwise $\mu_{ij}(t) = 0$. Also, we denote with $\mu_{ij}^d(t)$ the maximum possible communication traffic on the link (i,j) for destination d at time t. The matrix, $M(t)$, expresses collectively the maximum communication traffic for all the links at time slot t. Each source-node i associates its satisfaction for each one of its external flow rates λ_i^d, $\forall d : i \in S_r(d)$, with a corresponding strictly concave and increasing utility function denoted as $U_i^d(\lambda_i^d)$.

We use the term I_S to refer to the finite set of communication traffic matrices of all possible independent sets of the graph, i.e. sets of links that do not interfere with each other. Let us denote as I^1, I^2, ..., I^m the independent sets included in I_S, among which the scheduler chooses one, $I(t)$, to transmit at slot t. If the link $(i,j) \in I(t)$, then $\mu_{ij}(t) = \mu_{max}$. If π_1, π_2, ..., π_m is the fraction of time that each one of I^1, I^2, ..., I^m is scheduled ($\sum_l \pi_l = 1$), then we define the rate of service of the link (i,j) as $\mu_{ij} = \sum_{l:(i,j) \in I^l} \mu_{max} \pi_l$. The arrival and service rates are considered bounded, i.e. $0 < \lambda_{min} \leq \lambda_i^d(t) \leq \lambda_{max}$, $\forall i,d : i \in S_r(d), t$, $0 \leq \mu_{min} \leq \mu_{ij} \leq \mu_{max}$, $\forall i,j$.

The capacity region [7], Λ_G, is the set of all arrival rate matrices $[\lambda_i^d]$ with $\lambda_i^d \neq 0$ only if $i \in S_r(d)$, such that there exists a set of service rates $\{\mu_{ij}^d\}_{\forall (i,j),d}$ satisfying the following constraints:

- Efficiency constraints: $\mu_{ij}^d \geq 0$, $\mu_{ii}^d = 0$, $\mu_{dj}^d = 0$, $\sum_d \mu_{ij}^d \leq \mu_{ij}$, $\mu_{ij} = \sum_{l:(i,j) \in I^l} \mu_{max} \pi_l$, $\sum_l \pi_l = 1$, $\pi_l \geq 0$, $\forall l,i,d,j$.
- Flow constraints: $\lambda_i^d + \sum_j \mu_{ji}^d \leq \sum_j \mu_{ij}^d$, $\forall i,d : i \neq d$.

When there are specific routing constraints to be taken into consideration in the network's operation, these are included in the definition of the capacity region. We use the notation $\lambda_i^d \in \Lambda_G$, $\forall i,d : i \neq d$, to refer to the source rates lying inside the capacity region Λ_G. However, the flow constraints will be also specified separately. Note that the capacity region refers to a specific network topology, i.e. set of nodes and their locations and set of links, i.e. changes in the network's topology lead to a different capacity region and thus a different set of supportable input (source) rates by the network.

4   NUM Problem Formulation and Algorithmic Solution

In the NUM formulation, the network is seen as an "optimizer"of its users' satisfaction with respect to their appropriately defined elastic Quality of Service (QoS) demands. The users' satisfaction over the network's operation is expressed via their utility function, defined over their appropriate elastic QoS demands, i.e. the source rates. In the latter case, only elastic flows, i.e. flows with adjustable input rates, can be easily treated via the NUM framework so that their optimal source rates for the maximum attained network utility are obtained. From the above discussion, the basic, general NUM problem, denoted as **NUM$_G$**, takes the following form:

$$\max_{\Theta} \sum_{i,d:i\in S_r(d)} U_i^d(\lambda_i^d)$$

$$s.\, t. : \text{"Constraint Set"} \tag{1}$$

where Θ is the set of parameters or resources over which the maximization takes place. According to the above, the NUM problem formulation can be appropriately tuned, mainly via the constraint set, so as to realize a large diversity of resource allocation scenarios that span the protocol stack. A NUM problem, based on **NUM$_G$**, is formulated by defining the users' utilities and the constraint set of Eq. (1). In the sequel, primal and/or dual decomposition techniques are used to vertically modularize and potentially distribute horizontally the solution algorithm [10]. Further, proofs of stability, optimality and fairness (if applied) should follow. Also, it is important to note that in the NUM problem formulations, the deterministic fluid model approximation is used, meaning that all the variables correspond to mean values [10].

We examine a specific case of the **NUM$_G$** problem that targets at the cross-layer design and operation of the wireless multihop networks and it is formulated as follows (based on the terminology of Sect. 3):

$$\max F(\Lambda) = \sum_{i,d:i\in S_r(d)} U_i^d(\lambda_i^d)$$

$$s.\, t. : \quad \lambda_i^d + \sum_j \mu_{ji}^d \le \sum_j \mu_{ij}^d, \ \forall i,d : i \neq d \tag{2}$$

$$\lambda_i^d \in \Lambda_G, \ \forall i,d : i \neq d \tag{3}$$

i.e. the constraint set is constructed based on the definition of the capacity region of Sect. 3. This is the most common NUM problem applied in wireless multihop networks since it leads to algorithms performing congestion control, scheduling and routing. Other definitions of the constraint set (Eq. (1)) of the **NUM$_G$** problem (e.g. [4]) may assume predetermined routing and/or fixed scheduling, which however are more suitable for wired networks.

The solution algorithm for the latter problem has been studied extensively in literature [3,5–8]. We define a Lagrange multiplier, $q_i^d \ge 0$, for each flow

---

Algorithm 1. Cross-layer NUM problem solution based network operation at time slot t [6]

---

1 **for** *each directed link* (i,j) **do**
2   **for** *each destination* d **do**
3   *%Lagrange Multiplier Difference*
4        $P_{ij}^d(t) = (q_i^d(t) - q_j^d(t));$
5   *%Define the weight* $P_{ij}(t)$ *as follows* :
6     $P_{ij}(t) = \max_d P_{ij}^d(t);$
7     $d^*(i,j) = \arg\max_d P_{ij}^d(t);$
8 *%Choose the traffic matrix through the maximization* :
9 $M(t) = \arg_{M' \in I_S} \max \sum_{(i,j)} \mu_{ij}' P_{ij}(t);$
10 **for** *each directed link* (i,j) **do**
11   **if** $\mu_{ij}(t) > 0$ **then**
12   *the link* (i,j) *serves* $d^*(i,j)$ *with* $\mu_{ij}^{d^*}(t) = \mu_{ij}(t);$
13   *For* $d \neq d^*(i,j)$ *we set* $\mu_{ij}^d(t) = 0;$
14 *%Congestion Control & Lagrange Multipliers' Update*
15 **for** *each node* i **do**
16   **for** *each destination* d **do**
17   **if** i *is a source node for* d **then**
18   *% Renewal of the arrival data rates via the Dual Controller*
19           $\lambda_i^d(t+1) = \left\{ U_i^{d'-1}\left(\frac{q_i^d(t)}{K}\right) \right\}_{\lambda_{\min}}^{\lambda_{\max}};$
20   *% Renewal of the Lagrange multipliers*
21        $q_i^d(t+1) = \left\{ q_i^d(t) + \lambda_i^d(t) - \sum_j \mu_{ij}^d(t) + \sum_j \mu_{ji}^d(t) \right\}^+;$

---

conservation constraint of the NUM problem in Eq. (2), i.e. for each pair i, d : $i \neq d$, with value equal to $q_i^d(t) \geq 0$ at each time slot t (which coincides with the iteration number of the subgradient method applied for solving the dual of the NUM problem [6]). Then, the algorithmic cross-layer solution of the NUM problem for wireless multihop networks is as follows [6]:

In the dual decomposition, the Lagrange multipliers are updated following a subgradient approach, and the source rates are optimally determined at each iteration (time slot) via direct differentiation of the Lagrangian with respect to them. In Algorithm 1, $U_i^{d'-1}$ stands for the inverse of the derivative of the utility function and $\{\}_a^b$, $\{\}^+$ denote the projection to the sets $[a, b]$, $[0, +\infty)$ respectively. The dual controller ensures network stability and optimality with respect to the achieved source rates, although the latter depends on the chosen value of the constant $K > 0$. Higher values of K achieve source rates' values closer to the optimal ones [8,9]. We will examine the convergence behavior of Algorithm 1 under network churn as the latter is described in the following section.

5   Network Churn and Topology Evolution

As mentioned above, topology evolution is a norm for most networks of interest, and especially wireless communications ones, where users enter and leave
the network (node churn) and channel quality dictates if a communication link
can be used in practice or not (on-off channel model), nodes increase/decrease
their transmission power or add/delete/turn directional antennas (edge churn).
Network churn, consisting of node and edge churn, causes network topologies to
evolve and thus, we assume the most general case of network churn taking place
in wireless distributed networks.

The complete topology modification evolutionary process consists of four
partial processes associated with edge churn (edge addition and deletion) or
node churn (node addition and deletion), as described in Chapter 6 of [1]. We
assume that at each time step, only one of these processes takes place and they
may be formally described as:

- **Edge Addition:** With probability p, $0 \leq p \leq 1$, we add m_1 ($m_1 < N$) new
  links, one to each of the m_1 randomly selected nodes. The selected nodes
  increase their transmission range to a value R^{up} and add one point-to-point
  directed connection to a randomly selected node within this new range, while
  the latter connection can be deployed via (turning) a directional antenna
  attached to the first node.
- **Edge Deletion:** With probability r, $0 \leq r \leq 1$, we delete directionally
  m_2 ($m_2 < N$) randomly selected edges, where the edge deletion is realized in
  routing level or via turning directional antennas.
- **Node Addition:** With probability w, $0 \leq w \leq 1$, we add M_a new nodes in
  the network, close to randomly chosen existing nodes.
- **Node Deletion:** With probability v, $0 \leq v \leq 1$, M_d randomly selected nodes
  of the current network instance are deleted.

Given all the considered churn processes it should hold $p+r+w+v \leq 1$, i.e. the
vector $[p \ r \ w \ v \ 1 - p - r - w - v]^T$ is a probability distribution over the above
processes.

6   Numerical Results and Discussion

In this section, we study if and how network churn, consisting of relay nodes'
churn and edge churn, will affect the convergence behavior of Algorithm 1. Moreover, several emergent impacts on capacity region due to node and edge churn
are identified and discussed.

6.1   Design of Experiments

We consider a random geometric graph of $N = 200$ nodes each having a communication radius equal to $6 \, \text{m}$ over a square deployment area of side $100 \, \text{m}$. The
number of edges in the initial network instance is equal to $E = 1204$. We assume

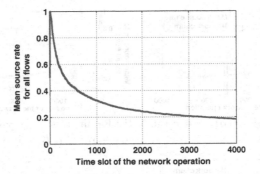

Fig. 1. Convergence behavior of the Algorithm 1 without network churn. Parameters: $p_{NC} = 0$ (no network churn), $K = 10$.

that one-hop neighboring nodes cannot transmit simultaneously (one-hop inter-ference model). In the numerical evaluations, we apply the Greedy Maximal Scheduling [13] instead of solving the NP-hard maximum weight matching prob-lem of line 9 in Algorithm 1 for computational efficiency. Note that I_S becomes also time varying as it is determined based on the changing network topology. There exist 50 flows in the network between randomly chosen pairs that are not direct neighbors. For the edge addition process, $R^{up} = 50$ m.

For the experiments that follow, Algorithm 1 is applied for 4000 time slots. At each time slot, except from applying Algorithm 1, we decide if we will perform network churn as described in Sect. 5, based on a probability denoted as p_{NC}. Thus, at each time slot, t, edge addition takes place with probability $p_{NC} \cdot p$, edge deletion takes place with probability $p_{NC} \cdot r$, node addition takes place with probability $p_{NC} \cdot w$ and finally node deletion takes place with probability $p_{NC} \cdot v$. The specific values of these probabilities and other parameters are given in the legends of the provided figures. Note that, node churn concerns only relay nodes, while the set of source-destination pairs remains invariant. In this case, the constraint set of the NUM problem in Eq. (2) evolves following the evolution of the network topology, while the summands of the objective function remain invariant. The choice of nodes and links for the network churn is performed under the restriction that the network will not be rendered disconnected.

Figure 1 depicts the convergence behavior of Algorithm 1, with respect to the mean source rate for all the flows, without network churn. This figure will serve as a benchmark behavior and will be compared with cases where node churn intervenes within the network operation.

6.2   Case of Node Churn

In this section, we evaluate the convergence behavior of Algorithm 1, when each one of the iterations corresponds to a time slot of the network operation and node churn takes place in the network topology. Figure 2 presents the corresponding

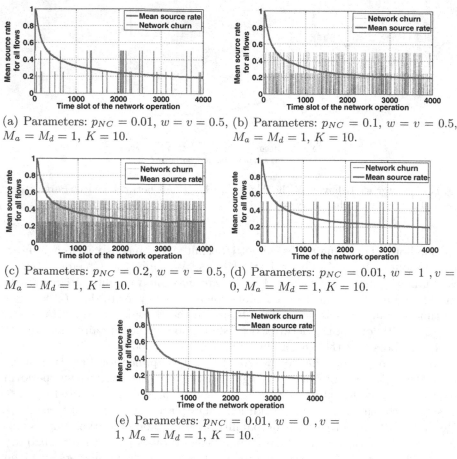

(a) Parameters: $p_{NC} = 0.01$, $w = v = 0.5$, $M_a = M_d = 1$, $K = 10$.

(b) Parameters: $p_{NC} = 0.1$, $w = v = 0.5$, $M_a = M_d = 1$, $K = 10$.

(c) Parameters: $p_{NC} = 0.2$, $w = v = 0.5$, $M_a = M_d = 1$, $K = 10$.

(d) Parameters: $p_{NC} = 0.01$, $w = 1$, $v = 0$, $M_a = M_d = 1$, $K = 10$.

(e) Parameters: $p_{NC} = 0.01$, $w = 0$, $v = 1$, $M_a = M_d = 1$, $K = 10$.

Fig. 2. Convergence behavior of Algorithm 1 under node churn (node addition and node deletion) (Color figure online).

results for diverse probabilities of network churn and parameters of node addition and deletion processes (Sect. 5).

Initially, Figs. 2(a), (b), (c) compare the convergence behavior of Algorithm 1 with respect to the mean source rate for all the flows when the expected number of nodes in the network topology remains invariant, i.e. the node addition/deletion processes take place with the same probability ($w = v = 0.5$) and the same number of nodes is added/deleted ($M_a = M_d = 1$) respectively. The vertical red lines indicate a node addition process for value equal to 0.5 and a node deletion one for value equal to 0.25. Since the expected total number of relay nodes remains invariant, node churn leads to a rearrangement of their deployment in the network area, a fact that leads to a change in the set of network links and can also affect the network's capacity region. Comparing Figs. 2(a), (b), (c) with Fig. 1, we observe that node churn does not impact the convergence behavior of Algorithm 1, although the value towards which it converges is slightly modified. In more detail, for the specific experiments, as the probability p_{NC} increases from 0.01 (Fig. 2(a)) to 0.1 (Fig. 2(b)) and then to 0.2 (Fig. 2(c)), the capacity region of the network seems to slightly increase as the final mean source rate value to which the algorithm converges slightly increases. This is mainly due to the fact that with higher probability new nodes are added close to the core of the network topology (since they are placed close to already existing nodes, as described in the node addition process in Sect. 5) than close to one specific network edge, and the same holds for node deletion. Similar behavior is obtained for the mean source rate when only node addition (Fig. 2(d)) and only node deletion (Fig. 2(e)) take place. Note also that the combination of node addition and deletion improves more the capacity region compared to the node addition process alone (comparison of Figs. 1, 2(c), (d)), since node deletion serves the purposes of interference reduction along with our previous observation for the locations from which the nodes are added/deleted. Therefore, a general observation from the conducted experiments is that Algorithm 1 ensures convergence even under node churn that has a faster rate than the one needed for Algorithm 1 to converge. Also, Algorithm 1 is able to track small changes in the capacity region.

Figure 3, shows the convergence behavior of Algorithm 1 for only one flow, and specifically for the flow with $ID = 50$, when the probability of network churn, p_{NC}, and the parameters of the node churn vary. Figure 3(a) depicts the convergence behavior of Algorithm 1 for the flow with $ID = 50$ without network churn. We observe that the convergence behavior of Algorithm 1 with respect to this specific flow is satisfactory even under network churn and it captures the possibilities of source rate improvements for the particular flow due to topology changes (e.g. Figure 3(d)). Similar results are also obtained for the other flows.

6.3   Case of Edge Churn

Similarly, in this section, we focus on edge churn, and we evaluate the convergence behavior of Algorithm 1, when each of the iterations corresponds to a time slot of the network operation and edge churn takes place over the network topology.

Figure 4 shows the convergence behavior of the Algorithm 1 with varying probability of network churn, p_{NC}, and varying parameters of the edge churn process. The convergence of Algorithm 1 is satisfactory and similar observations with the case of node churn arise. Comparing Figs. 4(a), (b), we further conclude that the capacity region of the network increases when the probability of the edge addition process increases, as the mean source rate value to which Algorithm 1 converges increases. Note that such a degree of capacity increase is not observed via the node addition process (e.g. Figure 2(d)). At this point, we should refer to the interference model under consideration, which is the one-hop interference model. Specifically, the node addition process may actually not improve the capacity region, as in the event that the newly added node transmits, all its neighboring nodes should remain silent. On the contrary, when a node to which

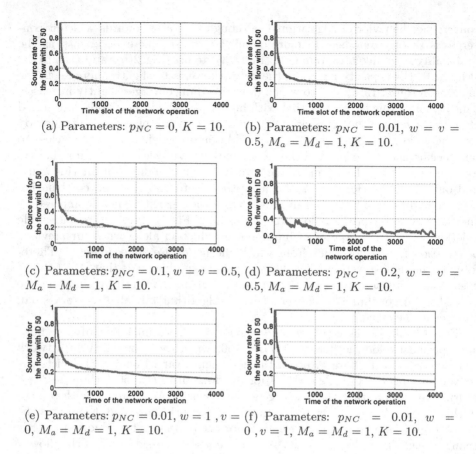

(a) Parameters: $p_{NC} = 0$, $K = 10$.

(b) Parameters: $p_{NC} = 0.01$, $w = v = 0.5$, $M_a = M_d = 1$, $K = 10$.

(c) Parameters: $p_{NC} = 0.1$, $w = v = 0.5$, $M_a = M_d = 1$, $K = 10$.

(d) Parameters: $p_{NC} = 0.2$, $w = v = 0.5$, $M_a = M_d = 1$, $K = 10$.

(e) Parameters: $p_{NC} = 0.01$, $w = 1$, $v = 0$, $M_a = M_d = 1$, $K = 10$.

(f) Parameters: $p_{NC} = 0.01$, $w = 0$, $v = 1$, $M_a = M_d = 1$, $K = 10$.

Fig. 3. Convergence behavior of Algorithm 1 under node churn for the flow with ID equal to 50.

an edge is added via a directional antenna transmits, only one more node should remain silent compared with the previous network instance (i.e. without the new link), i.e. the one to which the added link points at. Similarly, node deletion can increase the capacity region by reducing interference.

Figure 5 shows the convergence properties of Algorithm 1 for the flow with $ID = 50$, when the probability of network churn, p_{NC}, and the parameters of the edge churn vary. We observe that the fluctuations increase compared to the case without network churn (Fig. 5(a)), however, Algorithm 1 can track the possibilities of source rate increase for this particular flow due to new edges or interference reduction (due to edge deletion) and perform the appropriate adaptations (Figs. 5(b), (c)).

Briefly, the conclusions of our experimental study, can be summarized as follows:

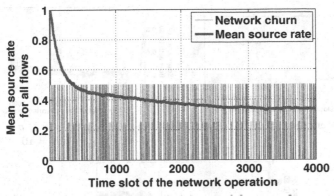

(a) Parameters: $p_{NC} = 0.1$, $p = 0.8$, $r = 0.2$, $m_1 = 3$, $m_2 = 2$, $K = 10$.

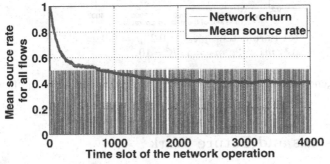

(b) Parameters: $p_{NC} = 0.2$, $p = 0.9$, $r = 0.1$, $m_1 = 3$, $m_2 = 1$, $K = 10$.

Fig. 4. Convergence behavior of Algorithm 1 under edge churn (edge addition and edge deletion).

– The convergence behavior of Algorithm 1 is satisfactory even when network churn takes place during its application for the network operation. Note that node churn applies only for the relay nodes (i.e. the flows remain invariant but their paths change). Also, Algorithm 1 tracks and responds to possible changes of the capacity region.
– Edge churn with increased edge addition, improves more the capacity region of the network than node churn does, both considered under the one-hop interference model.
– Generally, node churn over the relay nodes, with equal participation of the node addition and deletion processes, does not change in a significantly observable degree the capacity region. This is expected since the initial topology assumes a random node deployment (random geometric graph) that is randomly modified via node churn.

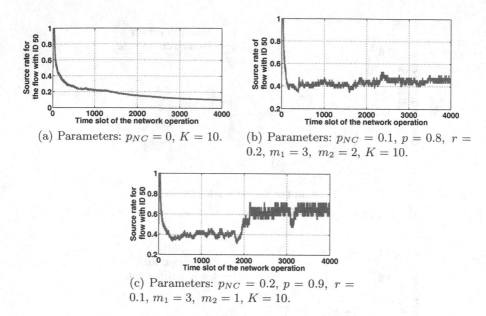

(a) Parameters: $p_{NC} = 0$, $K = 10$.

(b) Parameters: $p_{NC} = 0.1$, $p = 0.8$, $r = 0.2$, $m_1 = 3$, $m_2 = 2$, $K = 10$.

(c) Parameters: $p_{NC} = 0.2$, $p = 0.9$, $r = 0.1$, $m_1 = 3$, $m_2 = 1$, $K = 10$.

Fig. 5. Convergence behavior of Algorithm 1 under edge churn for the flow with ID equal to 50.

7   Conclusions and Future Work

In this work, we studied the behavior of cross-layer network problems in the form of network utility maximization when network topology evolves under the typical processes of network churn, namely node and edge churn. This is a very critical problem with numerous implications for practical applications, since most network topologies do not remain static with respect to time. Through simulations we showed that convergence can be satisfactory even when network churn takes place. In addition edge churn with increased edge addition improves more the capacity region of the network compared to node churn. Also, node churn with equal degree of node addition and deletion does not change the capacity region in a noteworthy manner.

In our future work, we plan to address such evolutionary problems in a more generic manner, based on algebraic methods, which will allow a more seamless study of the corresponding cross-layer optimization problems and obtaining less assumption-restrictive solutions in more efficient manners.

Acknowledgment. This research is co-financed by the European Union (European Social Fund) and Hellenic national funds through the Operational Program 'Education and Lifelong Learning' (NSRF 2007-2013).

References

1. Karyotis, V., Stai, E., Papavassiliou, S.: Evolutionary Dynamics of Complex Communications Networks. CRC Press - Taylor & Francis Group, Boca Raton (2013)
2. Stai, E., Karyotis, V., Papavassiliou, S.: Topology enhancements in wireless multi-hop networks: a top-down approach. IEEE Trans. Parallel Distrib. Syst. **23**(7), 1344–1357 (2012)
3. Lin, X., Shroff, N.: Joint rate control and scheduling in multihop wireless networks. In: Proceedings of the IEEE Conference on Decision and Control (CDC), vol. 2, pp. 1484–1489, December 2004
4. Chiang, M.: Balancing transport and physical layers in wireless multihop networks: jointly optimal congestion control and power control. IEEE J. Sel. Areas Commun. **23**(1), 104–116 (2005)
5. Stai, E., Papavassiliou, S., Baras, J.S.: Performance-aware cross-layer design in wireless multihop networks via a weighted backpressure approach. IEEE/ACM Trans. Netw. October 2014. doi:10.1109/TNET.2014.2360942
6. Chen, L., Low, S.H., Chiang, M., Doyle, J.C.: Cross-layer congestion control, routing and scheduling design in ad hoc wireless networks. In: Proceedings of IEEE INFOCOM, pp. 1–13, April 2006
7. Georgiadis, L., Neely, M.J., Tassiulas, L.: Resource allocation and cross layer control in wireless network. Found. Trends Netw. **1**(1), 1–144 (2006)
8. Eryilmaz, A., Srikant, R.: Joint congestion control, routing and mac for stability and fairness in wireless networks. IEEE J. Sel. Areas Commun. **24**(8), 1514–1524 (2006)
9. Huang, L., Neely, M.J.: Delay reduction via lagrange multipliers in stochastic network optimization. IEEE Trans. Autom. Control **56**(4), 842–857 (2012)
10. Chiang, M., Low, S.H., Calderbank, A.R., Doyle, J.C.: Layering as optimization decomposition: a mathematical theory of network architectures. Proc. IEEE **95**(1), 255–312 (2007)
11. Eswaran, S., Misra, A., La Porta, T.: Control-theoretic optimization of utility over mission lifetimes in multi-hop wireless networks. IEEE Trans. Netw. **20**(4), 129–141 (2012)
12. Ma, K., Mazumdar, R., Luo, J.: On the performance of primal/dual schemes for congestion control in networks with dynamic flows. In: Proceedings of the 27th IEEE Conference on Computer Communications (INFOCOM), April 2008
13. Leconte, M., Jian, N., Srikant, R.: Improved bounds on the throughput efficiency of greedy maximal scheduling in wireless networks. IEEE/ACM Trans. Netw. **19**(3), 709–720 (2011)

On the Problem of Resource Allocation and System Capacity Evaluation via a Blocking Queuing Model in D2D Enabled Overlay Cellular Networks

Georgios Katsinis[✉], Eirini Eleni Tsiropoulou,
and Symeon Papavassiliou

Network Management and Optimal Design Laboratory (NETMODE),
School of Electrical and Computer Engineering,
National Technical University of Athens (NTUA),
9 Iroon Polytechniou str. Zografou, 15773 Athens, Greece
{gkatsinis, eetsirop}@netmode.ntua.gr,
papavass@mail.ntua.gr

Abstract. We assume a D2D overlay enabled cellular network where two types of users are considered. D2D enabled users may communicate either via the Base Station (BS) or directly with each other, while cellular users can communicate only via the BS. In this type of network resource blocks (RBs) are split into two different categories, applicable for D2D access and cellular access respectively. We first introduce a blocking queuing model that can be used to effectively represent the process of allocation of resources to the different types of users. Subsequently we address the problem of fair resource assignment by formulating it as a multi-objective optimization problem of minimizing the maximum blocking probability experienced by any type of user in the system. We show that the optimal RBs assignment policy is of threshold type, while an adaptive algorithm to estimate the required parameters of the optimal policy is presented. Through modeling and simulation the achievable system capacity is evaluated and the superiority of the proposed optimal policy is demonstrated.

Keywords: Blocking queueing model · Capacity evaluation · Markov decision process · D2D overlay · Spectrum sharing · Resource blocks allocation

1 Introduction

Device to Device (D2D) Communication emerges as new add-on communication paradigm in the modern cellular networks towards enhancing network's performance and supporting new services, such as social applications triggered by user proximity, advertisements, local exchange of information, smart communication between vehicles, public safety communication in case of infrastructure damage, as well as proximity awareness. D2D communication can incorporate several benefits for the modern cellular networks both in spectrum and power efficiency via improving throughput, energy efficiency, delay and fairness. D2D communication provides the cellular system the possibility of exploiting proximity between the users and using resources more efficiently compared to

© Springer International Publishing Switzerland 2015
S. Papavassiliou and S. Ruehrup (Eds.): ADHOC-NOW 2015, LNCS 9143, pp. 76–89, 2015.
DOI: 10.1007/978-3-319-19662-6_6

the traditional cellular networks via establishing a direct communication between two mobile users without routing the data paths through a network infrastructure.

More specifically, in D2D enabled cellular networks there are two types of users. D2D enabled users (in the following we refer to as Type-I users) that can communicate either via the Base Station (BS) or directly with their receiver (i.e. acting as D2D users), if proximity and interference conditions permit it, and conventional cellular users that can communicate only via the BS (in the following we refer to as Type-II users). The cellular system operators have the possibility to apply different policies with respect to the RBs assignment between the cellular and the D2D users of the network [4, 5]. For example dedicated resources may be used explicitly for the D2D and the cellular communication mode (overlay scenario) or common RBs may be used for both of them (underlay scenario) [2]. The BS of the cellular network is in charge of allocating certain RBs to certain users. Moreover, in a D2D enabled network the BS can choose the optimal mode of communication for every connection which lies inside its communication range. Different approaches have been proposed in the recent literature to find the optimal mode of communication for every potential transmitter inside the network [9–11]. These approaches mainly take into account the interference levels that each transmitter faces, while however ignoring RBs availability issues.

In a D2D enabled overlay network, the BS is in charge of controlling the RBs allocated to the cellular connections as well as the D2D connections [4, 5]. For this reason the BS is in need of estimating the required RBs for both cellular and D2D connection requests in a dynamic way. It should be noted here that RBs allocated to D2D communicating pairs of users typically present higher degree of reusability within a cell, when compared to typical cellular type of communications where each RB can be used only once in a cell. The degree of RB reusability for RBs allocated to D2D communication depends on the traffic distribution within the cell as well as the interference conditions and the required Quality of Service (QoS). It is evident, that under such a consideration, a RB allocated to D2D users may multiply its effectiveness by the corresponding reusability factor, thus increasing the potential capacity of the system. That means that the higher number of RBs allocated to the D2D users the higher the achievable resource efficiency. However this may hamper the performance of the corresponding pure cellular users, who may suffer from higher blocking probability in the case where high portion of the RBs is allocated to D2D users. Therefore, the problem of choosing the optimal portion of RBs dedicated to the D2D requests along with the specific policies to be followed in assigning RBs to the users is still an open issue [6], of high research and practical importance.

This paper aims at filling exactly this gap that currently exists in the related literature. In a nutshell the key contributions and structure of the paper is as follows. We assume a D2D overlay enabled cellular network where two types of users are considered as mentioned before, (i.e. Type-I and Type-II users). Given this setting we first introduce and devise a blocking queuing model that can be used to effectively represent and model the process of allocation of RBs to the different users (Sect. 2), Subsequently we address the problem of fair RBs assignment to the different users. We formulate this problem as a multi-objective optimization problem of minimizing the maximum blocking probability experienced by any type of user in the system. The latter is formulated as a Markov Decision Process (MDP) optimization problem.

We show that for such a system the optimal resource assignment policy is of threshold type (Sect. 3), while an adaptive algorithm to estimate the required parameters of the optimal policy is presented (Sect. 4). Numerical results demonstrate that such a policy balances and minimizes the probability of RBs unavailability for both types of users (Sect. 5). Furthermore it achieves the maximum offered load that can be handled by the system under the performance requirements that the blocking probabilities of both types of users should be less or equal to some pre-specified value. Finally, Sect. 6 concludes the paper.

2   System Model

We assume a D2D enabled cellular network operating in the overlay mode, where D2D communications occur over dedicated resource blocks (RBs), subtracted from the cellular users. Let the total number of available RBs is N. Two types of users reside within its coverage area, i.e. Type-I users and Type-II users. Type-I users have the flexibility of either communicating via the BS or directly with their receiver (i.e. act as D2D users). In the latter case we refer to D2D communication. The cellular users (Type-II) can communicate exclusively via the BS. We assume that a certain portion of RBs is dedicated to cellular communication, while the rest of them are assigned to D2D communications. Assuming an average reuse factor x of the RBs allocated to the D2D users (that means that on average x pairs of D2D users can simultaneously communicate within a cell under this operation mode using the same physical RB), then the effective number of RBs that can be potentially used simultaneously in the system increases and becomes effectively larger than N. For example assuming that R_1 RBs are allocated for exclusive use under D2D communications and R_2 RBs are allocated for use under the cellular mode of operation $(R_1 + R_2 = N)$ then the total number of effective available RBs in the system becomes: $(R_1 \cdot x + R_2 > N)$.

We assume that users arrive in a Poisson stream with a certain rate λ users' requests per time unit. This stream is comprised of Type-I and Type-II users. As a result all users have the option to access the BS (cellular communication). However, based on users' distribution within the cell, we assume that a certain portion of them are located within the required proximity to their receiver and thus a D2D communication may be initiated instead of communicating via the BS [3]. Considering that the traffic is split in the two types of traffic according to a factor f, $0 < f < 1$, then the arrival of the different type of users' requests can be described by a Poisson process with rate λ_i requests per time unit, the corresponding service times of users' requests are independent and identically distributed (i.i.d.) exponential random variables with mean $1/\mu$ time units. Therefore based on the above the system is represented by the blocking queuing system shown in Fig. 1.

The state at time t is the two-dimensional vector $X(t) = (i(t), j(t))$ where $i(t)$ and $j(t)$ denote the total number of occupied RBs dedicated to cellular and D2D communication respectively. We use the notation $\{X(t), t \geq 0\}$ for the stochastic process which describes the evolution of the system. The state space of that process is $X = X^1 \times X^2$, where $X^1 = \{0, 1, \ldots, K_1\}$, $X^2 = \{0, 1, \ldots, K_2\}$, K_1 and K_2 is the total number of RBs dedicated to D2D and cellular communication respectively. Note that

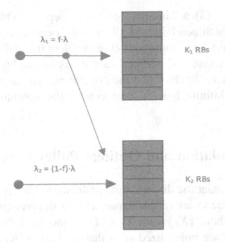

Fig. 1. Queueing System Model

K_1 refers to the effective number of RBs that can be used under D2D communications that essentially has taken into account the multiplication effect of the reusability factor as explained before. Transitions among the states in X are described in terms of the operators

$$l = 1, 2$$
$$A_l : X \rightarrow X,$$
$$D_l : X \rightarrow X,$$

defined as

$$A_1 : (i,j) = (i+1,j) \ \text{if} \ i < K_1$$
$$A_1 : (i,j) = (i,(j+1)^*) \ \text{if} \ i = K_1$$
$$A_2 : (i,j) = (i,(j+1)^*)$$
$$D_1 : (i,j) = ((i-1)^+,j)$$
$$D_2 : (i,j) = (i,(j-1)^+)$$

where $n^+ = \max\{n, 0\}$ *and* $m^* = \min\{m, K_2\}$.

The transition A_l represents an arrival of a user request (connection request) of type l. The D_l stands for the departure of a user request (connection release) for communication served by RBs belonging to group l of RBs. The arrival of a connection request does not imply necessarily a connection setup, due to the limited number of RBs which impose a blocking probability on users' requests.

Initially, it is argued that the original continuous-time system can be converted into an equivalent discrete-time system by sampling it at certain random instants of time. Now consider the original continuous time system sampled at the sequence of random times when either (1) an arrival occurs, (2) a "real" customer departs (a connection termination takes place) (3) a "dummy" customer departs. This sample system is a discrete-time system which is a faithful replica of the original continuous-time system. This procedure of converting a continuous-time problem into a discrete-time problem is well recorded in the literature and rigorous validations are readily available. Lippmanet et al. [7] and Varaiya et al. [8] show that the discrete and continuous-time problems are equivalent in that for infinite horizon cost criteria the optimal policies for the two formulations coincide.

3 Problem Formulation and Optimal Policy

In the following we consider the discrete time problem. Let $0 < t_1 < t_2 < \ldots < t_k < \ldots$ be the transition epochs (due to arrivals or departures) or the process $\{X(t)\}$. Consider the discrete time Markov chain $\{X_k\}$ with $X_k = X(t_k)$ and the following transition probabilities where $\lambda_1, \lambda_2, \mu$ are normalized such that $\lambda_1 + \lambda_2 + (K_1 + K_2) \cdot \mu = 1$.

$$P(X_{k+1}|X_K = (i,j), z_k) = \begin{cases} \lambda_1 & \text{if } X_{k+1} = A_1 X_k, \ z_k^1 = 0 \\ \lambda_1 & \text{if } X_{k+1} = X_k, \ z_k^1 = 1 \\ \lambda_2 & \text{if } X_{k+1} = A_2 X_k, \ z_k^2 = 0 \\ \lambda_2 & \text{if } X_{k+1} = X_k, \ z_k^2 = 1 \\ i\mu & \text{if } X_{k+1} = D_1 X_k \\ j\mu & \text{if } X_{k+1} = D_2 X_k \\ (K_1 + K_2)\mu - i\mu - j\mu & \text{if } X_{k+1} = X_k \end{cases}$$

Let us denote z_t^l the probability of blocking a user request of type l for connection at time t. A control policy z is any process $z = (z_1, z_2, \ldots)$, where $z_k = (z_k^1, z_k^2)$ and for every transition event ω, $z_k^i = 0$ if the transition $\omega \neq A_{1(2)}$ and $z_k^i \in [0, 1]$ if $\omega = A_{1(2)}$. This means that in the case of an arrival, the control policy z can take any value in the probability space [0,1] and in case of any other event (i.e. departure) the control policy takes the zero value. This approach provides the flexibility of applying a certain policy-action which results in a change of the current state of the system. Actually this policy determines the rules which govern the state transitions of the system. It is observed that this policy induces stochasticity in the decision process since it represents the probability of a user's request not being served by the network.

Let G denote the set of all possible policies of that form. The admission control policy determines whether a new user's request will be served by the system or not and it also indicates the desirable and feasible mode of communication for the Type-I users, i.e. D2D or cellular, via determining which type of RBs will be assigned to them. As mentioned above, each type of user, confronts the probability of not being served by the network due to the scarcity/finiteness of the available RBs. The cellular users (Type-II) have access only to a certain portion of RBs dedicated exclusively for communication between them and the BS. Type-I users have the option of either communicating directly with their receiver (making use of the RBs allocated for D2D communication) or communicating via the BS (making use of the RBs allocated for cellular communication). For this reason intuitively Type-I users face a lower blocking probability. This generates the need of applying a control policy which has to balance and minimize the blocking probability for both type of users simultaneously.

Let $P_{1,g}, P_{2,g}$ be the blocking probabilities experienced by the Type-I and Type-II users respectively. We can also define the long run average blocking cost associated with policy $g \in G$ for both type of users as follows.

$$P_{l,g}(x) = \lim_{n \to \infty} \sup \frac{1}{n} \cdot E_x^g \left[\sum_{k=0}^{n-1} z_k^l \right], x \in X$$

where $E_x^g[]$ denotes the expectation with respect to the probability measure induced by policy g, on the state process starting in state x.

The problem of fair admission control policy and optimal mode selection can be formulated as follows:

$$(P) : \min \{ \max (P_{1,g}, P_{2,g}) \},$$

over all possible RBs assignment policies g.

The formulation of the above problem is equivalent to the corresponding channel assignment policy in [1]. Therefore it can be shown that problem (P) has always a solution within a class of threshold policies [1]. This type of policies is as follows. When a user request of Type-I takes place and at least one RB of D2D RBs is free then the user request is forwarded there. If no D2D RBs are available then we accept the request if the number of busy RBs of the second type (cellular RBs) are less than a threshold T. While if exactly T RBs of the second type are occupied then we accept the request with a certain probability q and reject it with a probability $(1 - q)$. Problem (P) is formulated as a Markov Decision Process (MDP) optimization problem. It is shown that this problem is equivalent to a constrained MDP optimization problem with linear cost, where the analysis of the latter problem shows the existence of threshold type optimal policies [1].

4 Online Estimation of the Optimal Policy Parameters

The optimal policy proposed in Sect. 3 depends on the choice of the threshold T and the corresponding probability q. In this section we present an adaptive online algorithm which can be used to estimate those parameters. This algorithm relies on the measured levels of the blocking probabilities for both type of connection requests based on the past activity of the system (i.e. history of blocked requests).

The choice of the parameters values of T and the probability q should capture the expected influence that those parameters should have on the blocking probability of the two types of requests which are sent to the BS. The threshold T influences in a reverse way the blocking probabilities of the Type-I and the Type-II requests for communication.

The impact of the threshold T on the blocking probabilities of the users requests can be described as follows: an increase in the value of the threshold T makes more cellular

RBs accessible to the Type-I of requests which means that it is expected that the blocking probability of the Type-I requests decreases. On the other hand by increasing the participation of the type-I requests to the cellular RBs produces an additive burden for the second type of requests (cellular requests). Thus it is expected that this will lead to an increase in the corresponding blocking probability.

Inversely a decrease in the threshold T has a negative impact on the blocking probability for the Type-I users since it reduces the amount of cellular RBs which are also accessible to them. On the contrary it has a positive impact on the blocking probability of the Type-II since more cellular RBs become available to the users who have only cellular access to the network. Thus the choice of the parameter T plays a crucial role in the blocking probabilities for both type of requests.

Following the previous analysis we assume that the above parameters T and q are related to the blocking probabilities and thus a function f exists such as $h = T + q = f(P_1, P_2)$ where P_1 and P_2 are the blocking probabilities for the Type-I and Type-II users respectively.

We will propose an iterative online algorithm which at each step computes a new value for the parameters T and q. It updates the estimates of the threshold T and the randomized factor q at certain time slots which are determined by the network designer. The algorithm terminates when it converges to a value for both T and q (i.e. the relative absolute difference between two steps of the algorithm is less than some predefined value, e.g. *1* %).

We denote by T_n and q_n the control parameter at iteration n. By P_i^n we denote the estimate of the blocking probability for the i_{th} type of requests at iteration n which is estimated looking at the history of the previous w requests. It should be clarified that $T_n \in \{0, 1, 2, \ldots, K_2\}$ and $q_n \in [0, 1)$. The threshold T and the randomized factor q are represented both by the variable $h_n = T_n + q_n$, with $h_n \in [0, K_2]$. The threshold is the integer part of this variable while the randomized factor is the fractional part.

The iterative algorithm should have the following form

$$h_{n+1} = I[h_n + b_n]$$

$$T_{n+1} = \lfloor h_{n+1} \rfloor$$

$$q_{n+1} = h_{n+1} - \lfloor h_{n+1} \rfloor$$

where b_n is an estimate of the necessary adjustments on the variable $(T_n + q_n)$ based on the estimates of the blocking probabilities for both type of requests and $I[]$ is defined by

$$I[x] = \left\{ \begin{array}{ll} x & if\ 0 \leq x \leq K_2 \\ 0 & if\ x < 0 \\ K_2 & if\ x > K_2 \end{array} \right\}.$$

Therefore it remains to describe how we can estimate b_n, at each iteration as well as how the estimates of the blocking probabilities are updated. Since our target is to balance the blocking probabilities, a reasonable choice for our estimate b_n, which indicates the necessary adjustments should actually depend on the difference: $P_1^n - P_2^n$. Based on the above form of the iterative algorithm for the computation of b_n, we observe that the b_n reflects the required changes to the parameters of the optimal policy T and q in order to balance the differences in the two blocking probabilities.

We assume that the parameter h_n is a function of the difference between the blocking probabilities of the two types of requests at each time iteration as it was presented before. Therefore, a reasonable assumption is that the parameter b_n should have the following form which seems like the derivative of the function h_n with respect to the difference between the two types of blocking probabilities.

An iterative algorithm which captures this idea can be of the form

$$b_n = \frac{P_1^n - P_2^n}{\left|(P_1^n - P_2^n) - (P_1^{n-1} - P_2^{n-1})\right|} \Delta_{T,q}^n \tag{1}$$

where $\Delta_{T,q}^n = |(T_n + q_n) - (T_{n-1} + q_{n-1})|$.

Actually based on the fact that a change of on the factors T and q produces a change in the quantity: $\left|(P_1^n - P_2^n) - (P_1^{n-1} - P_2^{n-1})\right|$ on the difference between P_1 and P_2, b_n indicates the proposed change on T and q in order to balance the blocking probabilities of the two types of requests in the next step. Therefore the previous equations describe an adaptive algorithm for estimating the optimal values of T and q.

In the following, we investigate how often the algorithm should iterate and thus update the estimated parameter values of T and q. In order to answer this question we need to clarify how the blocking probabilities are calculated. Since we want the system to be able to keep track of the sudden changes on the traffic characteristics in the two types of requests, we have to decrease as much as possible the effect of the past of the system on the estimation of P_1^n and P_2^n. In the following we consider that the estimates of the blocking probabilities P_1^n and P_2^n are based on the last w requests for communication only.

In Fig. 2 we present the convergence of the above iterative algorithm for different initial values of T and q. Considering an initial change in the h_n function (i.e. in the values of T and q parameters), this produces a respective change on the two blocking probabilities. Thus starting with a pair of initial values for the parameters T, q and the corresponding values for the blocking probabilities we can compute the initial value for the b_n parameter according to Eq. (1). Then, we are able to update the h, T, q values based on the iterative algorithm. Thus, this algorithm proceeds the same way until the values of the parameters T and q converge to a certain value (i.e. the values of T, q do not differ from the values of the previous iteration by 1 %).

The simulated scenario assumes an average traffic of $\lambda = 80$ requests per time unit in a system with 50 D2D and 50 cellular RBs and a splitting factor of $f = 0.6$. The three different curves correspond to three different pairs of initial values of T and q.

Next we demonstrate that the above presented algorithm can converge in a limited number of iterations under multiple time varying traffic loads. In Fig. 3 we observe the convergence of the algorithm under four different traffic loads. We indicate via the 'x' markers the time instances at which the traffic load changes. Initially the average traffic load is 80 requests per time unit and changes to 50, 70 and 30 requests per time unit in a network with 50 D2D and 50 cellular RBs.

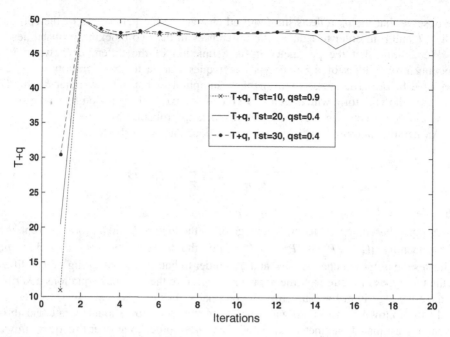

Fig. 2. Convergence of the iterative algorithm to the optimal T and q values under different initial values

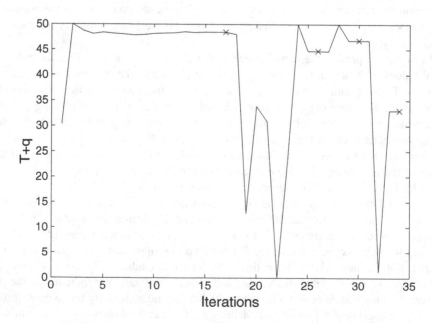

Fig. 3. Convergence of the algorithm under time varying loads

5 Performance Analysis

First we evaluate the performance of the system in terms of the blocking probability for both type of requests and on the average system blocking probability. We assume a fixed number of available RBs within a cell. As mentioned before, the total number of RBs is divided into two groups: those that are accessible for (a) D2D communication (D2D RBs) and (b) for the traditional cellular communication (cellular RBs). We show via an extensive set of simulations that there is a natural tradeoff of this split with respect to the corresponding blocking probabilities of D2D and cellular communications, and therefore an optimal division point of RBs dedicated to D2D communication and cellular communication respectively exists, which minimizes and balances the weighted sum of the blocking probabilities (system blocking probability) for both types of requests for service.

The simulated scenario under consideration assumes a network with *100* available RBs, average rate of incoming traffic $\lambda = 150$ requests per time unit and $f = 0.5$, while the degree x of resource reusability for resources allocated to D2D communications is assumed $x = 1$. We confirm in Figs. 4 and 5 that as expected the optimal point in terms of balancing and minimizing the blocking probabilities of the two types of requests, expressed as the ratio of D2D allocated resources to the allocated resources for usage under cellular communications, is very close to *1*.

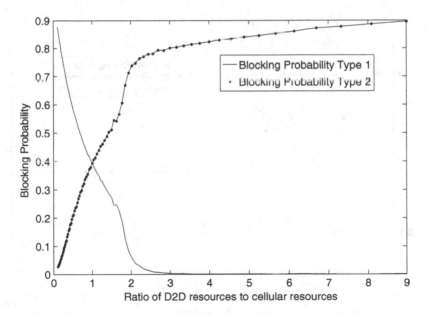

Fig. 4. Blocking probabilities of users' requests vs. the ratio of D2D RBs to the cellular RBs

Next we evaluate and demonstrate the benefits of the degree of RBs reusability for RBs allocated to D2D communications (in the following we refer to as reuse factor x) in terms of the blocking probability for both types of requests. Specifically in Fig. 6

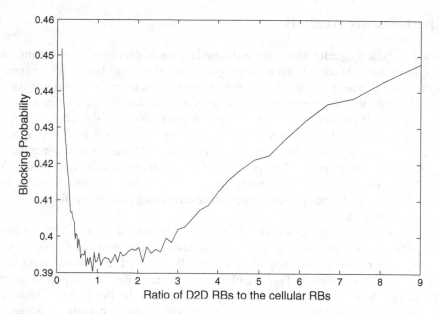

Fig. 5. System blocking probability vs. the ratio of D2D RBs to the cellular RBs.

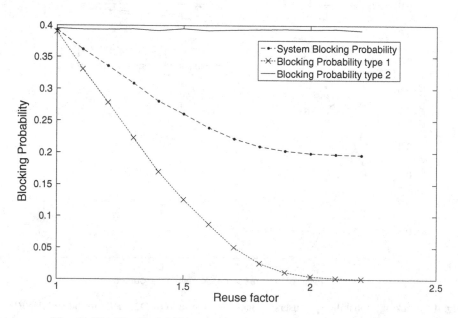

Fig. 6. Blocking Probabilities vs. the re-use factor of the D2D resources

we observe that the increase in the value of the reuse factor x improves both the system blocking probability, as well as the blocking probability of type I requests.

We also present some numerical results that demonstrate the superiority of the proposed optimal policy in terms of the achievable blocking probability for both types of request and for various amounts of traffic loads. We assume that 20 % of the available RBs are dedicated for D2D access and 80 % of them are allocated for cellular access. The mobile traffic is not known a priori and it may be changing due to time varying users requests. We assume that $f = 0.3$. Specifically, in Fig. 7 we show the evolution of the system blocking probability with respect to the offered load in the system for four different scenarios. The first one refers to a system where the threshold is set to zero, that is the Type-I users have exclusive access to the D2D RBs only, and the Type-II users have exclusive access to the cellular RBs, while in the second one the optimal policy is applied, assuming in both cases reuse factor $x = 1$, in order to evaluate the actual benefits achieved only by the application of the optimal policy. Furthermore, the last two scenarios refer to the application of two different reuse factors with respect to the D2D RBs, namely factors $x = 1.5$ and $x = 2$, combined with the optimal policy. Comparing the first two scenarios we clearly observe that the optimal policy produces significantly better results with respect to the average system blocking probability (in the case considered here the improvement is 20 %). It should be noted that the optimal policy not only reduces the overall average system blocking probability but also achieves to balance the blocking probabilities of the two different types of uses thus alleviating any inherent unfairness that may be introduced due to the static allocation/split of RBs to the D2D type of communication and to the cellular type of communication. Specifically for the case considered here the average difference between the two corresponding blocking probabilities of the two types of users is

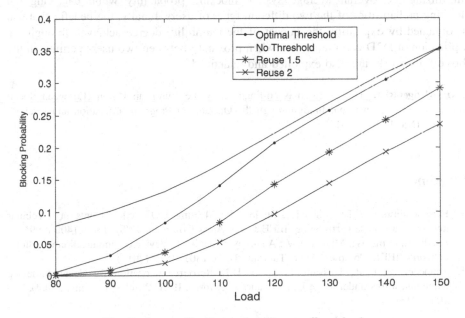

Fig. 7. System blocking probability vs. offered load

0.3407 for the no threshold policy and *0.0752* for the optimal policy. Finally, comparing the two cases of applying different reuse factors we observe that the system blocking probability can be further improved, specifically by *37.91* % and *56.47* % on average for the reuse factors of *1.5* and *2* respectively.

6 Conclusion

It has been argued in the literature that D2D enabled cellular networks can provide significant capacity improvements by taking advantage of the inherent characteristics and benefits provided by the D2D communication, capitalizing on the exploitation of the proximity between the users and thus using resources more efficiently compared to the traditional cellular networks. In this paper we considered the problem of fair resource blocks allocation in a D2D overlay enabled cellular network where enhanced users with the possibility of establishing either D2D communication or cellular type of communication co-exist with traditional users that can establish only conventional cellular communication. We provided a blocking queuing model that can be exploited to conveniently represent the allocation of RBs to the different types of users. Based on that we formulated the problem of fair RBs assignment to the different users as a multiobjective optimization problem of minimizing the maximum blocking probability experienced by any type of user in the system. We showed that in our setting the optimal RBs assignment policy is of threshold type and we provided an adaptive algorithm to estimate the required parameters of the optimal policy, without the need to apriori know the traffic details. Extensive numerical results were presented that demonstrate the benefits and superiority of the proposed optimal policy in terms of both minimizing the overall average system blocking probability while balancing the blocking probabilities of the two different types of users. Finally, the benefits that can be obtained by exploiting the impact of the reusability degree achieved through the application of D2D communication when proximity between two users permits, in the above framework, are also exploited and quantified.

Acknowledgement. This research is co-financed by the European Union (European Social Fund) and Hellenic national funds through the Operational Program 'Education and Lifelong Learning' (NSRF 2007-2013).

References

1. Papavassiliou, S., Tassiulas, L., Tandon, P.: Meeting QOS requirements in a cellular network with reuse partitioning. IEEE J. Sel. Areas Commun. **12**(8), 1389–1400 (1994)
2. Asadi, A., Qing, W., Mancuso, V.: A survey on device-to-device communication in cellular networks. IEEE Commun. Surv. Tutorials **16**(4), 1801–1819 (2014)
3. Phunchongharn, P., Hossain, E., Kim, D.I.: Resource allocation for device-to-device communications underlaying LTE-advanced networks. IEEE Wireless Commun. **20**(4), 91–100 (2013)

4. Fodor, G., Dahlman, E., Mildh, G., Parkvall, S., Reider, N., Miklós, G., Turányi, Z., Research, E.: Design aspects of network assisted device-to-device communications. IEEE Commun. Mag. **50**(3), 170–177 (2012)
5. Lei, L., Zhong, Z., Lin, C., Shen, X.: Operator controlled device-to-device communications in LTE-advanced networks. IEEE Wireless Commun. **19**(3), 96–104 (2012)
6. Lin, X., Andrews, J.G. and Ghosh, A.: Spectrum sharing for device-to-device communication in cellular Networks, September 2014. http://arxiv.org/abs/1305.4219
7. Lippman, S.: Applying a new device in the optimization of exponential queueing systems. Oper. Res. **23**, 687–710 (1975)
8. Varaiya, P., Rosberg, Z., Walrand, J.: Optimal control of service in tandem queues. IEEE Trans. Automatic Control **27**, 600–609 (1982)
9. Akkarajitsakul, K., Phunchongharn, P., Hossain, E., and Bhargava, V.K.: Mode Selection for Energy-Efficient D2D Communications in LTE-Advanced Networks: A Coalitional Game Approach. IEEE International Conference on Communication Systems, November 2012
10. Doppler, K., Yu, C.-H., Ribeiro, C.B., Jänis, P.: Mode selection for Device-to-Device Communication underlaying an LTE-Advanced Network. IEEE Wireless Communications & Networking Conference, April 2010
11. Liu, Z., Peng, T., Xiang, S., Wang, W.: Mode Selection for Device-to-Device (D2D) Communication under LTE-Advanced Networks", International Conference on Communications, June 2012

Localization, Sensor Deployment, and Mobility Management

Localization of a Mobile Node in Shaded Areas

Salvador Jauregui[1]($\boxtimes$), Michel Barbeau[2], Evangelos Kranakis[2],
Edson Scalabrin[3], and Mario Siller[1]

[1] Electrical Engineering and Computer Science Department,
Center of Research and Advanced Studies CINVESTAV,
45019 Zapopan, Jalisco, Mexico
{sajaureg,msiller}@gdl.cinvestav.mx
[2] School of Computer Science, Carleton University,
Ottawa, ON K1S 5B6, Canada
{barbeau,kranakis}@scs.carleton.ca
[3] Graduate Program in Computer Science,
Pontifical Catholic University of Parana,
Curitiba, Parana 80215-901, Brazil
scalabrin@ppgia.pucpr.br

Abstract. Many algorithms and applications use GPS as standard for
outdoor usage. But they cannot perform correctly for shaded areas such
as tunnels, canyons, and near large buildings. Other localization algorithms and reference information are then required for aiding GPS.
In this paper, we propose an architecture for tracking of a mobile node
considering: (i) a main node tracking system that is feasible enough for
non-shaded areas and (ii) a subsystem that supports the location during shaded areas. In (i), we propose the Probabilistic Random Mobility Model for simulating paths based on Center Turning Radius (CTR)
which is both an inherent vehicular feature and a vehicular displacement restriction. In (ii), Particle Filtering approach is used because it
is able to handle location uncertainty during shaded areas and improves
over time. The status and CTR of a mobile node are used to reduce
and adapt the space of uncertainty where particles are drawn. Finally, a
priority-selective control based on the suppress principle is employed for
choosing (ii) during shaded areas.

Keywords: Mobile localization · Particle filtering · Probabilistic Random Mobility Model · Center Turning Radius · Flexible space of uncertainty

1 Introduction

A Wireless Sensor Network (WSN) is made up of small devices (nodes) which
are capable of collecting information from the environment such as temperature, vibrations, humidity, sound, light, and motions. A WSN can be used in
military surveillance, environmental monitoring, intelligent spaces, habitat and
structural monitoring, and robotics, among others. Nowadays, there are many
Node Localization Algorithms (NLAs) with different accuracies and approaches.

© Springer International Publishing Switzerland 2015
S. Papavassiliou and S. Ruehrup (Eds.): ADHOC-NOW 2015, LNCS 9143, pp. 93–106, 2015.
DOI:10.1007/978-3-319-19662-6_7

Despite the large amount of algorithms, they can be classified into two wide groups: range-free and range-based NLAs. In the former, algorithms use only the content of received messages; a good example of this classification is the Centroid Localization (CL) [8]. In the latter, NLAs utilize distances or angles; e.g. RADAR [3] and GPS. There are also NLAs that belong to both classes because they use the content of the received messages and distances/angles. These are referred to as hybrid NLAs; e.g. the Triangular Centroid Localization Algorithm [15], Weighted Centroid Localization [6], and Improved Centroid Localization [20].

2   Background

2.1   Range-Free Node Localization

Range-free localization algorithms estimate the location of sensor nodes by, either, exploiting the sensing capabilities of each sensor, or exploiting the RF transceiver already on board among neighboring nodes. They are low-cost implementation since the need for extra per-node devices or additional infrastructure is not required. Range-free NLAs are usually less accurate than range-based NLAs and their performance deteriorates sharply in the presence of obstacles which can detour the shortest path between two nodes and enlarge the distance estimation between them. However, they can obtain reasonable performance such as the CL algorithm [8]. In the CL algorithm each unknown node estimates its physical position in a Cartesian space by averaging the positions of heard beacon nodes. The experimental results show different error performance for an outdoor and indoor scenario. For the former, the error never exceeded 2 m while for the latter it varied widely from 4.6 m to 22.3 m. According to the authors, the error fluctuation depends on: the number of walls, the node location, and the node orientation. The CL algorithm was developed assuming perfect spherical radio propagation and an identical transmission range, unlike the Approximate Point In Triangulation (APIT) algorithm [12]. The APIT algorithm uses the Degree Of Irregularity (DOI) equal to 0.1 and 0.2. The DOI is defined as an indicator of the ratio pattern irregularity, e.g. if $DOI = 0.3$, the radio range r in each direction takes a random value from $0.7r$ to $1.3r$.

2.2   Range-Based Node Localization

The range-based algorithms exploit distances or angle information between neighbor nodes, and then use the information to localize nodes. Range-based NLAs usually perform better than range-free but additional hardware is required and thus, they are very expensive to be used in large scale sensor networks. Another issue to consider is the calibration process and time that might be required before actual location estimation can be executed. For instance, the RADAR [3] is a radio-frequency based system for locating and tracking people in indoor environments; it uses the signal strength from three beacon nodes inside a building. Based on empirical measurements the user's location is inferred by triangulation. The authors consider the wall attenuation based on a floor attenuation factor propagation model. Nevertheless, RADAR requires an extensive effort and time for the generation of the signal database based on previous survey measurements. Another similar work is presented in [21] where a wireless signal strength map is generated by a ray-tracing approach in order to include absorption and reflection characteristics of various obstacles (home environment). The locations of users are computed using Bayesian Filtering on sample sets derived by Monte Carlo Sampling. Authors' results show a sub-room precision (~1 m). However the training effort requires some time because they have to measurement the wall, windows, and door length of all rooms as well as identification of major obstacles.

2.3 Hybrid Node Localization

Hybrid node localization algorithms take advantage of range-free and range based NLAs to adapt the solution based on application-specific factors, requirements of the application, coverage range, and the budget for the whole project. Hybrid NLAs use the content of the received messages and distance/angle such as Triangular Centroid Localization (TCL) [15], the Weighted Centroid Localization (WCL) [6], and EDIPS [19]. However, the integration of a large amount of factors into a NLA is not related directly to a high accuracy. In fact, the integration of all possible and suitable factors for a given real environment into a single NLA might not be feasible to achieve. In some cases, NLAs are designed for specific scenarios and only the most representative factors are considered such as in [3].

2.4 Mobile Node Tracking

According to Yaakov [5], tracking is the processing of measurements obtained from a target in order to maintain an estimate of its current state. In the Particle Filtering approach, external information is referred to as observations and they can be captured for example by GPS, video cameras, odometers, beacon nodes, and so forth. Observations are the only information that helps keep and weight particles, obviously the hardware quality plays an important roll, but also physical phenomena that affect the wave propagation and measurements. Information fusion [11] is commonly used as a solution to merge observations coming from several and diverse sources in order to improve the node localization estimation such as [4,9,16,18].

The general particle filtering approach includes three steps: (i) Initialization, (ii) Prediction, and (iii) Filtering. In (i), Particles are dropped from the proposal distribution. In (ii), the prior distribution is often used as importance function because it is easier to drop particles and perform subsequent importance weight calculations. That is, avoiding the situation that all but one of the importance weights are close to zero. The weight assigned to each particle represents the probability of occurrence. Hence, it is assumed that the particle with the highest weight is the most likely position for the mobile node. In (iii), filtering based on connection, exact range, bounded ranging and weighting are usually employed [2,10,13,14,17,21].

3 Localization of a Mobile Node in Shaded Areas

Our proposed mobile node architecture addresses the tracking of a mobile node considering: (i) a main node tracking system/algorithm that is feasible enough for non-shaded areas and (ii) a subsystem that supports the node tracking in shaded areas. A vehicle with a GPS on-board is a suitable example for (i); nevertheless our approach does not consider any particular main system/algorithm. Instead, our proposed Probabilistic Random Mobility Model (PRMM) generates sequential location points, trajectory, based on the Center Turning Radius (CTR) that in turn is both an inherent vehicular feature [1] and a vehicular displacement restriction. On this basis alone diverse mobile nodes can be represented, for instance a utility car with $CTR = 6.4\,\mathrm{m}$ or smaller values for robots. Our proposed model neither is limited to simulate vehicle trajectories nor tries to replicate the driving style of a person.

As secondary system/algorithm, we adopt the well-known particle filtering approach because it is able to handle the location uncertainty during a shaded area and improves the node location estimation over the time. We propose not only a solution for locating a mobile node in shaded areas but also a Probabilistic Random Mobility Model to generate paths, both based on CTR.

The proposed architecture for tracking a mobile node is made up of three main modules: (i) Probabilistic Random Mobility Model, (ii) Location Subsystem, and (iii) Priority Suppress as shown in Fig. 1. In (i), several parameters can

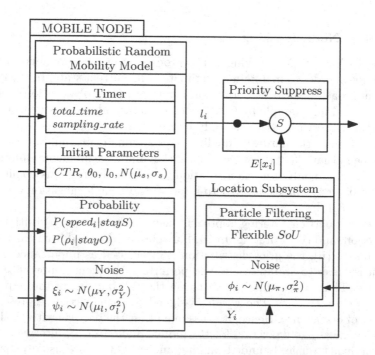

Fig. 1. General Architecture for mobile node tracking in shaded areas.

be established in order to obtain sequential location points (l_i) and observations (Y_i) at time i. The four submodules: Timer, Initial Parameters, Probability, and Noise are given in greater detail in Sect. 3.1. In (ii), Y_i is used as a unique input for generating an alternative estimated position $E[x_i]$ based on particle filtering. The Location Subsystem is given in greater detail in Sect. 3.2. In (iii), a selective process based on priority and availability is performed in order to select l_i when available and $E[x_i]$ during shaded areas such that the black dot over the line l_i represents the highest priority.

3.1 Probabilistic Random Mobility Model

Our Probabilistic Random Mobility Model generates sequential points, trajectories, of l_i and Y_i in a two dimensional Cartesian space (x, y) such that units are expressed in meters. The sequential points are generated from: (i) a bounded displacement based on the mobile node CTR, (ii) probability of turn every amount of seconds, (iii) probability of increase/decrease in speed every amount of seconds, and (iv) noise based on Normal Distribution. Algorithm 1 shows the procedure for the generation of l_i and Y_i. Let $sampling\_rate$ be the number of samples per second, i be a time slot s.t. $i = 1\ second/sampling\_rate$, and $total\_time$ be the total time of simulation. In line 1, the default initialization ($i = 0$) is contextualized as follows: A utility car ($CTR = 6.4$) starts its movement at the position $(0, 0)$ with an orientation θ_0 randomly chosen from 0 to 2π radians. The initial speed $speed_0$ and threshold thr are established starting from the assumption that $speed_i \sim N(\mu_s, \sigma_s^2)$ with a Normal distribution based on the central limit theorem. The use and definition of thr are addressed in line 10 of Algorithm 1.

The update frequency of PRMM outputs is based on the smallest time slot i. Hence, the generated amount of l_i and Y_i is determined by $total\_time \times sampling\_rate$, (line 2). The mobile node speed can be changed every $secondS$ seconds through the probabilistic function $A(\cdot)$ such that $secondS = 5$ is the default value, (lines 3–6). An increase or decrease in the mobile node speed have the same probability of occurrence and it is expressed in Eq. (1) such that $stayS$ represents the probability of maintaining the mean speed and its default value is 0.8.

$$P(speed_i | stayS) = \begin{cases} stayS, & speed_i = \mu_s \\ \frac{1-stayS}{2}, & otherwise \end{cases} \qquad (1)$$

The speed assignment by Eq. (2) is based on the three sigma rule such that j and i are two uniform random numbers in the interval [0,1] and $g = 1 - stayS$.

$$A(\cdot) = \begin{cases} \mu_s, & i = [0, stayS) \\ \mu_s + j\sigma_s^2, & i = \left[stayS, stayS + 0.341g\right) \\ \mu_s - j\sigma_s^2, & i = \left[stayS + 0.341g, stayS + 0.682g\right) \\ \mu_s + \sigma_s^2 + j\sigma_s^2, & i = \left[stayS + 0.682g, stayS + 0.818g\right) \\ \mu_s - \sigma_s^2 - j\sigma_s^2, & i = \left[stayS + 0.818g, stayS + 0.954g\right) \\ \mu_s + 2\sigma_s^2 + j\sigma_s^2, & i = \left[stayS + 0.954g, stayS + 0.976g\right) \\ \mu_s - 2\sigma_s^2 - j\sigma_s^2, & i = \left[stayS + 0.976g, 1\right] \end{cases} \qquad (2)$$

Algorithm 1. Generating of trajectories from the vehicle CTR.

Require: $i \geq 0$, $total\_time > 0$, $sampling\_rate > 0$
Ensure: $l_i, speed_i, \rho_i, \theta_i, Y_i$

1: $CTR \leftarrow 6.4$ or given, $l_0(x,y) \leftarrow (0,0)$ or given, $\theta_0 \leftarrow Random[0,2\pi)$ or given, $speed_0 \leftarrow \mu_s$, $thr \leftarrow \mu_s + \sigma_s^2/2$
2: **for** $i = 1$ **to** $total\_time \times sampling\_rate$ **do**
3:   **if** $i \bmod sampling\_rate \bmod secondS = 0$ **then**
4:     $speed_i \leftarrow A(speed_{i-1}, P(speed_i|\cdot), \mu_s, \sigma_s^2)$
5:   **else**
6:     $speed_i \leftarrow speed_{i-1}$
7:   **end if**
8:   $displacement_i \leftarrow speed_i/sampling\_rate$, $b_i(x) \leftarrow l_{i-1}(x) + displacement_i \times cos(\theta_{i-1})$, $b_i(y) \leftarrow l_{i-1}(y) + displacement_i \times sin(\theta_{i-1})$
9:   **if** $i \bmod sampling\_rate \bmod secondO = 0$ **then**
10:     $\rho_i \leftarrow B(CTR, P(\rho_i|\cdot), thr, displacement_i)$
11:   **end if**
12:   **if** $\rho_i \neq 0$ **then**
13:     $l_i \leftarrow C(|\rho_i|, l_{i-1}, b_i)$
14:   **else**
15:     $l_i \leftarrow b_i$
16:   **end if**
17:   $Y_i \leftarrow D(l_i, \mu_Y, \sigma_Y^2)$, $l_i \leftarrow E(l_i, \mu_l, \sigma_l^2)$
18:   **return** l_i, Y_i
19: **end for**

Based on the fact that if the mobile node speed is greater than zero a sudden displacement in the opposite direction might not be possible. On this basis alone, l_i can be established in a well-bounded space from CTR. Let Fig. 2 takes place in order to illustrate the bounded displacement of a mobile node based on its CTR as well as the rest of Algorithm 1. The dotted arc represents all possibilities $l_i(x,y)$ for a mobile node based on its CTR. If the mobile node (black square) goes in a straight line, it would stay at $b_i(x,y)$ which is determined from $l_{i-1}, displacement_i$, and θ_{i-1}, (line 8). If it turns, the position $b_i(x,y)$ is then displaced over the dotted arc ρ radians. The mobile node turning can be changed every $secondO$ seconds through the probabilistic function $B(\cdot)$ such that $secondO = 7$ is the default value, (lines 9–11). A left or right turn have the same probability of occurrence and $stayO$ represents the probability of non-turning. Hence, the turning probability is $1 - stayO$ and this is expressed by Eq. (3), with $stayO = 0.8$ as the default value.

$$P(\rho_i|stayO) = \begin{cases} stayO, & \rho_i = 0 \\ \frac{1-stayO}{2}, & otherwise \end{cases} \qquad (3)$$

The maximum turning of a mobile node is bounded by the angle λ_i which is derived from CTR, $l_{i-1}(x,y)$, and $b_i(x,y)$. The angular velocity w_i can be simplified as $w_i = \frac{displacement_i}{CTR}$ and $displacement_i = speed_i/sampling\_rate$. The angle α_i can be then easily estimated by $alpha_i = \frac{\pi - w_i}{2}$. Besides, the angle λ_i

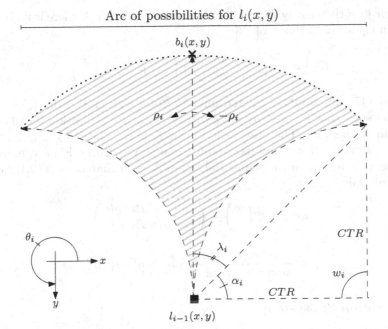

Fig. 2 Arc of possibilities for $l_i(x,y)$ for a mobile node ■ based on the vector rotation (l_{i-1}, b_i) such that its magnitude is $\sqrt{2CTR^2}$.

is reduced as the mobile node increases its speed in order to avoid skidding or overturning. The safe range of turning can be calculated considering mobile node mass, friction, sensor measurements, and forces involved, among others. However, this calculation is beyond the scope of this article. Instead, the angle λ_i is gradually reduced while the mobile node speed is greater than the threshold thr as shown in Eq. (4) such that $lambda_i = \pi/2 - \alpha_i'$ and $thr = \mu_s + \sigma_s^2/2$ is given as the default value, (lines 1).

$$\alpha_i' = \begin{cases} \frac{(speed_i - thr)(\frac{\pi}{2} - \alpha_i)thr}{\mu_s + \sigma_s^2} + \alpha_i, & speed_i > thr \\ \alpha_i, & otherwise \end{cases} \tag{4}$$

The value of ρ_i is assigned by function $B(\cdot)$ as shown in Eq. (5) such that i and j are two uniform random numbers in the interval $[0,1]$, (line 10).

$$B(\cdot) = \begin{cases} 0, & i = [0, stayO) \\ j\lambda_i, & i = \left[stayO, \frac{stayO+1}{2} \right) \\ -j\lambda_i, & i = \left[\frac{stayO+1}{2}, stayO \right] \end{cases} \tag{5}$$

If there is a turning i.e., $\rho_i \neq 0$ the mobile node position l_i is then displaced ρ_i radians over the dotted arc as shown in Fig. 2. The turning is determined by the function $C(\cdot)$ in Eq. (6) such as $dx = l_i(x) - l_{1-i}(x)$ and $dy = l_i(y) - l_{1-i}(y)$.

The angle θ_i is then easily determined by using l_i and l_{i-1}. If there is no turning, l_i is then equal to b_i, (lines 14–16).

$$C(\cdot) = \begin{pmatrix} \cos \rho_i & -\sin \rho_i \\ \sin \rho_i & \cos \rho_i \end{pmatrix} \begin{pmatrix} dx \\ dy \end{pmatrix} + l_{i-1} \begin{pmatrix} x \\ y \end{pmatrix} \qquad (6)$$

In line 17, function $D(\cdot)$ determines the observation Y_i from l_i as shown in Eq. (7) such that $\xi_i \sim N(\mu_Y, \sigma_Y^2)$ with a Normal distribution based on the central limit theorem. The observation error ξ_i is calculated by Eq. (8), where $Rangle = Random[0, 2\pi)$ and its default value is $N(0.3, 0.1)$. Finally, in line 17, the noise ψ_i is added to the sequential points l_i such that $\psi_i \sim N(0.1, 0.05)$ as the default value.

$$D(\cdot) = l_i \begin{pmatrix} x \\ y \end{pmatrix} + \begin{pmatrix} \xi_i \times \cos Rangle \\ \xi_i \times \sin Rangle \end{pmatrix} \qquad (7)$$

$$\xi_i = \frac{speed_i \times \mu_Y}{sampling\_rate} + Random[-\sigma_Y^2, \sigma_Y^2] \qquad (8)$$

3.2   Location Subsystem

We define Space of Uncertainty (SoU) as the space where a mobile node might stay at a given time instance. We also define Flexible SoU as the SoU resizing over time. In Fig. 2, suppose that $|b_i - l_{i-1}|$ is the maximum mobile node displacement. The rising tiling pattern represents then SoU at the time instance i. In other words, the arc of possibilities for $l_i(x, y)$ is forward projected regarding the maximum mobile node displacement. Hence, the pseudo-cone shape is maintained despite variations in speed. The Flexible Space of Uncertainty can be expressed as a 4-tuple $(E[x_{i-1}], \theta_i, CTR, maxD)$ such that $E[x_{i-1}]$ is the expected value in the previous instance, θ_i is the standard angle, and $maxD$ is the maximum mobile node displacement $(\mu_s + \sigma_s \times 3)/sampling\_rate$. As far as we know, in the-state-of-the-art works the space of uncertainty of a mobile node is represented by a circle or square. However as stated earlier, SoU can be reduced significantly considering the mobile node status and its inherent features. For example, in a noise-free environment let a circle with radius equal to 6.4 m be the uncertainty space. The area where a mobile node might stay is $128.68\, m^2$ and the area of SoU is $13.99\, m^2$. In other words, there is a reduction of 89.13 % in the area, which can be even greater as the mobile speed increases after the threshold thr. In the Location Subsystem module, we have successfully introduced the Flexible SoU into Particle Filtering. Our proposed algorithm in this module adopts the Sequential Importance Resampling general framework in order to approximate the posterior probability distribution $p(x_i|Y_{0:i})$ by a weighting set of N particles. The calculation of the distribution is performed recursively using a Bayes filter and assuming that the Markov property is held.
 •   Algorithm 2 shows the procedure to compute the expected mobile node position $E[x_i]$. Initially, the Location Subsystem has no knowledge of its position. Hence, N particles are drawn randomly around the expected value $E[x_0]$ i.e.,

---

Algorithm 2. Particle Filtering + Flexible SoU.

Require: Y_i
Ensure: $E[x_i], x_i^p$
    {INITIALIZATION, $i = 0$}
 1: Draw particles N around expected value $E[x_0]$
 2: Uniform weighting. $w_i^p = 1/N$, $E[x_i] \leftarrow \sum_{p=1}^{N} x_i^p \times w_i^p$
 3: **loop**
 4:     {PREDICTION, $i > 0, \forall p$}
 5:     $\pi(x_i^p | E[x_{i-1}^p], Y_i, \phi_i)$, $w_i^p \leftarrow P(Y_i | x_i^p) = 1 - \frac{|Y_i - x_i^p|}{\sum |Y_i - x_i^p|}$, $E[x_i] \leftarrow \sum_{p=1}^{N} x_i^p \times w_i^p$
    {FILTERING}
 6:   **while** $x_i^p \notin SoU_i$ **do**
 7:     Resampling($x_i^p | E[x_i], SoU_i$)
 8:   **end while**
 9:   $w_i^p \leftarrow 1 - \frac{|E[x_i] - x_i^p|}{\sum |E[x_i] - x_i^p|}$, $\tilde{w}_i^p \leftarrow w_i^p \times w_{i-1}^p$, $w_i^p \leftarrow \frac{\tilde{w}_i^p}{\sum_{p=1}^{N} \tilde{w}_i^p}$
    { ESTIMATION }
10:   $E[x_i] \leftarrow \sum_{p=1}^{N} x_i^p \times w_i^p$
11:   **return** $E[x_i]$
12: **end loop**

---

at any possible location. In line 2, the probability for each particle w_0^p is uniformly established because there is no prior information. The expected value $E[x_0]$ is determined by a centroid estimation because all particles have the same weight. In line 5 x_i^p is computed using x_{i-1}^p and Y_i since x_i^p reflects all previous observations. Hence, the Probabilistic Random Mobility Model plus noise ϕ_i is used as the importance function π such that $\phi_i \sim N(0.3, 0.1)$ is given as default value. The assigned weights w_i^p are normalized in the range $[0,1]$ based on the distance between Y_i and x_i^p. Finally, the expected mobile node position $E[x_i]$ is determined by all particles and their corresponding weights such that $\{(x_i^p, w_i^p) | p \in [1, N], \sum_{p=1}^{N} = 1\}$. Through lines 6–8, particles x_i^p with negligible weights are replaced by new particles with higher weights around $E[x_i]$ and within SoU_i. In line 9, the particle weights are updated to $\tilde{w}_i^p$ based on previous valid possible locations and $\tilde{w}_i^p$ is normalized to w_i^p. In line 10, the estimated mobile node position, $E[x_i]$, is calculated based on the posterior distribution which is represented by the weighted set (x_i^p, w_i^p). Finally, $E[x_i]$ is provided to the Priority Suppress module.

3.3 Priority Suppress

We successfully introduced the suppress principle of the subsumption architecture [7] into our proposed general architecture for mobile node tracking in shaded areas. The principle has been widely employed in robotics and software agents because complex behaviors can be divided into simple modules that in turn are organized into layers. Information of an upper layer can subsume information of lower layers and this is graphically represented by the symbol "s", as shown in Fig. 1. In our proposed architecture the upper layer is identified by the black dot over line l_i. In other words, the outputs l_i of the main node tracking system subsumes the outputs $E[x_i]$ of Location Subsystem during non-shaded areas.

4   Experimental Results

In this section, we describe the experiments we conducted to measure the effectiveness of our proposed architecture for a mobile node localization in shaded areas. In particular, we are interested in the Location Subsystem evaluation considering the space of uncertainty as a circle and pseudo-cone shape. The former alludes to the most used form in the node tracking based on Particle Filtering and the latter refers to our proposed Flexible SoU. We use the term *General PF* to refer to the use of a circle as space of uncertainty and *PF + Flexible SoU* to refer to our proposal.

Hereafter, all simulation results are based on the default values, unless indicated otherwise: $total\_time = 18000$ s, $sampling\_rate = 4$ /s, $CTR = 6.4$ m, $\theta_0 = Random[0\pi, 2\pi)$, $l_0 = (0,0)$, $speed_i = N(8.33, 2.77)$, $P(speed_i|stayS) = stayS = 0.8$, $secondS = 5$ s, $P(\rho_i|stayO) = stayO = 0.8$, $secondO = 5$ s, $\xi_i = N(0.3, 0.1)$, $\psi_i = N(0.1, 0.05)$, $\phi_i = N(0.3, 0.1)$, $N = 25$ particles, and Markov chain size $= 50$ states. In Fig. 3a is shown the particle behavior of General PF and PF + Flexible SoU such that $N = 50$, $\theta_0 = 0$, and $i = 4$. The mobile node is represented by the large circle where arrows indicate its trajectory. The General PF particles are grouped around the mobile node at the current position (3.33, 0). However, 20% of the particles are behind the mobile node at the position $i-1$. Based on the fact that if the mobile node speed is greater than zero a sudden displacement in the opposite direction might not be possible. Hence, only 80% of the particles must be considered as valid. Moreover, considering the vehicle's CTR even more particles might be removed. On the other hand, the particles of our proposed PF + Flexible SoU are drawn regarding the bounded-forward displacement of the mobile node and they are grouped so that a pseudo-cone shape is formed. Besides, there is no particle behind the mobile node at the position $i - 1$.

The pseudo-cone form is observed even for a higher speed as long as it is less than the threshold thr. For example, Fig. 3b depicts the particle behavior when the mobile node speed is greater than, thr which changes the opening of the pseudo-cone form making it seem as a pseudo line. The update of $E[x_i]$ is mainly determined by the sampling rate which can increase or decrease the localization error of Location Subsystem module. We established $sampling\_rate = 4$ as the default value because the average estimation error is kept below of one meter for both General PF and PF + Flexible SoU, as shown in Fig. 4a. A higher value for $sampling\_rate$ involves a higher execution frequency of Location Subsystem module and therefore higher energy consumption. However, it is not a guarantee that the error will be less than one meter. For example, a higher mobile node speed increases the distance separation among measurement points so that the error for $sampling\_rate = 4$ and $speed = 41.66$ is similar to $sampling\_rate = 1$ and $speed = 11.11$ as shown in Fig. 4b where (n) represents

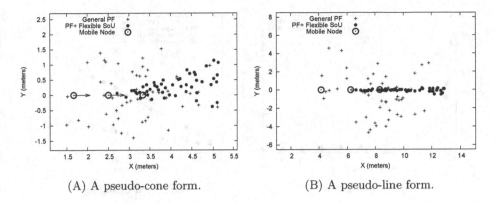

(A) A pseudo-cone form.   (B) A pseudo-line form.

Fig. 3. The behaviour of a set of particles under different speed.

the value for *sampling_rate*. The value of σ_s^2 was kept to 2.77 m/s and μ_s was gradually increased in units of 2.77 m/s staring from 2.78 m/s. Obviously, the probability of changing the speed also affects the accuracy, but it is a PRMM property used in combination with other parameters to generate diversity in trajectories. Hence, the probability is not considered as an element of the Location Subsystem module but rather the recorded states of the mobile node using recursively a Bayes filter and assuming that the Markov property is held.

On the other hand, the number of particles is another key factor to be considered. Maintaining a large number of particles can improve the accuracy, but requires additional memory and increases the execution time because the particle matrix size is determined by the Markov chain size (states) and number of particles. Figure 5a shows the Location Subsystem module accuracy under different number of particles, where (n) represents the value for *sampling_rate*. The error is quickly reduced as the number of particles increases, but it is fairly stable after 25 particles. The accuracy improvement for $25 < N < 501$ is negligible.

The particle movement from the previous to current state is carried out by the importance function π in which the noise ϕ allows particles to effervesce. But, a high noise can make particles go far away and causes a lot of resampling. We found that $\phi \sim N(0.3, 0.1)$ has the best performance and PF + Flexible *SoU* dampens better high values in μ_π than General PF as show in Fig. 5b, where (n) represents the value for *sampling_rate*.

Based on our experiments, the most relevant accuracy elements we found are: sampling rate, speed, number of particles, and noise (ξ, ψ, and ϕ). Both General PF and PF + Flexible *SoU* are fairly stable for different values in CTR, $P(\rho_i|stayO)$, and $P(speed_i|stayS)$. The change frequency in the mobile orientation, *secondO*, does not have an important impact in the accuracy because there is a bounded-forward displacement based on CTR. On the other hand, the change frequency in the mobile speed, *secondS*, affects the location accuracy from one time instance to another, but the average error tends to be almost the same for both a low and high value in *secondS*.

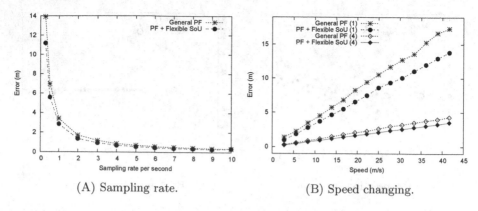

(A) Sampling rate.   (B) Speed changing.

Fig. 4. The Location Subsystem performance over a different sampling rate and speed.

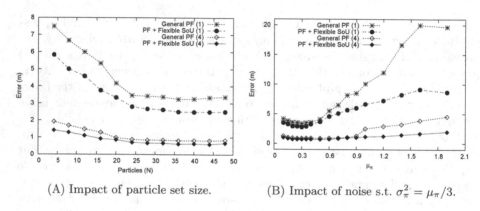

(A) Impact of particle set size.   (B) Impact of noise s.t. $\sigma_\pi^2 = \mu_\pi/3$.

Fig. 5. The Location Subsystem performance over a different particle set size and Noise.

5   Conclusion

We not only presented a solution for locating a mobile node in shaded areas but also a Probabilistic Random Mobility Model for the generation of paths, both based on CTR. Our Mobility Model is able to generate mobile node paths including the probability of change both the speed and orientation in a bounded-forward displacement. It is believed that better paths can be achieved by incorporating more kinetic elements in the Model, but it is beyond the scope of this article. Our main goal is to evaluate the Location Subsystem module performance under diversity of scenarios considering the space of uncertainty as a circle and pseudo-cone shape (Flexible SoU). The latter is used as part of our solution for mobile localization in shaded areas, which reduces the space at least 89.13 %. Nevertheless, the space reduction is not kept in the same proportion for accuracy reduction between General Particle Filtering and our proposed PF

+ Flexible *SoU* because there are many independence factors in the Mobility Model and Location Subsystem modules.

The major contributions of this paper are: (i) an architecture for locating a mobile node during shaded areas. (ii) We successfully introduced the suppress principle of the subsumption architecture [7] into our proposed architecture as a priority-selective control between l_i and $E[x_i]$. (iii) A Probabilistic Random Mobility Model for the generation of paths of any vehicle [1]. (iv) We successfully introduced the Flexible *SoU* upon the area of Particle Filtering and generation of paths. (v) The use of the CTR and status of a mobile node in combination to reduce and adapt the space of uncertainty where particles are drawn.

Acknowledgment. Research of M. Barbeau and E. Kranakis supported in part by NSERC grants and S. Jauregui by Emerging Leaders in the Americas Program (ELAP) and the National Council of Science and Technology (CONACYT) while the author was visiting the School of Computer Science, Carleton University.

References

1. American association of state highway and transportation officials
2. Baggio, A., Langendoen, K.: Monte carlo localization for mobile wireless sensor networks. Ad Hoc Netw. **6**(5), 718–733 (2008)
3. Bahl, P., Padmanabhan, V.: Radar: an in-building rf-based user location and tracking system. In: INFOCOM 2000 Proceedings of IEEE Nineteenth Annual Joint Conference of the IEEE Computer and Communications Societies, vol. 2, pp. 775–784 (2000)
4. Bancroft, J.B., Lachapelle, G.: Data fusion algorithms for multiple inertial measurement units. Sensors **11**(7), 6771–6798 (2011)
5. Bar-Shalom, Y.: Tracking and Data Association. Academic Press Professional Inc, San Diego, CA (1987)
6. Blumenthal, J., Grossmann, R., Golatowski, F., Timmermann, D.: Weighted centroid localization in zigbee-based sensor networks. In: Proceedings of IEEE International Symposium on Intelligent Signal Processing, WISP 2007, pp. 1–6 (2007)
7. Brooks, R.: A robust layered control system for a mobile robot. IEEE J. Robot. Autom. **2**(1), 14–23 (1986)
8. Bulusu, N., Heidemann, J., Estrin, D.: Gps-less low-cost outdoor localization for very small devices. IEEE Pers. Commun. **7**(5), 28–34 (2000)
9. Choi, B.S., Lee, J.J.: Sensor network based localization algorithm using fusion sensor-agent for indoor service robot. IEEE Trans. Consum. Electron. **56**(3), 1457–1465 (2010)
10. Fu, Y.J., Lee, T.H., Chang, L.H., Wang, T.P.: A single mobile anchor localization scheme for wireless sensor networks. In: proceedings of IEEE 13th International Conference on High Performance Computing and Communications, HPCC 2011, pp. 946–950 (2011)
11. Hall, D., Llinas, J.: An introduction to multisensor data fusion. Proc. IEEE **85**(1), 6–23 (1997)
12. He, T., Huang, C., Blum, B.M., Stankovic, J.A., Abdelzaher, T.: Range-free localization schemes for large scale sensor networks. In: Proceedings of the 9th annual international conference on Mobile computing and networking, MobiCom 2003, pp. 81–95 (2003)

13. Hu, L., Evans, D.: Localization for mobile sensor networks. In: Proceedings of the 10th Annual International Conference on Mobile Computing and Networking, MobiCom 2004, pp. 45–57 (2004)
14. Huang, R., Záruba, G.V.: Monte carlo localization of wireless sensor networks with a single mobile beacon. Wireless Netw. **15**(8), 978–990 (2009)
15. Jauregui-Ortiz, S., Siller, M., Ramos, F.: Node localization in wsn using trigonometric figures. In: Proceedings of 2011 IEEE Topical Conference on Wireless Sensors and Sensor Networks (WiSNet), pp. 65–68 (2011)
16. Lee, D., Son, S., Yang, K.W., Park, J., Lee, H.: Sensor fusion localization system for outdoor mobile robot. ICCAS-SICE **2009**, 1384–1387 (2009)
17. Mirebrahim, H., Dehghan, M.: Monte carlo localization of mobile sensor networks using the position information of neighbor nodes. In: Ruiz, P.M., Garcia-Luna-Aceves, J.J. (eds.) ADHOC-NOW 2009. LNCS, vol. 5793, pp. 270–283. Springer, Heidelberg (2009)
18. Tseng, Y.-C., Kuo, S.-P., Lee, H.-W., Huang, C.-F.: Location tracking in a wireless sensor network by mobile agents and its data fusion strategies. In: Zhao, F., Guibas, L.J. (eds.) IPSN 2003. LNCS, vol. 2634, pp. 625–641. Springer, Heidelberg (2003)
19. Vera, R., Ochoa, S.F., Aldunate, R.G.: Edips: an easy to deploy indoor positioning system to support loosely coupled mobile work. Pers. Ubiquit. Comput. **15**(4), 365–376 (2011)
20. Yu, L., Xiao, Y., You, H.: A novel centroid localization for wireless sensor networks. Int. J. Distrib. Sens. Netw. **2012**, 8 (2012). Article ID 829253
21. Záruba, G.V., Huber, M., Kamangar, F.A., Chlamtac, I.: Indoor location tracking using rssi readings from a single wi-fi access point. Wireless Netw. **13**(2), 221–235 (2007)

CAMS: Consensus-Based Anchor-Node Management Scheme for Train Localisation

Adeel Javed[1]([✉]), Zhiyi Huang[1], Haibo Zhang[1], and Jeremiah D. Deng[2]

[1] Department of Computer Science, University of Otago,
Dunedin, New Zealand
{adeel,zhuang,haibo}@cs.otago.ac.nz
[2] Department of Information Science, University of Otago,
Dunedin, New Zealand
jeremiah.deng@otago.ac.nz

Abstract. Train localisation is important to railway safety. Using Wireless Sensor Networks (WSNs) in train localisation is a robust and cost effective way. A WSN-based train localisation system contains anchor nodes that are deployed along railway tracks and have known geographic coordinates. However, anchor nodes along the railway tracks are prone to hardware and software deterioration such as battery outage, thermal effects, and dislocation. Such problems have negative impacts on the accuracy of WSN-based localisation systems. In order to reduce these negative impacts, this paper proposes a novel Consensus-based Anchor-node Management Scheme (CAMS) for WSN-based localisation systems. CAMS can assist WSN-based localisation systems to exclude the input from the faulty anchor nodes and eliminate them from the system.

The nodes update each other about their opinions on other neighbours. Each node uses the opinions to develop consensus and mark faulty nodes. It can also report the system information such as signal path loss. Moreover, in CAMS, anchor nodes can be re-calibrated to verify their geographic coordinates. In summary, CAMS plays a vital role in the life of the WSN-based localisation systems and in their ability to accurately estimate the train location. We have evaluated CAMS with simulations and analysed its performance based on real data collected from field experiments. To the best of our knowledge, CAMS is the first protocol that uses consensus-based approach to manage anchor nodes in train localisation.

1 Introduction

Railway systems have provided an important means of transport over the past hundred years, with significant investments having been made in safety infrastructure. In recent years, real-time train localisation is becoming more essential in serving the need of safety. Even though many GPS-based approaches have been designed and deployed for localisation and tracking applications, drawbacks such as limited coverage and sophisticated infrastructures have prevented them from being used for train localisation. For example, trains in subway systems primarily operate underground. Even in the above-ground railway transportation, trains

© Springer International Publishing Switzerland 2015
S. Papavassiliou and S. Ruehrup (Eds.): ADHOC-NOW 2015, LNCS 9143, pp. 107–120, 2015.
DOI:10.1007/978-3-319-19662-6_8

may frequently pass through GPS dark territories such as tunnels and hilly regions [2]. Other technologies like WLAN, RFIDs and bluetooth also have certain limitations for the train localisation scenario due to either high cost of infrastructure installation and complicated protocol stack, or limited transmission range and lack of collision avoidance with limited computational capabilities. The need of heavy investment and highly trained staff make such systems less feasible for real-time tracking of trains. There is an increasing desire to be capable of tracking trains in real-time and cost-effective way in order to improve efficiency while ensuring safety.

WSNs are cost effective and useful technology that can be used for train localisation. In the past years, many researches were conducted on using WSNs for localisation [12,17]. In WSN-based train localisation, anchor sensor nodes that have known geographical coordinates are deployed along the railway track to detect the incoming train. Upon detection, these anchor nodes report their geographical coordinates to the gateway node that is installed on the train. Then the gateway node estimates the transmission distance based on Received Signal Strength (RSS) using the path loss model [18] and localisation algorithms such as particle filter [7].

The anchor nodes along the railway track may suffer from the location errors caused by software or hardware bugs. Therefore, they need to be re-calibrated in terms of their geographic coordinates and the path loss of the signals sent by the anchor nodes. Moreover, the presence of faulty nodes [6] in the system can also deteriorate the accuracy of the location estimation. All these issues should be addressed in the WSN-based train localisation system. Manually sorting out such problems by human beings incurs significantly high cost. The management and maintenance of the anchor nodes with the help of each other play an important role in the stability of the whole localisation system.

Therefore, the need of a management scheme comes into play which can enable anchor nodes to detect the faulty nodes among themselves. The faults should be reported to the gateway node for further analysis. Furthermore, the management scheme should assist anchor nodes to estimate the path loss ratio of their signals, which depends very much on the surrounding environment and affects directly the distance estimation based on RSS. Such management scheme can clearly improve the accuracy of train localisation by excluding the faulty nodes and re-calibrating the parameters of the anchor nodes like path loss ratio.

In this paper, we propose a management scheme called CAMS (Consensus-based Anchor-node Management Scheme) for our WSN-based train localisation system. CAMS allows anchor nodes to share their opinions about trustworthiness of their neighbour nodes and develop consensus to detect the faulty nodes. The anchor nodes can be automatically re-calibrated in terms of path loss ratio and geographical coordinates. The main contributions of this work are summarised as follows.

- We propose the novel CAMS for management and maintenance of anchor nodes in WSN-based train localisation systems. As far as we know, CAMS is the first scheme that uses a consensus-based approach to manage anchor nodes in train localisation. Additionally our consensus algorithm uses history data as well as the current data to reduce false detection ratio of faulty nodes and increases the accuracy of the re-calibrated path loss ratio.

- CAMS is implemented in a simulated environment using MATLAB. The simulation is based on the real data collected from field experiments in various environments such as open fields, train stations and tunnels. These data make our simulation very close to the real-world implementation.
- From the results collected from the simulation, we find that CAMS can effectively detect the presence of faulty nodes in the system. Our results show that, with the re-calibration of the path loss ratio of the anchor nodes, the accuracy of train localisation can be improved up to 15%.

The rest of the paper is organised as follows. Section 2 discusses the related work. Section 3 describes the system model. In Sect. 4, we present the proposed CAMS, which is then evaluated in Sect. 5. Finally, Sect. 6 concludes the paper and sheds light on future work.

2 Related Work

Management of sensors in WSNs has become one of the major focuses of research in the recent decade. It is an important research issue for the stability of the system. In particular, safety related applications such as train localisation make sensor management increasingly more important. Sensor management includes the calibration of anchor nodes' location, calibration of path loss ratio, and detection of faulty nodes in the system.

Marti et al. [9] proposed two techniques to monitor the misbehaving nodes in the ad-hoc networks: watchdog, and pathrater. In the scheme, watchdog locates the malicious nodes and pathrater helps to establish routing path by avoiding those nodes. Both help improve the system performance. However, it is a centralised approach and does not use the consensus approach as we do in CAMS which is more robust.

Cooperative sensing [5,6] is an important technique to observe the phenomenon of interest in WSNs. However, it raises new concerns for the reliability and the security as expressed by Mishra et al. [11]. The work in [13,20] also discusses the drawbacks of cooperative sensing for large scale networks such as need of synchronisation and more energy consumption. These problems are properly addressed in CAMS, as to be shown in the following sections.

Srinivasan et al. [15] used an interesting idea of majority voting scheme, in which each beacon node with known location computes the reputation of other beacon nodes and casts the votes upon request by sensor node to judge the trust factor of that beacon. However, if a beacon node pretends to be trusted sensor for some time until it gains positive reputation, it may not be detected as malicious and later it can start to spread mis-information. Moreover, it is not clear how the scheme deals with ID forgery. However, in CAMS, the inclusion of the neighbour node's opinion makes it harder for a malicious anchor node to achieve its purpose and in case of ID forgery, system will be notified about the peculiar behavior due to the different nodes with the same ID and will mark the duplicate ID as a faulty node.

Srinivasa et al. [14] proposed a technique which takes empirical and analytical distribution of received mean power to estimate the path loss ratio by comparing the empirical and theoretical distribution. In [8] Cayley-Menger determinant was proposed to determine the geometric constraints which are used to estimate the path loss ratio. However, in CAMS, the path loss ratio is estimated based on consensus. Each anchor node sends its path loss ratio to other anchor nodes and calculates the estimated path loss ratio based on the trust factors of the anchor nodes. This method makes the estimation of path loss ratio smoother and more trustworthy.

Zhao et al. [19] propose da technique to monitor energy levels using local energy level aggregations. Each individual node scans its energy and reports the range of residual energy to the gateway which aggregates the results and sends them to the servers. In [10], the authors presented a technique to generate the energy map in which sink node uses local information received from sensors to update energy level based on the activity performed at each sensor. In CAMS, anchor nodes adopt the technique in [10] to report its battery level to the gateway node, and mark anchor nodes under acceptable threshold as faulty nodes.

3   System Models and Problem Statement

This section discusses the system models used in our WSN-based train localisation system and defines a list of problems which will be addressed in the following sections.

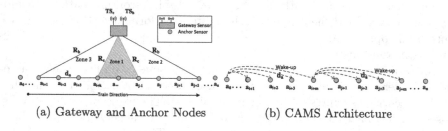

(a) Gateway and Anchor Nodes   (b) CAMS Architecture

Fig. 1. WSN-based train localisation system.

3.1   System Models

In our network model, there are a set of anchor nodes and one gateway node, as shown in Fig. 1(a). The set of n anchor nodes $\{a_0, a_1, \ldots, a_n\}$ are uniformly deployed along the track with equal distance d_a between any two consecutive anchor nodes. Each anchor node is equipped with a single radio transceiver with transmission range of R_c. We assume that each anchor node is hard-coded with its geographic coordinates before deployment. The gateway node is installed on the train. It is equipped with two radio transceivers: TS_c and TS_b. TS_c is used to communicate with the anchor nodes that fall into its transmission range, and TS_b is used to continually broadcast beacon packets to activate the anchor nodes before they approach the transmission range of TS_c. The transmission range for TS_c and TS_b is R_c and R_b, respectively. We assume that R_b is larger than R_c. To avoid interference, we assume TS_c and TS_b operate on two non-overlapping channels ch_c and ch_b, respectively. Each anchor node operates on both channels, that is, uses ch_b during duty-cycling and switches to ch_c to communicate with TS_c when it is within the transmission range of TS_c. As shown in Fig. 1(a), zone 1 is the region covered by TS_c, and zone 1, zone 2 and zone 3 are the regions covered by TS_b.

In the duty-cycling model of our WSN-based train localisation, anchor nodes operate in an asynchronous duty-cycling mode in which each anchor switches between sleep and wake-up states independently without synchronisation among the nodes. Each anchor node periodically wakes up to detect beacon packets, called Clear Channel Assessment (CCA) checks. Upon detection of beacon packets, it stays in active state until the scheduled communication with the gateway node (i.e. the train) is completed; otherwise, it switches back to sleep state and repeats another duty-cycle.

In our localisation model, as the train moves, the gateway node continually broadcasts beacon packets through its TS_b transceiver. These broadcasted beacon packets contain the information regarding train such as current train speed, and latest computed location. Upon receiving a beacon during CCA checks, an anchor node wakes and communicates with the gateway node. Using algorithms like particle filter, the gateway node computes its new location based on the received location information from the anchor nodes and the RSS information from the communication with the anchor nodes.

In our train localisation system, A BWS scheme [3,4] is used to wake-up the anchor nodes. This scheme has three stages: duty cycling sleep, communication with the gateway node, and duty cycling resumption. During communication with the gateway node, there are multiple rounds of information exchange between the gateway and each anchor node.

3.2   Problem Statement

In this paper, we consider a set of m single hop anchor sensor nodes denoted as $\{a_1, a_2, a_3, \ldots, a_m\}$ We model the network as an undirected graph $G = (V, E)$, where V is the set of anchor nodes and E the set of communication links between the nodes. An edge exists between any two nodes that are in each other's communication range. As the anchor nodes in the network follow asynchronous duty-cycling without any knowledge of sleep schedules of neighbouring anchor nodes, they must wake up to perform the faulty node detection and calibration. Each anchor node sleeps for t_{sleep}^{ub} time which is computed as in [3]. In CAMS, periodically, any anchor node that wakes up continually broadcasts beacons for t_{sleep}^{ub}

to wake up its neighbour nodes to detect faulty nodes and to calibrate. Once awoken, the anchor nodes develop opinions about their neighbour nodes. The opinion of a neighbour node includes the evaluation of the claimed location of the node. If the claimed location of the neighbour node is found to be at the estimated distance according to RSS with acceptable threshold, the opinion score of the neighbour increases; otherwise opinion score decreases. The acceptable threshold depends upon the radio chipsets noise range which is $\pm 6\,dB$ for CC2420 radio chipset [16]. Each anchor node stores opinions about its neighbours in Neighbour Opinion Table (NOT).

The first problem we address is "how to make anchor nodes to communicate and develop opinions about their neighbour nodes in the presence of asynchronous duty-cycling". The opinions are important to analyse the trustworthiness of neighbour nodes. The untrusted anchor nodes, e.g. faulty nodes, must be reported to the gateway node to eliminate their false input which may affect the accuracy of the WSN-based localisation system. To calculate opinion scores about the neighbour nodes, they must be in wake-up state to communicate. In CAMS, BWS [3] is used to enable anchor nodes to wake up.

The second problem is "how to evaluate their NOT in order to detect the faulty nodes based on consensus of the received individual opinions". Anchor nodes are required to report NOT to each other. Each anchor node develops consensus based on received NOTs and marks and eventually eliminates the faulty anchor nodes from the system. The detection of such anchor nodes allows the gateway node to ignore the inputs of the faulty anchor nodes and reduce the effects of their biased opinions.

The third problem we address is "how to assist the gateway node to estimate the consensus-based re-calibration of path loss ratio estimated by the individual anchor nodes". The consensus-based re-calibration of path loss ratio helps the gateway node to improve the accuracy of train localisation.

3.3   System Assumptions

CAMS is proposed under the following assumptions, though these assumptions can be relaxed with slight modification of our existing system.

- Each anchor node is anchored at a fixed location along the railway track. They cannot move without physical intervention. If a node is maliciously removed to a different location, it will be treated as a faulty node in CAMS.
- The geographic coordinates of each node are hard-coded before deployment. However, the location information could be different later from the real location due to node malfunction or malicious dislocation.
- The ID of an anchor node is unique and encrypted with the message using a shared key. In the case of intrusion where the shared key is cracked and the ID is forged, the node with the forged ID will be detected by CAMS as a faulty node due to its peculiar behaviour such as incorrect claimed location. All reports on faulty nodes will be sent urgently to the management center for immediate human response.

4   CAMS: Consensus-based Anchor Node Management Scheme

CAMS enables anchor nodes to work in their consensus-based management. The following sections discuss CAMS in detail.

4.1   Impacts of Faults in Anchor Nodes

Anchor nodes provide gateway node with information such as geographic coordinates, and track health status. Therefore, to infer meaningful conclusions with the received information, its quality must be ensured. The potential occurrence of faults in the sensor network can affect the data integrity which can lead to wrong estimation of the train's location. There are several features which lead to faults in the sensor nodes such as transceiver input-output detection range, sensor age, battery state, noise, sensor response hysteresis, and dislocation. Input from anchor nodes with aforementioned faults can affect the accuracy of WSN-based train localisation systems.

Over time, anchor nodes may not remain consistent with their hard-coded geographic coordinates due to change in their position by physical unauthorised intervention or due to node malfunction. This inconsistency leads to dissemination of mis-information. Moreover, low battery states can also make anchor node's parameters such as path loss ratio deviate from the expected range. Similarly, environmental effects such as weather can change those parameters to an unacceptable range.

CAMS allows anchor nodes to communicate with each other to detect such faulty nodes based on consensus and assist the gateway node to neglect their inputs. In addition, it allows anchor nodes to re-calibrate their path loss ratio to improve the accuracy of WSN-based train localisation. It is worth noting that the assumption of encrypted ID keeps the system safe from the intrusion of malicious nodes. However, if a malicious node successfully breaks into the system with a forged duplicated ID, it can be detected as a faulty node as soon as it starts to transmit wrong location information.

4.2   Computation of Opinion Score

Periodically, anchor nodes perform calibration and exchange their opinion about each other. When an anchor node wakes up to perform CAMS tasks, it continually broadcasts beacons for t_{sleep}^{ub} time to wake up its neighbour anchor nodes as given in BWS protocol [3]. Once its single-hop neighbours are awoken and broadcast their location information, it computes and stores the opinion scores of its neighbour nodes in its NOT table. Initially, the opinion score computed by an anchor node about any of its neighbour is zero as it knows nothing about the neighbour.

Suppose a neighbouring node a_i broadcasts a packet to a node a_j. The receiver node a_j develops its opinion about a_i after examining the location information Loc_{a_i} sent by a_i. Let d be the distance between a_j's location Loc_{a_j} and

a_i's claimed location Loc_{a_i}, and d' be the distance estimated according to the log-distance path loss model in Eq. 1.

$$RSS(d') = P_{tx} - PL(d_0) - 10\eta \log_{10} \frac{d'}{d_0} - X,$$ (1)

The path loss model in Eq. 1 is a well-known radio propagation model [18] that predicts the path loss a signal encounters over distance, and it has been widely used for distance estimation. In Eq. 1, $RSS(d')$ is the received signal strength in dBm at a given distance d' from the transmitter, P_{tx} is the power in dBm of the transmitted signal, $PL(d_0)$ is the path loss at a reference distance d_0, and η is the path loss ratio. The term $P_{tx} - PL(d_0)$ refers to the base power at reference distance. X in Eq. (1) is a random variable that reflects the noise in the signal strength due to different environmental factors such as reflection and fading.

Node a_j computes its opinion about a_i as follows. If the difference between the estimated distance and the claimed distance of a_i is within acceptable threshold (θ_1), a_j gives a positive vote ($v = 1$) to a_i; otherwise, a_j casts a negative vote ($v = 0$) to a_i. The acceptable threshold depends upon the radio chipsets noise range which is $\pm 6\,dB$ for CC2420 radio chipset [16]. The opinion is computed using Eq. 2. It depends on the historic opinion as well.

$$O'_{j\rightarrow i} = pO_{j\rightarrow i} + (1-p)(1 - \frac{|d - d'|}{\theta_1})v,$$ (2)

where, $O'_{j\rightarrow i}$ is the update of the opinion about the trustworthiness that a_j develops about a_i, p is the weight assigned to the historic opinion. The term $(1 - \frac{d-d'}{\theta_1})$ decreases with the increase of deviation between d and d' if the deviation is within the acceptable threshold. If the deviation is larger than the threshold θ_1, v is zero and the opinion is gradually decreased to zero.

The above process is carried out in each node periodically. However, there is a tradeoff between energy saving and the frequency of the process. Assuming the nodes are not faulty very often, the process could be less frequent, say once a day. However, the process could also be triggered by the gateway on the train if the gateway finds the deviation of a node's claimed location and its estimated location based on Eq. 1 is larger than the threshold.

4.3   Consensus-based Faulty Nodes Detection

After the information exchange between the anchor nodes, each anchor node updates its NOT. Then each anchor node broadcasts its location information, opinion about its neighbours (NOT), transmission power (P_{tx}), date of deployment, and residual battery level (which can be estimated as given in [19]). Each node then develops consensus according to the received opinions from other anchor nodes using Eq. 3, where CS'_{a_i} is the consensus score about the anchor node a_i. It includes the averaged received opinions from other nodes and also incorporates historic consensus score CS_{a_i} with a weight q.

$$CS'_{a_i} = qCS_{a_i} + (1 - q)\frac{\sum_{j=1}^{m} O_{j \to i}}{m} \tag{3}$$

After the anchor nodes communicate and share their opinions with each other, if the concensus score of a node is below a threshold (θ_c), that node is identified as a faulty node and will be excluded from the trusted node list. Similarly, each anchor node also marks the anchor nodes with residual battery power under the acceptable threshold (θ_2) as faulty anchor nodes. The age of anchor nodes are computed from the date of deployment and their recommended operational period. Each anchor node receives the date of deployment of other nodes and computes their age. A list of anchor nodes with their expiry date under a threshold (θ_a) is reported back to the gateway node for possible replacement.

Algorithm 1. Detection of faulty anchor node by a_j

---

1 **for** *Each anchor node a_i* **do**
2   **if** $|d - d'| \le \theta_1$ **then**
3        $O'_{j \to i} = pO_{j \to i} + (1 - p)(1 - \frac{d - d'}{\theta_1})$
4   **else**
5        $O'_{j \to i} = pO_{j \to i}$

6 **for** *Each anchor node a_i* **do**
7     $CS'_{a_i} = pCS_{a_i} + (1 - p)\frac{\sum_{j=1}^{m} O_{j \to i}}{m}$
8   **if** $CS_{a_i} < \theta_c$ **then**
9   Enlist a_i as faulty anchor node
10   **else**
11   a_i is a trustworthy anchor node

4.4 Consensus-based Calibrated Path Loss Ratio Estimation

The anchor nodes can also calibrate the path loss ratio (η) in Eq. 1 to assist the gateway node in the WSN-based train localisation. Each anchor node broadcasts its transmitting power (P_{tx}) along with geographic coordinates which help the receiving anchor nodes to estimate the attenuation rate of signal strength, called path loss ratio. Equation 4 defines the simplified equation for an anchor node a_j to calculate η_{a_i}:

$$\eta_{a_i} = \frac{P_{tx}^{do} - P_{rx}^{d'}}{10 log d'}, \tag{4}$$

where $P_{rx}^{d'}$ is the receiving power. Each anchor node receives the calculated η_{a_i} from other anchor nodes and calculates the consensus η'_{cs} as shown in Eq. 5. The equation considers the weighted averages of the received η according to the consensus scores of the corresponding nodes. This gives more weight to the trusted anchor nodes' estimation. In addition, an anchor node also considers the historic path loss ratio η_{cs} when updating η'_{cs} and r is the weight assigned to the historic path loss ratio.

$$\eta'_{cs} = r\eta_{cs} + (1-r)\frac{\sum_{i=1}^{n} \eta_{a_i} CS_{a_i}}{\sum_{i=1}^{n} CS_{a_i}} \tag{5}$$

4.5   Reporting to Gateway

Each anchor node updates the gateway node about the new path loss ratio η, and a list of faulty nodes when train passes. The gateway node later on updates the human resource involved in the management of the anchor nodes to undertake the necessary steps in rectification of faults or removal of faulty sensors. In the long run such management improves the accuracy and lifetime of the system, as shown in the next section.

5   Performance Evaluation

In this section, we discuss the implementation of our CAMS scheme to assist the overall maintenance of train localisation system. In our simulation, we implemented CAMS scheme which enables the anchor nodes to coordinate and mark the faulty nodes among them, notify the faults, and calculate the path loss ratio to assist the gateway node to increase the accuracy of the train localisation.

The radio characteristics of the anchor and gateway nodes used in our analysis are taken from CC2420 chipset data sheet [16]. All simulations are run independently and their results are averaged for all iterations. The performance metrics we evaluated are: consensus score, opinion score, path loss ratio estimation, and distance estimation error. The other simulation parameters are: deployment density, average number of received packets, weights for historic opinion and consensus score, and weights for historic path loss ratio.

5.1   Simulation Setup

Our simulation has been done on a MATLAB simulator. In our simulation, we use real data collected from field experiments in railway representative environments such as open fields, railway stations, and tunnels. Our experiments are based on several Maxfor's MTM sensor platforms [1] such as MTM-CM3300, MTM-CM5000, and MTM-CM4000 with transmission ranges of 800 m, 150 m, 150 m, respectively. Moreover, in our simulation, we deployed anchor nodes in several deployment density settings to collect RSS measurements. In our simulation, we present the results based on single representative dataset due to limitation of space. The simulation results show how different variables shape the opinion and consensus score of anchor nodes which assist each anchor node to mark the faulty anchor nodes. The consensus threshold $\theta_c = 0.4$ is calculated based on average difference between consensus scores of trusted and faulty node over 500 iterations. The detailed configurations for the simulation parameters are given in Table 1.

Table 1. Simulation parameters

Parameters	Values	Parameters	Values
No. of 1-hop anchor nodes	25	Voltage	3.0 v
Average received packets	25	θ_1	5 m
p	0.5	q	0.5
r	0.5	θ_2	5 %
θ_c	0.4	θ_a	95 %

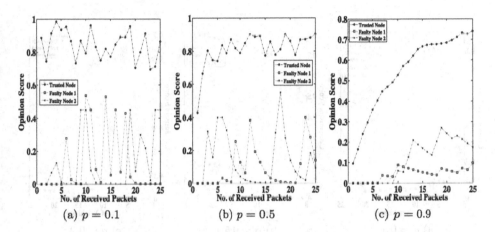

(a) $p = 0.1$ (b) $p = 0.5$ (c) $p = 0.9$

Fig. 2. Opinion score computed by an anchor node.

5.2 Detection of Faulty Anchor Nodes

Each anchor node develops opinion about its neighbour anchor nodes by receiving packets. The opinion score categorises anchor nodes as trusted or faulty based on difference between the claimed and estimated distance. The number of received packets also effects the computation of opinion score about sending anchor node. In the first set of results, we have 2 faulty anchor nodes in the system. Figure 2 shows the opinion scores of both faulty anchor nodes and a trusted anchor node calculated by a trusted anchor node. It can be seen that the opinion score fluctuates a lot when the weight of the historic opinion is small such as $p = 0.1$ as shown in Fig. 2(a). However, opinion about other anchor nodes become more stable with the increase in the weight assigned to historic opinion to 0.5 and 0.9, respectively, as shown in Fig. 2(b), (c). This takes more number of packets by sender to develop opinion score about it and thus takes longer time to pass the threshold of trustworthiness (θ_c).

Each anchor node compiles the consensus score from opinions received from other anchor nodes. In our simulation, the number of neighbouring anchor nodes varies from 5 to 25 which is called deployment density. In Fig. 3, we have shown consensus score computed by a trusted node about 3 anchor nodes: 1 trusted anchor node, and 2 faulty anchor nodes. In Fig. 3(a)–(c), it shows the impact of weight factor given to the past consensus score which is 0.1, 0.5, and 0.9, respectively. It can be seen that if the consensus score, CS_{a_i} relies least on the historic data the score is significantly high for trusted anchor node even if the claimed location deviates to the far extent within the acceptable margin. However, if we increase the weight percentage to 0.5 and 0.9 it takes more opinions from multiple anchor nodes to develop the consensus score and it takes more time to see the decline in the consensus score of a trusted node which turns to be faulty later on.

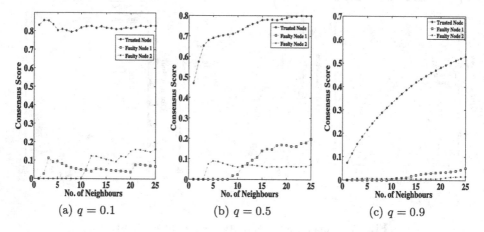

Fig. 3. Consensus score to mark the faulty anchor node.

5.3 Estimation of Path Loss Ratio

The estimation of consensus based path loss ratio represents the real time signal attenuation rate in particular environment. Such parameter assist the gateway node to improve the accuracy of WSN-based train localisation. The consensus based path loss ratio based on calibrated path loss ratio estimated by each anchor node and its impact on the accuracy of the localisation system are shown in Fig. 4(a), (b), respectively. Figure 4(a) presents the path loss ratio calibrated by three trusted anchor nodes. It can be seen that due to signal reflection, each anchor node's path loss ratio fluctuates significantly which is based on RSS. However we can see that the consensus based path loss ratio is relatively more stable and it improves with the increase in the number of packets received. The impact of consensus based path loss ratio in terms of localisation error is shown in Fig. 4(b). The localisation error stays from 0.22 m to 0.04 m which drops down to 0.06 m–0.001 m while using consensus based path loss ratio. This improves the localisation accuracy from almost 5 %–15 %.

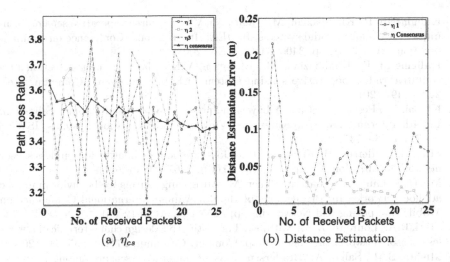

(a) η'_{cs} (b) Distance Estimation

Fig. 4. Impact of Consensus-based path loss ratio estimation.

6   Conclusions and Future Work

In this paper we presented a novel CAMS scheme to assist the train localisation system for management of anchor nodes deployed along the track. CAMS works on the mutual cooperation and consensus based theory to detect the faulty anchor nodes, report the faults, and assist the gateway node in the estimation of path loss ratio. We have shown through simulation that the proposed scheme is robust in dense networks and can detect the presence of faulty sensor nodes.

Though CAMS is capable of detecting malicious nodes as faulty nodes up to some extent but it cannot cope with a large number of intruding nodes with forged ID. As a future work, we will investigate how to defend the system from intruding of a large number of malicious nodes to make the train localisation system more secure. We would conduct more exhaustive simulations to confirm the robustness of our scheme in train localisation systems. We also plan to implement our scheme on hardware and test its performance in real scenario to compare with other schemes.

References

1. MTM specifications, March 2012. http://www.maxfor.co.kr/datasheet/MAXFOR_Digital_Brochure.pdf
2. Acharyaa, A., Sadhu, S., Ghoshala, T.: Train localization and parting detection using data fusion. Transp. Res. Part C: Emerg. Technol. 19(1), 75–84 (2011)
3. Javed, A., Zhang, H., Huang, Z., Deng, J.: BWS: beacon-driven wake-up scheme for train localization using wireless sensor networks. In: 2014 IEEE International Conference on Communications (ICC), pp. 276–281, June 2014
4. Javed, A., Zhang, H., Huang, Z.: Performance analysis of duty-cycling wireless sensor network for train localization. In: Proceedings of Workshop on Machine Learning for Sensory Data Analysis, p. 43. ACM (2013)

5. Kaligineedi, P., Khabbazian, M., Bhargava, V.K.: Secure cooperative sensing techniques for cognitive radio systems. In: IEEE International Conference on Communications, ICC 2008, pp. 3406–3410. IEEE (2008)
6. Kaligineedi, P., Khabbazian, M., Bhargava, V.K.: Malicious user detection in a cognitive radio cooperative sensing system. IEEE Trans. Wirel. Commun. **9**(8), 2488–2497 (2010)
7. Klepal, M., Pesch, D., et al.: A bayesian approach for rf-based indoor localisation. In: 4th International Symposium on Wireless Communication Systems, ISWCS 2007, pp. 133–137. IEEE (2007)
8. Mao, G., Anderson, B., Fidan, B.: Path loss exponent estimation for wireless sensor network localization. Comput. Netw. **51**(10), 2467–2483 (2007)
9. Marti, S., Giuli, T.J., Lai, K., Baker, M.: Mitigating routing misbehavior in mobile ad hoc networks. In: Proceedings of the 6th Annual International Conference on Mobile Computing and Networking, pp. 255–265. ACM (2000)
10. Mini, R.A., Loureiro, A.A., Nath, B.: The distinctive design characteristic of a wireless sensor network: the energy map. Comput. Commun. **27**(10), 935–945 (2004)
11. Mishra, S.M., Sahai, A., Brodersen, R.W.: Cooperative sensing among cognitive radios. In: IEEE International Conference on Communications, ICC 2006, vol. 4, pp. 1658–1663. IEEE (2006)
12. Ren, H., Meng, M.: Power adaptive localization algorithm for wireless sensor networks using particle filter. IEEE Trans. Veh. Technol. **58**(5), 2498–2508 (2009)
13. Song, C., Zhang, Q.: Sliding-window algorithm for asynchronous cooperative sensing in wireless cognitive networks. In: IEEE International Conference on Communications, ICC 2008, pp. 3432–3436. IEEE (2008)
14. Srinivasa, S., Haenggi, M.: Path loss exponent estimation in large wireless networks. In: Information Theory and Applications Workshop, pp. 124–129. IEEE (2009)
15. Srinivasan, A., Teitelbaum, J., Wu, J.: DRBTS: distributed reputation-based beacon trust system. In: 2nd IEEE International Symposium on Dependable, Autonomic and Secure Computing, pp. 277–283. IEEE (2006)
16. Texas Instruments: 2.4 GHz IEEE 802.15.4/ZigBee-ready RF transceiver (2003). http://www.ti.com/lit/ds/symlink/cc2420.pdf
17. Vijayakumar, J., Zhang, H., Huang, Z., Javed, A.: A particle filter based train localization scheme using wireless sensor networks. In: 2013 IEEE 11th International Conference on Dependable, Autonomic and Secure Computing (DASC), pp. 269–274, December 2013
18. Xu, J., Liu, W., Lang, F., Zhang, Y., Wang, C.: Distance measurement model based on RSSI in WSN. Wirel. Sens. Netw. **2**(8), 606–611 (2010)
19. Zhao, Y.J., Govindan, R., Estrin, D.: Residual energy scan for monitoring sensor networks. In: Proceedings of the IEEE Wireless Communications and Networking Conference, WCNC 2002, vol. 1, pp. 356–362. IEEE (2002)
20. Zhou, X., Ma, J., Li, G.Y., Kwon, Y.H., Soong, A.C.: Probability-based combination for cooperative spectrum sensing. IEEE Trans. Commun. **58**(2), 463–466 (2010)

Delay Analysis of Context Aware Mobility Management Systems Addressing Multiple Connectivity Opportunities

Adamantia Stamou[1,2]($\boxtimes$), Nikos Dimitriou[1], Kimon Kontovasilis[1], and Symeon Papavassiliou[2]

[1] National Centre for Scientific Research "Demokritos", Institute of Informatics and Telecommunications, Agia Paraskevi, Athens, Greece
{stamouad,nikodim,kkont}@iit.demokritos.gr
[2] NETwork Management and Optimal DEsign Laboratory, National Technical University of Athens, Athens, Greece
papavass@mail.ntua.gr

Abstract. The support of context aware Vertical Handover Operations (VHO) for optimally exploiting multiple connectivity opportunities is addressed by considering a typical generic architecture, reminiscent of relevant frameworks, such as IEEE 802.21. The paper develops a novel modeling methodology that can capture concisely, but also effectively, all important factors that have an impact on the system's performance, including the intensity of handover requests, network topological and availability characteristics, various sources of signaling overhead and the congestion points imposed by the architectural components. The model is validated by simulation and is employed for investigating the impact of the various parameters on the mean delay required for the VHO preparation phase. The principles of the proposed modeling methodology could be exploited for the future study of additional more decentralized solutions, towards addressing distributed mobility management scenarios, including ad-hoc and mesh network topologies.

Keywords: Mobility management · Context awareness · Delay analysis

1 Introduction

The requirement for mobility and global wireless broadband coverage has been a key research issue in the last ten years. Moreover, the Future Internet (FI) vision is expected to extend the 'always best connected' (ABC) notion, enabling end-users to attain any service using any available network, within a heterogeneous environment consisting of open, intelligent and collaborative wireless and wire-line access networks. "Heterogeneity" is being addressed in terms of the development of different Radio Access Technologies (RATs), such as Wireless Fidelity (WiFi), Worldwide Interoperability for Microwave Access (WiMAX), Long Term Evolution of 3rd Generation Partnership Project (3GPP LTE), etc., but also in different cell types of the same RAT (i.e. 3GPP Femto-, Pico-, Micro- and Macro-Cells). The latter can be seen also as multi-tier (or multi-layer) deployment, which can be implemented as an overlay of cells of different

© Springer International Publishing Switzerland 2015
S. Papavassiliou and S. Ruehrup (Eds.): ADHOC-NOW 2015, LNCS 9143, pp. 121–133, 2015.
DOI: 10.1007/978-3-319-19662-6_9

sizes using a single RAT. For instance, outdoor terminals may be served by a combination of micro and macro LTE cells [1].

Furthermore, other connectivity paradigms, such as Machine to Machine (M2 M) or Device to Device (D2D) have emerged, introducing the implementation of next-generation ad-hoc networks and particularly MANETs (Mobile Ad-hoc Networks). The spontaneous, ad-hoc wireless communication between devices, allows devices to establish communication, anytime and anywhere without the aid of a central infrastructure [2]. More specifically, M2 M virtual cell concepts can be implemented in low power devices, where the device is always connected to the same virtual cell during movement. In addition, direct M2 M communication may achieve better spatial reuse of transmission resources in a cellular network; in this context, it has been referred that in a local area cellular network overall network throughput may increase up to 65 per cent [3].

Based on the above, it is clear that a unified approach to manage heterogeneous networks is essential to achieve seamless broadband connectivity addressing the increasing user requirements. It is therefore of paramount importance to study in detail the enabling mechanisms with which seamless mobility among different access networks will be possible. Thus, a context aware vertical handover management framework is needed, in order to optimize the handover preparation procedure, ensuring service continuity and adapting to changes. The unified heterogeneous network management solution will have to cover all connectivity opportunities (RATs and tiers) and has currently been part of the subject of working groups within major standardization bodies [4, 5], introducing the concept of the information server as a key component that could be used in the direction of context-awareness.

As a contribution along this objective, this paper provides a model-based efficiency study of Vertical Handover Operations (VHO) management architectures and the associated signaling. The study addresses a typical architecture reminiscent of the IEEE 802.21 framework [4]. Context awareness is assumed to be provided through a centralized network Information Server (IS), an entity responsible for maintaining information relevant to the various Radio Access Networks (RANs) serving an area of interest. It is noted, however, that the modeling methodology is suitably generic, so that it can also be readily applied to the study of more decentralized solutions that might be of future relevance.

The model takes into account, in concise terms, many important factors that have an impact on the system's performance, including the intensity of handover requests, network topological and availability characteristics, various sources of signaling overhead and the congestion points imposed by the architectural components. With respect to performance metrics, emphasis is given on the calculation of the mean latency from the handover trigger to the completion of the VHO preparation phase. This is one of the most critical phases to control during the whole VHO process, due to its complexity [6]. Moreover, the impact of the various considered factors on congestion is further quantified by calculating the amount of the signaling load occurring in key architectural components.

The effectiveness of the model is validated extensively, by means of NS-2 simulations for an appropriately rich set of relevant parameters. It is demonstrated that, despite its simplicity, the model succeeds in capturing with precision the effect of all involved factors on the system's performance.

The rest of the paper is organized as follows: Sect. 2 discusses related work. Subsequently, Sect. 3 describes in detail the novel modeling methodology previously outlined. Section 4 provides numerical and simulation results, illustrating the system's performance under various conditions and providing validation of the modeling methodology. Lastly, Sect. 5 concludes the paper, by highlighting the main outcomes and providing insights for future work.

2 Related Work

According to the IEEE 802.21 standard [4], context information is collected from both client side (Mobile Nodes (MN)) and network side. Collectively, the so called, Media Independent Information Service (MIIS) provides a framework and the related mechanisms by which a corresponding entity may discover and obtain network information existing within a geographical area to facilitate the handovers. Information provided by the MIIS includes static link layer parameters such as channel information, a list of available networks and their associated operators, roaming agreements between different operators, costs for using the network, operator charging policies, network security and quality of service (QoS) mechanisms, as well as specific information about the different Points-of-Attachment (PoAs) for each one of the available access networks, including the channel parameters to optimize link layer connectivity [6].

Another standard similar to IEEE 802.21, defined by the 3GPP, is the Access Network Discovery and Selection Function (ANDSF) component [5], which offers support for, and mobility between 3GPP and non-3GPP systems (e.g., 802.11, 802.16, etc.). Furthermore, the ANDSF provides information to the MN about connectivity to 3GPP and non-3GPP access networks, as it assists the MN to discover the access networks and to provide policies for prioritizing and managing connections to these networks. More specifically, the ANDSF provides a list of access networks available for the MNs, and exchanges discovery information and policies according to operator requirements.

We now review relevant prior research works that draw upon the frameworks just outlined, with a focus on VHO performance: Publication [7] introduced an enhanced centralized information server (EIS) in the core network and proposed an improved vertical handover procedure that estimates the wireless channel conditions by using spatial and temporal locality at the EIS. According to this approach, time consuming channel scanning procedures were omitted, by exploiting periodic reports from each MN via the serving RAN air interface, regarding the MN's location and measurements. Therefore, the EIS was assumed capable of being informed about the MN's link conditions, towards determining the best candidate RAN for the MN. Although [7] discussed performance aspects, this was done in terms of a very simple model employing constant values for the delays between the MN and the Serving Network (SN), as well as between SN and IS. No attempt was made to quantify complex congestion phenomena, as targeted here.

With a similar orientation, [6] pursued an evaluation of the vertical handover mechanisms based on the IEEE 802.21 framework and compared the base framework with a proposed enhancement of the IS, named CIS (Context-aware Information

Server). The CIS was assumed capable of providing information (including the available radio resources for each one of the candidate RANs) by using only the Information Request – Response messages, without having to individually query all the available candidate RANs after their discovery. Comparative performance evaluation demonstrated that the differences between the static IS and the CIS were significant, due to the fact that the MNs did not need to query every possible candidate RAN. However, the simulation-based evaluation methodology in [6] was again simple and did not attempt to consider complex interactions and the associated congestion phenomena.

Finally, in [8] the authors presented a framework for Vertical Handovers (VHO) among collaborative wireless networks based on the principles of the IEEE 802.21 and IEEE 1900.4 standards. The paper also performed an experimental delay assessment considering a VHO preparation and execution between IEEE 802.11 and IEEE 802.16 showing the validity of the proposed framework, again without analyzing further the congestion mechanisms contributing to the VHO preparation delay, which is the focus of the current paper.

3 VHO Preparation Delay Analysis

We now proceed to discuss the model for the delay analysis of the VHO preparation phase. As already mentioned, the relevant signaling is assumed to employ a message sequence based on the IEEE 802.21 framework, including a static and centralized network information server in the core network. Moreover, the Mobile-Initiated HandOver (MIHO) case is considered, in which the handover trigger is originated by the MN. (The model can also be adapted for the complementary Network Initiated Handover (NIHO) scenario.)

According to the MIHO scenario, as soon as the handover is triggered, the MN sends an information request to the static IS in the core network, through its current serving network (SN). In the information request, the MN may include information about its subscription profiles, user preferences, current location and velocity, power status, as well as the MN's running applications demands. Thus, the IS can determine a sorted list of available networks according to the MN needs, based on available static information about the supported data rate of each network technology, coverage, pricing, energy consumption, etc. Once the MN receives the information response with the sorted candidate networks list, it scans the specific area of the band and queries the resources (QR) of the first candidate. If the response is positive, the VHO is executed. If the first candidate network is not available, the SN sends again a 'query resources' message, to the next candidate, until a target network selection succeeds or the candidate networks in the list have been exhausted.

The main entities in the abovementioned scenario are a number N_{RAN_s} of RANs serving the area under study, a number N_{MN} of MNs in each RAN, and the IS. Note that each RAN acts both as the SN of its own MNs and as a candidate network (CN) for (at least some of) the MNs in other SNs. Moreover, the signaling message exchanges that will be discussed shortly occur over communication links between the entities just mentioned. In particular, communication between a MN and its SN occurs over the wireless link provided by the corresponding RAN. Communication between different RANs and between a RAN and the IS occur over wired links of the network backbone.

In the interest of keeping the presentation simple, the following developments assume a homogeneous setup, where all RANs serve the same number of MNs and have the same wireless link characteristics and where links in the wired backbone are assumed to have the same capacity. Moreover, the MNs served by any given RAN and being subject to handover are assumed to be able to use all $N_{RAN_s} - 1$ other RANs as candidates for the handover target. The model can be readily extended, along the same principles, for addressing a heterogeneous setup, at the expense of somewhat more complicated expressions for the results and some extra notation to express the asymmetries.

The calculation of the overall mean delay for completing the VHO preparation phase involves various delay components, which will be enumerated in detail shortly. All of these components can be classified in three different types: The first type refers to transmission delays of signaling messages over a wired link in the backbone, or the wireless link of the SN. The corresponding mean delay values will be denoted as D_L and D_{WL}, for a transmission over a wired and a wireless link, respectively. (These mean delays are further expressed in terms of the link bandwidths and the mean lengths of the packets involved, later on.) The second type of a delay component is associated with the processing of a query. There are two such cases, queries at the IS, requiring a mean processing delay equal to D_{IS}, and queries for the availability of resources at a candidate network, requiring on the average time equal to D_{CN}. The third type of a delay component corresponds to the queueing delays experienced by signaling messages passing through their serving entities. Indeed, each SN acts as a gateway managing a certain signaling load, because of its role as a mediator between the MNs and the IS, and between the MNs and CNs, as well. For these reasons, the gateways associated with the SNs are modeled in the following as queues that serve all the incoming requests and responses, both from the network and from the MN sides. Service of the signaling messages at the queues just mentioned is assumed to occur in a First Come First Served manner.

It is noted that it would have been readily possible to also explicitly consider queueing phenomena at the IS too. The reason this has not been pursued in the current model is that the IS is assumed centralized, serving many more network entities beyond those whose performance is considered in the model. For a properly dimensioned IS in such as setting, traffic (or other parameter) changes relating to the networks under examination would not have a significant impact to the magnitude of the queueing delay experienced at the IS. Consequently, this delay has been incorporated into the overall IS-related "processing delay" (of mean D_{IS}, as already mentioned).

We now proceed to review all the steps involved in the VHO preparation phase. With reference to Fig. 1, the handover initiation generates an Information Request message (IR) sent from the MN to the SN. This involves a mean transmission delay D_{WL} over the wireless link between the MN and the SN. Then, the IR message passes through the SN queue, experiencing a queueing delay, followed by an additional mean delay D_L for the transmission over the wired link towards the IS. Subsequently, the IS takes, on the average, a processing time D_{IS} to retrieve the required information and sends back to the SN its response. After an additional time D_L, the SN receives the IR response (IRR) packet, and forwards it to the MN. These last two steps involve another queueing delay and a transmission delay of mean D_{WL}.

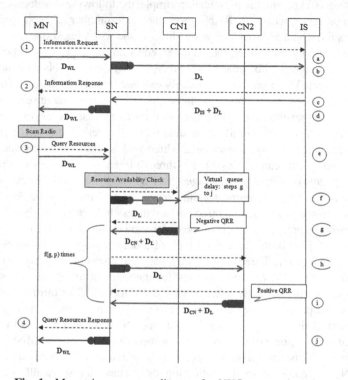

Fig. 1. Messaging sequence diagram for VHO preparation phase

On the basis of the IRR information, the MN sends a Query Resources (QR) message to the SN, which is received there after time D_{WL}. At this time, the SN starts to query the resources of the candidate networks, by forwarding the QR message to the first CN in the list. This involves a queueing delay through the SN queue plus a D_L transmission delay from the SN to the first CN. The availability of resources there is determined and a QR response (QRR) is sent back to the SN, after a queueing delay at the CN, a processing delay D_{CN} for determining the response and a transmission delay D_L. If the QRR is negative, the same procedure is repeated, until the SN gets a positive response from a CN, or until all CNs have been queried, and none has been found having sufficient available resources. In this second case, the last CN's response triggers a negative QRR that is sent back to the MN, after a final queueing delay through the SN's queue and a further time D_{WL} for transmission to the MN. A positive QRR involves the same delay overhead.

In order to proceed further in putting the various time components together towards computing the overall delay for the whole message sequence in Fig. 1, one must first express the mean transmission delays over the wireless or the wired links, in terms of system parameters. For the wired links this is easy. Given a link bandwidth BW_L and a mean packet length P, one simply has

$$D_L = P/BW_L. \tag{1}$$

The situation is slightly more complicated for the wireless links, because we must also consider how the MNs share the link. Here we assume that all MNs in each SN share equally the available wireless link bandwidth BW_{WL}, so each user has access to bandwidth BW_{WL}/N_{MN} and

$$D_{WL} = N_{MN}P/BW_{WL}. \tag{2}$$

Clearly, (2) can be adjusted to reflect other more sophisticated sharing policies (e.g., a proportionally fair link access).

The next step is to address the queueing delays. For this, one must determine the pattern of occurrence for the various signaling messages. From the standpoint of individual MNs subject to handover, the messaging sequence of Fig. 1 is initiated by the corresponding handover trigger. Such triggers can be reasonably assumed to occur randomly and without correlation across MNs, following a Poisson pattern. This pattern, however, is not maintained across all messages in the sequence, because there is causality between successive steps. On the other hand, taking an aggregate look over many MNs from the standpoint of the network, messages belonging to different sequences do appear to occur in a random and uncorrelated fashion, especially given that the timing characteristics of a sequence are independent from that of other sequences. In view of this fact, the arrivals of signaling messages in all queues are taken to occur according to corresponding Poisson processes.

The relevant Poisson rates depend on the mean rate of handover triggers, assumed equal to λ triggers/s from each mobile node. In view of the homogeneous setup considered, the total rate of handover requests originating at a SN is equal to

$$\lambda_1 = \lambda N_{MN}. \tag{3}$$

Given that many of the messages in the sequence of Fig. 1 (including the IR, IRR, QRR) occur once per trigger, the rate of such messages occurring from an SN is also equal to λ_1 (see Fig. 2). However, this equivalence is not maintained for the QR messages, because, in general, more than one CNs are queried per handover trigger. In order to determine the mean number of CNs queries, consider a total number of CNs g, and let the CN resources availability probability (i.e. the probability of finding a sufficient amount of available resources in a queried CN) be denoted as p. Assuming independence between different attempts, a number n of total QR attempts occurs with probability $(1-p)^{n-1}p$ for $1 \leq n < g$. When $n = g$, the sequence may also end unsuccessfully, so the probability becomes $(1-p)^{g-1}p + (1-p)^g = (1-p)^{g-1}$. Therefore, the mean number of QR attempts per trigger is equal to

$$f(g,p) = \sum_{n=1}^{g-1} np(1-p)^{n-1} + g(1-p)^{g-1} = \frac{1-(1-p)^g}{p}. \tag{4}$$

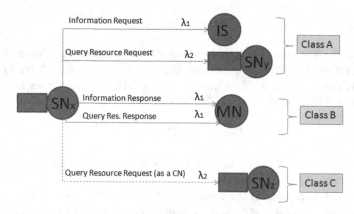

Fig. 2. Signaling flows for the queueing model

In the homogeneous setup considered, each MN served by any given SN has all other RANs as candidates, so $g = N_{RANs} - 1$ and the mean number of QR attempts from each SN is equal to $f(N_{RANs} - 1, p)$. Therefore, the average rate of QR requests originating from each SN is equal to

$$\lambda_2 = \lambda_1 f(N_{RANs} - 1, p) = \lambda N_{MN} f(N_{RANs} - 1, p) \tag{5}$$

where the second equality makes use of (3). The same rate λ_2 applies also to the occurrence of QR requests on a candidate network, considering that these messages having originated from all other RANs. This is because these $N_{RANs} - 1$ RANs collectively generate a rate of $(N_{RANs} - 1)\lambda_2$ QR requests, evenly distributed between $N_{RANs} - 1$ CNs.

The previous results about the occurrence rates of the various signaling messages are illustrated in Fig. 2.

In view of the Poisson arrivals of the signaling messages to the queues, the mean queueing delay can be calculated by using results for M/G/1 systems. As we shall see shortly, different types of messages require different service times, so a multiclass M/G/1 queue must be employed. Consider first a general system of this type, with K classes of customers arriving with rates λ_k and having service requirements with means $E(S_k)$ and second moments $E(S_k^2)$, for $k = 1, .., K$. The class-specific traffic intensities are equal to $\rho_k = \lambda_k E(S_k)$, for a total load $\rho = \sum_{k=1}^K \rho_k$. The queue is stable exactly when $\rho < 1$. Customers from all classes experience the same mean waiting time (see, e.g., [9]), equal to

$$E(W) = \frac{1}{1-\rho} \sum_{k=1}^K \frac{\lambda_k}{2} E(S_k^2) = \frac{1}{1-\rho} \sum_{k=1}^K \rho_k E(S_k) \frac{1 + C_V^2(S_k)}{2}, \tag{6}$$

where the second equality expresses the second order properties through the coefficients of variation $C_V^2(S_k)$. As is well known, for exponentially distributed service times $C_V^2(S_k) = 1$, while for deterministic services $C_V^2(S_k) = 0$. Given $E(W)$, the overall mean delay (waiting plus service) for a given class follows simply as

$$E(R_k) = E(W) + E(S_k), \quad k = 1, \ldots, K. \tag{7}$$

The specific application of interest here employs three classes of customers, related to the three types of service times required by different signaling messages. Specifically, the class A messages are those directed to the wired link (towards the IS or the CNs) with service time $E(S_A) = D_L$, the class B messages are those directed to the wireless link (towards the MNs) with service time $E(S_B) = D_{WL}$, and the class C messages are the QR requests processed at a CN. In this case there is a processing time of mean D_{CN}, followed by a wired transmission for returning the reply to the querying SN, so the mean service time $E(S_C) = D_{CN} + D_L$. The Poisson rates corresponding to the three classes are determined easily by inspection of Fig. 2, as

$$\lambda_A = \lambda_1 + \lambda_2 = \lambda N_{MN}(1 + f(N_{RANs} - 1, p)), \tag{8}$$

$$\lambda_B = 2\lambda_1 = 2\lambda N_{MN}, \tag{9}$$

$$\lambda_C = \lambda_2 = \lambda N_{MN} f(N_{RANs} - 1, p). \tag{10}$$

The total load at the SN queue, encompassing all three classes is

$$\begin{aligned}\rho &= \lambda_1((1 + f(N_{RANs} - 1, p)E(S_A) + 2E(S_B) + f(N_{RANs} - 1, p)E(S_C)) \\ &= \lambda N_{MN}((1 + f(N_{RANs} - 1, p))E(S_A) + 2E(S_B) + f(N_{RANs} - 1, p)E(S_C)),\end{aligned} \tag{11}$$

while the total rate of signaling message arrivals at a queue can be seen equal to

$$\lambda_A + \lambda_B + \lambda_C = \lambda N_{MN}(3 + 2f(N_{RANs} - 1, p)).$$

It is now a simple matter to calculate the total mean delay for the handover preparation phase, by simply accumulating the time components of the messaging sequence depicted in Fig. 1, in line with the discussion in the beginning of this section. It is reminded that delays relevant to the processing of QR requests spent at the SN and the CN must be multiplied by the factor $f(N_{RANs} - 1, p)$, to account for the multiple attempts involved. After following the various steps of this calculation, one obtains a

$$\begin{aligned}\text{Total delay} &= D_{WL} + E(R_A) + D_{IS} + D_L + E(R_B) \\ &\quad + D_{WL} + f(N_{RANs} - 1, p)[E(R_A) + E(R_C)] + E(R_B) \\ &= 4D_{WL} + D_{IS} + (2 + 2f(N_{RANs} - 1, p))D_L \\ &\quad + f(N_{RANs} - 1, p)D_{CN} + (2f(N_{RANs} - 1, p) + 3)E(W).\end{aligned} \tag{12}$$

Equation (12) clearly illustrates how much of the mean total VHO preparation delay is spent due to transmission over communication links, processing, and queueing. The magnifying effect on the delay when candidate networks are less likely to be available is expressed through $f(N_{RANs} - 1, p)$, calculated according to (4). The congestion phenomena are expressed through $E(W)$, as in (6). Further remarks on the impact of various parameters on the total delay will be provided in the following section.

4 Results

In order to assess the applicability and accuracy of the analytical VHO preparation delay model that was presented in Sect. 3, a number of simulation experiments were performed in the NS-2 simulator. In these experiments we measured the handover preparation delay and the resulting signaling load, by considering the mobility management model, the involved nodes and the interactions between them. The simulations considered a number of MNs and RANs that acted both as SNs and as CNs, and an IS. The parameters used for the simulations and the analytical model are depicted in Table 1.

Table 1. Parameters used in analytical and simulation models.

Parameter	Value	
Number of RANs: N_{RANs}	5	
Number of MNs (per SN): N_{MN}	10	
Rate of VHO triggers per mobile node: (mean) λ (triggers/sec)	In range [0.01, 0.1]	
Wired Link Bandwidth: BW_L (Mbps)	1000	
Wireless Link Bandwidth: BW_{WL} (Mbps)	10	
Packet Length: P (bits)	12000 (1500 × 8)	
Mean IS Delay: D_{IS} (sec)	0.01	
Mean Process Delay (CN): D_{CN} (sec)	0.030	0.300

According to the derived simulation scenario each MN -within the area of a specific SN- sent VHO Requests with mean rate λ. All the traffic directed to and from the MNs passed through the SN, where packets were enqueued and dequeued. Transmission delays over the wired and the wireless links (corresponding to S_A and S_B, respectively) were taken exponentially distributed (implying exponentially distributed message lengths). The same distribution was also used for the processing times of QR requests at the CNs.

The simulation results confirmed the analytical model that was presented in the previous section. In Fig. 3, the mean VHO preparation delay (on the left plot), is depicted, considering different arrival rates per MN, while the CN resources availability probability was set to $p = 0.3$ and assuming $D_{CN} = 0.3 \, sec$ (i.e. service delay in the CN). Note that the total load at the SN queue is depicted on the right plot of Fig. 3, signifying that even with very high loads at the SN queue (up to 70 %), the analytical model outputs are still in agreement with the simulation results.

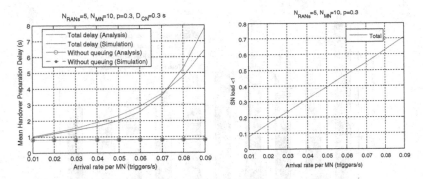

Fig. 3. Mean VHO preparation delay (on the left), and total load at the SN queue (on the right), under different arrival rates per MN, considering $D_{CN} = 0.3$ s and $p = 0.3$

Furthermore, the rate λ of VHO requests by each MN plays an important role on the total signaling load and also on the total mean VHO preparation delay, causing a considerable increase in the queueing delay, as it can be seen from the difference between the total delay (red and blue lines without markers) and the delay without queueing (red and blue lines with markers). Note that in reality only the initial arriving rates (i.e. the arriving rates generated by the MNs) follow the Poisson distribution, while the subsequent arrival rates are smoother, which is the only approximation included in the analytical model (through $E(W)$ that was calculated considering all packet arrivals on the M/G/1 queue to follow the Poisson distribution), and this is the reason that a slight difference can be observed between simulation and analytical results.

In Fig. 4, the total packet arrival rate at an SN is depicted, considering $\lambda = 0.1$ triggers/s. It can be seen that the total arrival rate increases as the CN resources availability probability p is decreasing, and that it is actually being doubled as the value of the CN resources availability probability drops from $p = 1.0$ (considering that the first queried CN is always available) down to $p = 0.1$ (considering that each CN will answer positively only at 10 % of the queries). Also, the figure depicts the packet arrival rate related to the traffic that is generated in all stages for the MNs that are being served by the SN (blue-colored bar), the packet arrival rate regarding the traffic that is generated from the other SNs that send QR request packets to the specific SN (green-colored bar) and their contribution to the total load. It can be seen that as the CN resources availability probability decreases (i.e. it is harder to find available resources during the QR phase and thus more CNs have to be queried sequentially) more traffic is generated throughout the wired network segments and this affects the incoming packet rates observed at the serving network, both concerning its own MNs and regarding the queries from the other networks that consider the SN as a candidate for their own MNs.

In addition, in Fig. 5, the mean VHO preparation delay is depicted considering $\lambda = 0.1$ triggers/s and $D_{CN} = 0.03$ s. It can be seen that a lower CN resources availability probability leads to higher mean VHO preparation delay. Also, the figure shows the mean delay caused by the IS, the mean delay caused by transmission and service times, as well as, the resulting total queueing delays, caused by the SN queue. From both figures, it can be seen that the simulation results confirm the analytical results, and

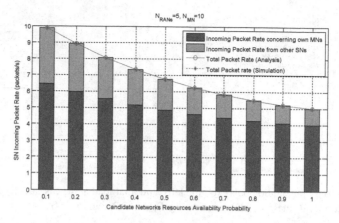

Fig. 4. SN total incoming packet rate under varying CN resource availability probability, considering $\lambda = 0.1$ triggers/s

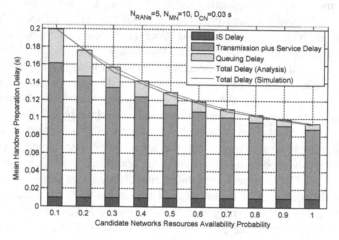

Fig. 5. Mean VHO preparation delay under varying CN resources availability probability considering $D_{CN} = 0.03$ s and $\lambda = 0.1$ triggers/s

also that p plays an important role at the total signaling load and total delay. The same trend is also observed regarding the background users, such as other MNs of the same SN, as well as, other RANs querying for resources the current SN that cause considerable queueing delays.

5   Conclusions

The paper presented a novel modeling methodology that can capture concisely, but also effectively, all important factors that have an impact on the system's performance, including the intensity of handover requests, network topological and availability characteristics, various sources of signaling overhead and the congestion points imposed by the architectural components. The model was validated by simulation and was employed for investigating the impact of the various parameters on the mean delay required for the VHO preparation phase. The principles of the proposed modeling methodology can be used to motivate the study of decentralized solutions towards addressing distributed mobility management scenarios, including ad-hoc and mesh network topologies.

Acknowledgements. The work of the first three authors was undertaken in the context of the "SYNAISTHISI" project, which is co-financed by the Greek General Secretariat for R&T, Ministry of Education & RA and the European RDF of the EC under the Operational Program "Competitiveness and Entrepreneurship" (OPCE II), in the action of Development Grants For Research Institutions (KRIPIS).

References

1. Landström, S., Furuskär, A., Johansson, K., Falconetti, L., Kronestedt, F.: Heterogeneous networks–increasing cellular capacity. Data Boom: Oppor. Chall. 4(1) (2011)
2. Frodigh, M., Johansson, P., Larsson, P.: Wireless ad-hoc networking-the art of networking without a network. Ericsson Rev. 4(4), 249 (2000)
3. Doppler, K., Rinne, M., Wijting, C., Ribeiro, C.B., Hugl, K.: Device-to-device communication as an underlay to LTE-advanced networks. Commun. Mag. IEEE 47(12), 42–49 (2009)
4. IEEE Standard for Local and Metropolitan Area Networks- Part 21: Media Independent Handover, IEEE Std 802.21-2008, pp. c1–301, 21 January 2009. doi:10.1109/IEEESTD. 2009.4769367
5. Architecture enhancements for non-3GPP accesses, 3GPP TS 23.402 (2007)
6. Neves, P., Soares, J., Sargento, S., Pires, H., Fontes, F.: Context-aware media independent information server for optimized seamless handover procedures. Comput. Netw. 55(7), 1498–1519 (2011)
7. Kim, Y., Pack, S., Kang, C.G., Park, S.: An enhanced information server for seamless vertical handover in IEEE 802.21 MIH networks. Comput. Netw. 55(1), 147–158 (2011)
8. Dimitriou, N., Sarakis, L., Loukatos, D., Kormentzas, G., Skianis, C.: Vertical handover (VHO) framework for future collaborative wireless networks. Int. J. Netw. Manage 21, 548–564 (2011)
9. Haviv, M.: Queues: A Course in Queueing Theory. Springer, New York (2013)

AdaMap: Adaptive Radiomap
for Indoor Localization

Zhiqi Yang, Yongcai Wang[✉], and Lei Song

Institute for Interdisciplinary Information Sciences (IIS), Tsinghua University,
Beijing 100084, People's Republic of China
wangyc@tsinghua.edu.cn

Abstract. In wireless networks, radiomap (also known as fingerprint-
ing) based locating techniques are commonly used to cope the diverse fading signatures of radio signal, in which probabilistic or static radiomaps
are trained in offline phase. A challenging problem of radiomap locating is that the radiomap can be outdated when environments change.
Reconstruction of radiomap is time consuming and laborious. In this
paper, we exploit the inter-beacon radio signal strength (RSS) to construct adaptive radiomap (AdaMap) by an online self-adjusted linear
regression model. The distinct feature of AdaMap is that not only the
radio signatures at the training locations vary with the online inter-
beacon RSS measurements, but also the coefficients of the model are self-
adjusted when the environments change significantly, so that AdaMap
is highly adaptive to the environment changes. The proposed schemes
are evaluated by extensive simulations, with comparisons to the state of
art of the radiomap wireless localization methods. The results showed
that AdaMap presented dramatical advantages in preserving positioning
accuracy when the environments changed over time.

1   Introduction

Nowadays, considerable research efforts were dedicated to develop wireless networks as indoor location infrastructure. Access points (AP) or other wireless
nodes which have been widely deployed in the indoor environments are treated
as reference points of locating, called *beacons*. A target to be located measures
the received signal strength (RSS) from the beacons to infer its own location.
Propagation model regarding distance and the *signature of signal strength* are two
ways of utilizing the RSS. The former way is to infer distance by RSS, which is
generally inaccurate for the complex fading nature of RF signals in the physical environments; the latter one is to adopt a patten-matching based approach
[1,2], which contains two phases. In the offline phase, the RSS signatures of a
set of training positions are measured to construct a database, which is called
radiomap. In the online phase, a target takes online RSS measurement to search
in the radiomap to find a position that has the least RSS-distance (distance in
RSS space) to the measurement, which will be treated as the position estimation
of the target.

© Springer International Publishing Switzerland 2015
S. Papavassiliou and S. Ruehrup (Eds.): ADHOC-NOW 2015, LNCS 9143, pp. 134–147, 2015.
DOI:10.1007/978-3-319-19662-6_10

Radiomap based locating method can achieve meter level positioning accuracy when the radiomap is trained to be fine-grained (for example in 1 m resolution). But fine-grained radiomap positioning method is challenged by two key problems: (1) training the radiomap of fine-grained locations is laborious, especially when the area of interest (AOI) is large; (2) the laboriously trained radiomap is easily outdated when the environment changes. The variations of weather, temperature, movement of objects all can deviate the real-time radio signatures from the offline trained radiomap. But reconstruction the radiomap is time-consuming and laborious.

To overcome these challenges, designing adaptive radiomap mechanism to update the radiomap online is therefore critically important for the radiomap method to be valuable in practice. Existing works have investigated this problem mainly from two ways: (1) using additional reference points, and (2) designing adaptive radio-map model using inter-beacon RSS measurements.

(1) *Using additional reference points:* LANDMARC [3] used the active RFID and training data learned from online sources to perform indoor localization. The main advantage of LANDMARC is that it improves locating accuracy by utilizing the reference tags. LEMT [4] used additional reference points (with known positions) to online collect radio signatures and proposed model-tree method to construct piece-wise linear regression radio-map which can be online calibrated. Reference [5] used radio signatures learned from reference points to online calculate the channel models constants to adapt the radio-map model. Reference [6] proposed to use RFID sensors and environment sensors to help the location systems to adapt to the environment dynamics. These methods all need additional reference points or sensors to provide online assistance for radio-map adaptation.

(2) *Adaptive radio-map model by measuring inter-beacon RSS:* To relieve the needs for additional reference points or sensors, the closest related work in [7] proposed to use the beacons themselves as reference points, because most of current beacons (such as WiFi or sensor nodes) have both transmitting and receiving capabilities. Linear regression type radiomap was trained for each training location by using the inter-beacon RSS as a reference. We will introduce the details in Sect. 2. Another important approach is to use hybrid method to improve the reliability of radio-map based locating [25]. We focus on pure radio-map based solutions.

However, in existing adaptive radiomap model as stated in [7], the coefficients of linear regression model will remain constant after training without further adapting to the environments online. The principle of radiomap adaptivity is only to substitute the online measured inter-beacon RSSs into the static functions to infer the RSS signatures of specific locations. Since the function coefficients are non-adaptive, such model is partially adaptive, because when environments change dramatically, the linear regression model, i.e., the trained coefficients may become out-of-date, which fails to capture the changes of radio signatures. Therefore, how to design adaptive radio-map model to make not only the radio signatures at the training locations vary with the online inter-beacon

RSS measurements, but also to online adjust the coefficients of the trained functions autonomously is critically important.

To achieve this, this paper presents AdaMap, which uses the inter-beacon RSSs to online justify the disparity between the trained model and the online RSS measurements. It updates the coefficients of the model when noteworthy disparity between the trained model and the online measurements is detected. Algorithms are designed to check the disparity and to update the coefficients adaptively. We further investigated the distance and angle metrics for selecting reference beacons to make the online model updating more efficient and more accurate. We show by extensive simulations and hardware experiments that AdaMap provides adaptivity to the environment changes, which helps to provide dramatically better online positioning performances than the other radiomap locating methods.

The remainder of this paper is organized as follows: in Sect. 2, we introduce radiomap model and related works. In Sect. 3, we present our methodologies of AdaMap. Extensive simulation results are shown in Sect. 4. Final section is conclusion and future work.

2   Background and Problem Model

2.1   Radiomap Locating Method

Radiomap locating is essentially a *pattern-matching* based approach, which offline learns the RSS signatures of the a set of locations to construct a *radio-map* and online searches in the radio-map to find the location whose RSS fingerprint best matches the measured RSS at the location to be determined [1,2,9]. In offline phase, k training locations are selected in the sensing field, which are denoted by $\mathbf{L} = \{l_1, l_2, ..., l_k\}$. Suppose there are n beacons (WiFi APs or wireless sensors) in the sensing field, which are denoted by $\mathbf{B} = \{b_1, b_2, ..., b_n\}$. In the training phase, the RSS values of all beacons at each training location l_i will be measured over a period of time. So that a signal signature vector of location l_i is constructed as $\mathbf{r}_i = \{r_{i,1}, r_{i,2}, r_{i,n}\}$, where $r_{i,j}$ is inferred deterministically or probabilistically by the RSS values measured from beacon b_j at location l_i. When only the mean value of RSS is considered, $r_{i,j}$ represents the average RSS value [9]. When signature distribution is considered [10–12], $r_{i,j}$ can be probabilistic density function (pdf) of RSS. The signature vectors of all training locations are stored as a database, called *radiomap*, denoted by $\mathbf{R} = \{r_1, r_2, ..., r_k\}$.

In the online positioning phase, a mobile target measures its current RSS vector $\mathbf{s} = s_1, s_2, ..., s_n$ and finds the best match (Euclidean distance in signal space) of $\mathbf{s}$ in $\mathbf{R}$ to estimate its location. In mean value type radio-map, matching can be conducted by Nearest Neighbor algorithm [9]. In probabilistic radio-maps, maximum likelihood estimate and Bayesian estimation were applied to improve the positioning accuracy [11].

2.2 Radio-Map Adaptation by Inter-Beacon Signatures

The radiomap can be outdated when environment changes, but re-training radiomap will be time-consuming and laborious. Existing work investigated radiomap adaptation methods [3,4,7,13]. Most of these methods use additional reference points (with known locations), sensors or other resources to help handle the RSS dynamics. Instead of using the additional reference points or additional sensors, [7] proposed to use the beacons themselves as reference points to construct linear regression model, which can relieve the additional hardware costs. This method explores the facts that some off-the-shelf WiFi stations can measure RSSs from other beacons. We introduce this method briefly before presenting our solution.

Let's consider there are n beacons $b_1, b_2, \cdots, b_n$ in the network and let l be a training location. In offline phase, to train the signal signature of beacon i at location l, i.e., $d_{i,l}$, the inter-beacon RSS vector $\{r_{i,1}^t, r_{i,2}^t \cdots, r_{i,n}^t\}$ are measured over a period of time, where $r_{i,j}^t, j \neq i$ represents the RSS of beacon i measured by beacon j at time t. Then a linear regression model is built by assuming the RSS at location l, i.e., $d_{i,l}^t$ is a linear combination of the inter-beacon RSSs vector:

$$d_{i,l}^t = \alpha_{i,1}^l r_{i,1}^t + \alpha_{i,2}^l r_{i,2}^t + \cdots + \alpha_{i,n}^l r_{i,n}^t + \alpha_{i,n+1}^l. \tag{1}$$

So that by the measurements from 1 to T, the following equation can be constructed to calculate the coefficient vector. Least square estimation can be applied and the equation can be made overdetermined by improving T.

$$\begin{bmatrix} r_{i,1}^{(1)} & \cdots & r_{i,n}^{(1)} & 1 \\ \vdots & \cdots & \vdots & \vdots \\ r_{i,1}^{(T)} & \cdots & r_{i,n}^{(T)} & 1 \end{bmatrix} \begin{bmatrix} \alpha_{i,1}^l \\ \vdots \\ \alpha_{i,n}^l \\ \alpha_{i,n+1}^l \end{bmatrix} = \begin{bmatrix} d_{i,l}^1 \\ \vdots \\ d_{i,l}^T \end{bmatrix} \tag{2}$$

So that the RSS fingerprint at each location l is no longer a static value or a vector, but an adaptive value calculated by the linear regression function. In online phase, the inter-beacon RSS $\{r_{i,1}^t, r_{i,2}^t, \cdots, r_{i,n}^t\}$ are measured, and are substituted into the linear regression function to calculate the RSS value at location l, which is denoted by $d_{i,l}^c$. Since the inter-beacon RSS values $\{r_{i,1}^t, r_{i,2}^t, \cdots, r_{i,n}^t\}$ change adaptively with the environments, the changes are embodied by the linear regression model to make $d_{i,l}^c$ be adaptive to the environment.

However, in this linear regression model (LRM), once the linear coefficients are calculated in offline phase, the coefficients will not change in online phase. When environment changes slowly, the self-adaptivity of LRM could keep the effectiveness of the radio-map model to provide reasonable locating accuracy. While in case striking environment change happens, the linear regression model may lose the correctness in modeling the RSS signatures. We call this *partial-adaptivity* problem.

Figure 1 uses an example to show the *partial-adaptivity* problem, i.e., ineffectiveness of the linear regression model in case the environment changes dramatically. We simulate a block effect on the experimental environment, i.e., a big block is placed in the sensing field. Any signal that passes through the block would have 20db signal attenuation.

Figure 1 presents online calculated radiomap by LRM with same measured radiomap while facing two different environment scenarios. Compared to the same measured radiomap from Fig. 1(a), which is accurate, Fig. 1(b) shows that when environment doesn't change, the self-adaptivity of LRM can ensure reasonable locating accuracy. While Fig. 1(c) shows that when facing block effect, which means striking environment changes, LRM fails to perform locating and the calculated radiomap differs much from measured radiomap in Fig. 1(a).

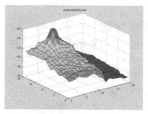 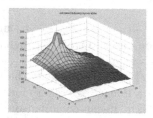 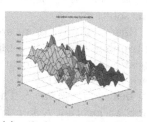

(a) Measured Radiomap   (b) Online Calculated Radiomap, when environment doesn't change   (c) Online Calculated Radiomap, when environment changes by block effect

Fig. 1. Block effect on LRM model locating

3 Adaptive Radiomap Model

3.1 Overview

To solve above problems, we present AdaMap, an efficient and flexible online radiomap adaptation model. It not only trains the linear regression model to make the radio signatures be adaptive to environment dynamics, but also can online update the coefficients of the linear model by judging whether the calculated inter-beacon radiomap differs much from the measured one. The overview of the AdaMap model is shown in Fig. 2.

In offline phase, we apply LRM to calculate offline coefficients by solving equation (2). In online phase, the trained offline coefficients are utilized to calculate the RSS signature of each location. These calculated signatures are then compared with the measured RSS signature of the target. The location whose signature best matches the measured RSS signature of the target will be determined as the real-time position of the target.

Different from existing approaches, Adamap designs a judgment step to determine whether the trained coefficients by LRM are outdated. If the coefficients are detected to be outdated, efficient algorithms to update the coefficients of LRM model were designed. The updated coefficients make the LRM be adaptive

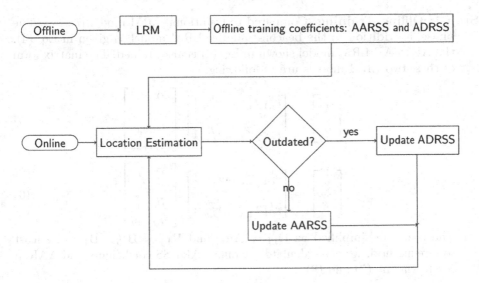

Fig. 2. Flowchart of AdaMap.

to the environment changes. The whole process ensures that the calculated RSS signatures at each location be adapt to the environment dynamics by two ways: (1) LRM model; (2) coefficients updating scheme of LRM model.

To present the coefficient updating scheme of Adamap clearly, we define AARSS as short for AP-to-AP RSS:

$$r_{i,j}^c = \sum_{v=1,v\neq i}^{n} \beta_{i,j,v} r_{i,v}^m + \beta_{i,j,n+1} \cdot 1 \tag{3}$$

$r_{i,j}^c$ is the calculated RSS value from beacon i to beacon j. It is represented by the linear combination of the measured RSS values from beacon i to other beacons. β is the coefficient vector, and the superscript c and m means calculated and measured RSS respectively.

We define ADRSS as short for AP-to-Device RSS, which indicates the beacon to device's location RSS values.

$$d_{i,l}^c = \sum_{j=1,j\neq i}^{n} \alpha_{i,j,l} r_{i,j}^m + \alpha_{i,n+1,l} \cdot 1 \tag{4}$$

$d_{i,l}^c$ is the calculated RSS value at location l of beacon i, which is represented by the linear combination of inter-beacon RSS values. α in (4) represents the coefficients. The superscript c and m represents calculated and measured value respectively.

3.2 AdaMap Framework

AdaMap is designed based on above two linear regression models. It contains two phases: (1) offline training and (2) online adapting.

Step 1: Offline Training. Compared to traditional LRM model [7], in offline phase, in addition to the beacon-to-device LRM model as given in Eq. (4), the AP-to-AP LRM model shown in Eq. (3) is also trained. The matrix form of these two LRM models are as following:

$$
\begin{bmatrix} d_{i,l}^1 \\ \cdot \\ d_{i,l}^T \end{bmatrix} = \begin{bmatrix} r_{i,1}^1 \ r_{i,2}^1 \ \cdots \ r_{i,n}^1 \ 1 \\ \cdot \\ r_{i,1}^T \ r_{i,2}^T \ \cdots \ r_{i,n}^T \ 1 \end{bmatrix} \begin{bmatrix} \alpha_{i,1,l} \\ \cdot \\ \alpha_{i,n,l} \\ \alpha_{i,n+1,l} \end{bmatrix} \tag{5}
$$

$$
\begin{bmatrix} r_{i,j}^1 \\ \cdot \\ r_{i,j}^T \end{bmatrix} = \begin{bmatrix} r_{i,1}^1 \ r_{i,2}^1 \ \cdots \ r_{i,n}^1 \ 1 \\ \cdot \\ r_{i,1}^T \ r_{i,2}^T \ \cdots \ r_{i,n}^T \ 1 \end{bmatrix} \begin{bmatrix} \beta_{i,j,1} \\ \cdot \\ \beta_{i,j,n} \\ \beta_{i,j,n+1} \end{bmatrix} \tag{6}
$$

The could be simplified as $\mathbf{D_{i,l}} = \mathbf{A}\alpha_{i,l}$ and $\mathbf{R_{i,j}} = \mathbf{B}\beta_{i,j}$. By using least square method, we can calculate the offline ADRSS coefficients and AARSS coefficients as (7) and (8).

$$
\alpha_{i,l} = (\mathbf{A^T A})^{-1}\mathbf{A^T D_{i,l}} \tag{7}
$$

$$
\beta_{i,j} = (\mathbf{B^T B})^{-1}\mathbf{B^T R_{i,j}} \tag{8}
$$

Note that $\alpha_{i,l}$ is the ADRSS coefficient vector of beacon i at location l. $\beta_{i,j}$ is the AARSS coefficient vector from beacon i to beacon j.

Step 2: Online Adapting. In online phase, $r_{i,j}^c$ is calculated by (3) and $r_{i,j}^m$ is measured in real-time. Then we judge whether $r_{i,j}^c$ is close to $r_{i,j}^m$ to decide whether to update the coefficient vector or not.

(1) **If $r_{i,j}^c$ is close to $r_{i,j}^m$** ($\|r_{i,j}^c - r_{i,j}^m\| \leq H$), where H is a threshold, we consider the ADRSS model (4) is still effective, which don't need to be updated. We can calculate the RSS signature at location l of beacon i by (4).

(2) **If $r_{i,j}^c$ differs greatly from $r_{i,j}^m$** ($\|r_{i,j}^c - r_{i,j}^m\| > H$), which means that the trained LRM model is outdated. Then we firstly collect the inter-beacon RSS measurements to update the inter-beacon coefficients β by (8). The updated coefficient vector is denoted by β'. Then we show how to update α.

$$
d_{i,l}^c = \sum_{j=1,j\neq i}^{n+1} \alpha_{i,j,l}\left\{ \sum \beta'_{i,j,v}r_{i,v}^m \right\} = \sum_{j=1,j\neq i}^{n+1} \alpha'_{i,j,l}r_{i,j}^m \tag{9}
$$

By simple mathematical derivation, we can conclude the updated coefficients α' as shown in (10).

$$
\alpha'_{i,j,l} = \sum_{k=1,k\neq i,k\neq j}^{n+1} \alpha_{i,k,l}\beta'_{i,k,j} \tag{10}
$$

From the derivation, we can see clearly that the ADRSS coefficients α can be updated online.

Step 3: Device Location Estimation. Finally we can calculate the estimated location $d_{i,l}^c$ with updated coefficients and inter-beacon RSS values, as in the traditional methods.

3.3   Algorithm Properties

The adaptive model solves the problem that linear regression model coefficients cannot adapt online. The algorithm pseudo code is written as Algorithm 1. The computation complexity of Algorithm 1 can be analyzed as follows. Solving the least square estimation has complexity $O(n^3)$, n is beacon number. In online phase, the outloop has complexity $O(n^2)$, updating coefficients inside the loop has complexity $O(nm)$, m is training location number. Therefore, the overall complexity of this algorithm is $O(n^2(n^3 + nm))$. It is affordable, since n is generally not large.

Input: $\{r_{i,j}^m\}$ and $\{d_{i,l}^m\}$ at running times $t = 1, 2, \cdots, T$, AP number n and location number k, threshold H
Use $\{r_{i,j}^m\}$ and $\{d_{i,l}^m\}$ to construct $B_{i,j}$, A_i, and $R_{i,j}$ within training time T
Calculate AARSS training coefficients $\beta = (B^T B)^{-1} B^T R$
Calculate ADRSS training coefficients $\alpha = (A^T A)^{-1} A^T D$
while *(system is running)* **do**
   for $i - 1; i \leq n; i + +$ **do**
      for $j = 1; j \leq n, j \neq i; j + +$ **do**
         Calculate $r_{i,j}^c$ with $B_{i,j}$ and $\beta_{i,j}$
         if $\left\| r_{i,j}^c - r_{i,j}^m \right\|_2 < H$ **then**
          | update ADRSS coefficients $\alpha_{i,l}$
         end
         else
           Construct $B_{i,j}$ and $R_{i,j}$, *both at time* t, and update AARSS
           coefficients: $\beta_{i,j} = (B_{i,j}^T B_{i,j})^{-1} B_{i,j}^T R_{i,j}$
         end
      end
      Calculate location estimation $d_{i,l}^c$ with updated coefficients
   end
end

Algorithm 1. Algorithm for AdaMap Model

4   Performance Evaluation

Simulations were conducted to evaluate the performance of AdaMap compared with simple linear regression model (LRM). More specifically, the locating accuracy and robustness of AdaMap locating against block effect and temperature effect were evaluated in this section.

4.1 Simulation Settings

We conducted simulation in MATLAB environment. Area is partitioned into grids, and a number of beacons are deployed randomly in the area, they all possess the same initial transmission power, and environment noise follows a gaussian distribution. The target moves in this area, and its transmission power can be received by beacons. In our experiment, the parameters are set as Table 1. We use 20*20 area size, the beacon number is set as 10, initial transmission power is set as 150. The gaussian noise variance is set as 2. The target follows a sine wave path in the area. Besides, in order to verify the performance convincingly, the 10 APs are deployed with random location generation algorithm in the area.

Table 1. Setting of simulation

Parameters	Value
Area size	20*20
Beacon number	10
Transmission power	150
Noise variance	2
Target trajectory	Sine wave in the area

4.2 Environment Dynamics

To verify our adaptive algorithm performance on environment dynamics, we consider the 3 following cases:

(1) **Free Space:** environment doesn't change, the signal transmission follows the propagation model [9]. The model is shown as Eqn (11).

$$r = p_t - p_0 - 20 * \log(d) + \sigma \tag{11}$$

In this formula, p_t is transmitter's initial power, p_0 is constant term, d is the distance between transmitter and receiver, σ is environment noise which follows normal gaussian distribution, r is the power received from transmitter.

(2) **Block Effect:** we simulate a big block moving in a constant velocity inside the sensing area. The block would cause specific amount of signal attenuation for signals passing through the block, and we use *blockeffect* to define this value loss;

(3) **Temperature Effect:** the temperature would cause the transmission powers of all the beacons decrease slowly, and we use *powerdecrease* to express the decrease value.

   In the latter two cases, AdaMap and LRM model would be applied to compare the locating performance.

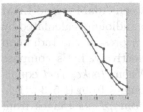 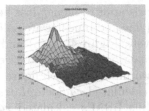 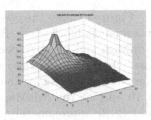

(a) Tracking performance   (b) Measured Radiomap   (c) Online Calculated Radiomap

Fig. 3. When environment doesn't change

4.3   Simulation Results

Environment Doesn't Change. Fig. 3 presents LRM model locating performance when environment doesn't change. In Fig. 3(a), the red line is the real path, and blue line is estimated path. In our simulation, the real path is a sine wave in the area, in each time slot, we apply LRM model to calculate the most possible estimated location. We can see that the tracking performance is quite good when the environment doesn't change, Fig. 3(b) and(c) shows that the online calculated radio map matches the measured radiomap quite well, which means that LRM performs reasonably well in free space.

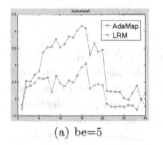 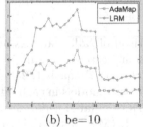 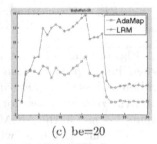

(a) be=5   (b) be=10   (c) be=20

Fig. 4. Block effect radiomap loss: AdaMap vs. LRM

Block Effect and Temperature Effect. We define *radiomapLoss* to present the similarity between measured radiomap and online calculated radiomap by AdaMap and LRM models at each time slot. Suppose measured radiomap is denoted as 2D matrix **A**, online calculated radiomap is denoted as 2D matrix **B**, the *radiomapLoss* is defined as (12).

$$radiomapLoss = \frac{sum(abs(A - B))}{row * column} \tag{12}$$

The definition shows that if *radiomapLoss* is smaller, the measured radiomap and online calculated radiomap is more similar, which means more locating accuracy.

Figure 4 shows the comparison of radiomap loss between AdaMap and LRM when facing block effect. Red line and blue line are online radiomap calculated by AdaMap and LRM respectively. Figure 4(a), (b) and (c) present the radiomap loss when *blockeffect(be)* equals 5, 10 and 20 *dB* respectively. Let's compare the *radiomapLoss* for the two models in each figure. When *blockeffect* equals 5, 10 and 20, the AdaMap's calculated *radiomapLoss* ranges from [1.5, 3], [1, 5] and [2, 8] approximately, while LRM's calculated *radiomapLoss* ranges from [1.5, 4.5], [1, 8] and [2, 14] approximately. As the *blockeffect* increases, both LRM and AdaMap's *radiomaoLoss* increases, LRM model can partially adapt to the increasing block effect, so that the range is not that wide, while AdaMap's range is less than LRM in all scenarios, which means that when facing block effect, AdaMap performs better locating accuracy when compared to LRM.

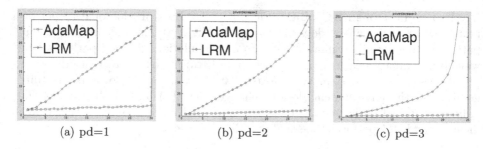

(a) pd=1   (b) pd=2   (c) pd=3

Fig. 5. Power transmission decrease radiomap loss: AdaMap vs. LRM

Figure 5 shows the comparison of radiomap loss between AdaMap and LRM when facing power transmission decrease. Figure 5(a), (b) and(c) present the radiomap loss when *powerdecrease(pd)* equals 1, 2 and 3 respectively. Let's compare the *radiomapLoss* for the two models in each figure. When *powerdecrease* equas 1, 2 and 3, the AdaMap's calculated *radiomapLoss* ranges from [1.7, 3.6], [1.7, 5.7] and [1.7, 6.5], while LRM's calculated *radiomapLoss* ranges from [1.7, 31.1], [1.7, 89.7] and [1.7, 235.0] approximately. The power transmisson decrease effect obviously affect LRM's locating accuracy, the *radiomapLoss* increases almost exponentially, which means the failure of locating by LRM, while AdaMap's adaptivity presents more stable(only increases from 3.6 to 6.5 with worst case) and less radiomap loss when facing power transmission decrease, which means our AdaMap model successfully achieves better locating accuracy.

Figure 6 shows the locating performance comparison between AdaMap model and LRM model. Block effect CDF errors are presented in Fig. 6(a). The solid lines and dotted lines are AdaMap CDF errors and LRM CDF errors respectively, the comparison shows that as *blockeffect* increases, AdaMap error increases smaller than LRM and achieve better locating accuracy. Temperature effect CDF errors are presented in Fig. 6(b). The result shows that although *powerdecrease* decreases AdaMap locating accuracy because of inherent effect, AdaMap CDF errors are still smaller than LRM.

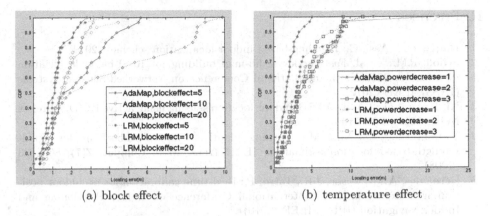

(a) block effect   (b) temperature effect

Fig. 6. AdaMap model vs. LRM model

In summary, these simulation results verified the locating efficiency of AdaMap model. They show that better satisfactory locating accuracy can generally be obtained by AdaMap model compared with LRM model.

5 Conclusion and Future Work

Considerable research efforts were dedicated to develop wireless networks as indoor location infrastructure, we have investigated current locating methods, in which radiomap based locating method can achieve meter level positioning accuracy when the radiomap is trained to be fine-grained. However, the challenging problem is that the radiomap would be outdated when environment changes. The reconstruction would be time-consuming and laborious. In this paper, we present AdaMap, a novel radio map model to deal with environment changes. We exploit the inter-beacon RSS to construct adaptive radiomap based on an online self-adjusted linear regression model. AdaMap not only trains the linear regression model to make the radio signatures be adaptive to environment changes, but also can online update the coefficients according to online inter-beacon RSS measurements. The updating algorithms ensures the efficiency and accuracy of AdaMap locating. The simulation results that shows our proposed method can efficiently make the trained radiomap model online adapt to the environment dynamics and the signal dynamics. It can help improve the locating accuracy of radiomap based algorithms without the pain of recalibration of the radio maps. In the future, work should focus on the online adaptive algorithm optimization and complete model performance evaluation in real systems.

Acknowledgments. This work was supported by in part by National Natural Science Foundation of China Grant 61202360, 61073174, 61033001, 61061130540, the Hi-Tech research and Development Program of China Grant 2006AA10Z216, and the National Basic Research Program of China Grant 2011CBA00300, 2011C-BA00302.

References

1. Haque, I.T., Assi, C.: Profiling-based indoor localization schemes (2013)
2. Scholl, P.M., et al. Fast indoor radio-map building for RSSI-based localization systems. In: 2012 Ninth International Conference on Networked Sensing Systems (INSS), IEEE (2012)
3. Ni, L.M., et al.: LANDMARC: indoor location sensing using active RFID. Wireless Netw. **10**(6), 701–710 (2004)
4. Yin, J., Yang, Q., Ni, L.M.: Learning adaptive temporal radio maps for signal-strength-based location estimation. IEEE Trans. Mob. Comput. **7**(7), 869–883 (2008)
5. Bernardos, A.M., Casar, J.R., Tarro, P.: Real time calibration for rss indoor positioning systems. In: 2010 International Conference on Indoor Positioning and Indoor Navigation (IPIN), IEEE (2010)
6. Chen, Y.-C., et al.: Sensor-assisted wi-fi indoor location system for adapting to environmental dynamics. In: Proceedings of the 8th ACM International Symposium on Modeling, Analysis and Simulation of Wireless and Mobile Systems. ACM (2005)
7. Lo, C.-C., Hsu, L.-Y., Tseng, Y.-C.: Adaptive radio maps for pattern-matching localization via inter-beacon co-calibration. Pervasive Mob. Comput. **8**(2), 282–291 (2012)
8. El-Kafrawy, K., et al.: Propagation modeling for accurate indoor WLAN RSS-based localization. In: 2010 IEEE 72nd Vehicular Technology Conference Fall (VTC 2010-Fall), IEEE (2010)
9. Bahl, P., Padmanabhan, V.N.: RADAR: An in-building RF-based user location and tracking system. In: Proceedings Nineteenth Annual Joint Conference of the IEEE Computer and Communications Societies (INFOCOM 2000), Vol. 2, IEEE (2000)
10. Roos, T., et al.: A probabilistic approach to WLAN user location estimation. Int. J. Wirel. Inf. Netw. **93**, 155–164 (2002)
11. Dieter, F., et al.: Bayesian filtering for location estimation. IEEE Pervasive Comput. **2**(3), 24–33 (2003)
12. Youssef, M.A., Agrawala, A., Udaya Shankar, A.: WLAN location determination via clustering and probability distributions. In: Proceedings of the First IEEE International Conference on Pervasive Computing and Communications, (PerCom 2003). IEEE (2003)
13. Yin, J., Yang, Q., Ni, L.: Adaptive temporal radio maps for indoor location estimation. In: Third IEEE International Conference on Pervasive Computing and Communications, (PerCom 2005). IEEE (2005)
14. Roberts, B., Pahlavan, K.: Site-specific RSS signature modeling forWiFi localization. In: IEEE Global Telecommunications Conference, GLOBECOM. IEEE (2009)
15. Atia, M.M., Noureldin, A., Korenberg, M.J.: Dynamic online-calibrated radio maps for indoor positioning in wireless local area networks. IEEE Trans. Mob. Comput **12**(9), 1774–1787 (2013)
16. Pahlavan, K., Levesque, A.H.: Wireless Information Networks, vol. 95. Wiley, New York (1995)
17. Sharma, P., et al.: KARMA: Improving WiFi-based indoor localization with dynamic causality calibration. In: 2014 Eleventh Annual IEEE International Conference on Sensing, Communication, and Networking (SECON), IEEE (2014)

18. Pan, S.J., et al.: Adaptive localization in a dynamic wifi environment through multi-view learning. In: Proceedings of the National Conference on Artificial Intelligence, 22(2). AAAI Press, MIT Press, London, Cambridge (1999, 2007)
19. Liu, H., et al.: Survey of wireless indoor positioning techniques and systems. IEEE Trans. Syst. Man Cybern. Part C Appl. Rev. **37**(6), 1067–1080 (2007)
20. Gu, Y., Lo, A., Niemegeers, I.: A survey of indoor positioning systems for wireless personal networks. Commun. Surv. Tutorials IEEE. **11**(1), 13–32 (2009)
21. Yang, Z., Wu, C., Liu, Y.: Locating in fingerprint space: wireless indoor localization with little human intervention. In: Proceedings of the 18th Annual International Conference on Mobile Computing and Networking. ACM (2012)
22. Rai, A., et al.: Zee: zero-effort crowdsourcing for indoor localization. In: Proceedings of the 18th Annual International Conference on Mobile Computing and Networking. ACM (2012)
23. Sujak, B., et al.: Indoor propagation channel models for WLAN 802.11 b at 2.4 GHz ISM band. In: Asia-Pacific Conference on Applied Electromagnetics, APACE 2005. IEEE (2005)
24. Ji, Y., et al.: Impact of building environment on the performance of dynamic indoor localization. In: IEEE Annual Wireless and Microwave Technology Conference, WAMICON 2006. IEEE (2006)
25. Geng, X., Wang, Y., et al.: Hybrid radio-map for noise tolerant wireless indoor localization. In: 2014 IEEE 11th International Conference on Networking, Sensing and Control (ICNSC), 233–238, April 2014
26. Wu, C., et al.: WILL: Wireless indoor localization without site survey. IEEE Trans. Parallel Distrib. Syst. **24**(4), 839–848 (2013)
27. Ji, Y., Player, R.: A 3-D indoor radio propagation model for WiFi and RFID. In: Proceedings of the 9th ACM International Symposium on Mobility Management and Wireless Access. ACM (2011)

On the Displacement for Covering a Square with Randomly Placed Sensors

Rafał Kapelko[1]([✉]) and Evangelos Kranakis[2]

[1] Faculty of Fundamental Problems of Technology, Department of Computer Science,
Wrocław University of Technology, Wrocław, Poland
`rafal.kapelko@pwr.edu.pl`
[2] School of Computer Science, Carleton University, Ottawa, ON, Canada

Abstract. Consider n sensors placed randomly and independently with the uniform distribution in a unit square. The sensors have identical sensing range equal to r, for some $r > 0$. We are interested in moving the sensors from their initial positions to new positions so as to ensure that the unit square is completely covered, i.e., every point in the square is within the range of a sensor. If the i-th sensor is displaced a distance d_i, what is a displacement of minimum cost? As cost measure for the displacement of the team of sensors we consider the a-*total movement* defined as the sum $M_a := \sum_{i=1}^{n} d_i^a$, for some constant $a > 0$. We assume that r and n are chosen so as to allow full coverage of the square and $0 < a \leq 4$. The main contribution of the paper is to show the existence of a tradeoff between the square sensing radius and a-total movement and can be summarized as follows:

1. If the *square sensing radius* is equal to $\frac{1}{2\sqrt{n}}$ and n is the square of a natural number we present an algorithm and show that in expectation the a-total movement is in $O(n^{1-a/4})$.
2. If the *square sensing radius* is greater than $\frac{2\sqrt{3}}{\sqrt{n}}$ and n is natural number then we present an algorithm and show that in expectation the a-total movement is in $O(n^{1-a/2}(\ln n)^{a/4})$.

Therefore this sharp decrease from $O(n^{1-a/4})$ to $O(n^{1-a/2}(\ln n)^{a/4})$ in the a-total movement of the sensors to attain complete coverage of the square indicates the presence of an interesting threshold on the square sensing radius when it increases from $\frac{1}{2\sqrt{n}}$ to $\frac{2\sqrt{3}}{\sqrt{n}}$. In addition, we simulate our algorithms above and discuss the results of our simulations.

Keywords: Displacement · Random · Sensors · Unit square

1 Introduction

A key challenge in utilizing effectively a group of sensors is to make them form an interconnected structure with good communication characteristics. For example, one may want to establish a sensing and communication infrastructure for

R. Kapelko—Research supported by grant nr S40012/K1102.

E. Kranakis—Research supported in part by NSERC Discovery grant.

S. Papavassiliou and S. Ruehrup (Eds.): ADHOC-NOW 2015, LNCS 9143, pp. 148–162, 2015.
DOI:10.1007/978-3-319-19662-6_11

robust connectivity, surveillance, security, or even reconnaissance of an urban environment using a limited number of sensors. For a team of sensors initially placed in a geometric domain such a robust connectivity cannot be assured a priori e.g., due to geographic obstacles (inhibiting transmissions), harsh environmental conditions (affecting signals), sensor faults (due to misplacement), etc. In those cases it may be required that a group of sensors originally placed in a domain be displaced to new positions either by a centralized or distributed controller. The main question arising is what is the cost of displacement so as to move the sensors from their original positions to new positions so as to attain the desired communication characteristics?

A typical sensor is able to sense a limited region usually defined by its sensing radius, say r, and considered to be a circular domain (disc of radius r). To protect a larger region against intruders every point of the region must be within the sensing range of at least one of the sensors in the group. Moreover, by forming a communication network with these sensors one is able to transmit to the entire region any disturbance that may have occurred in any part of the region. This approach has been previously studied in several papers. It includes research on (1) *area coverage* in which one ensures monitoring of an entire region [11,15], and (2) on *perimeter* or *barrier coverage* whereby a region is protected by monitoring its perimeter thus sensing intrusions or withdrawals to/from the interior [2,3,5, 6,14]. Note that although barrier coverage is less expensive (in terms of number of sensors) than area coverage also the former cannot be considered a substitute of the latter.

1.1 Related Work

Assume sensors of identical range all initially placed on a line. It was shown in [5] that there is an $O(n^2)$ algorithm for minimizing the max displacement of a sensor while the optimization problem becomes NP-complete if there are two separate (non-overlapping) barriers on the line (cf. also [4] for arbitrary sensor ranges). If the optimization cost is the sum of displacements then [6] shows that the problem is NP-complete when arbitrary sensor ranges are allowed, while an $O(n^2)$ algorithm is given when all sensing ranges are the same. Similarly, if one is interested in the number of sensors moved then the coverage problem is NP-complete when arbitrary sensor ranges are allowed, and an $O(n^2)$ algorithm is given when all sensing ranges are the same [16]. Further, [7] considers the algorithmic complexity of several natural generalizations of the barrier coverage problem with sensors of arbitrary ranges, including when the initial positions of sensors are arbitrary points in the two-dimensional plane, as well as multiple barriers that are parallel or perpendicular to each other.

An important setting in considerations for barrier coverage is when the sensors are placed at random on the barrier according to the uniform distribution. Clearly, when the sensor dispersal on the barrier is random then coverage depends on the sensor density and some authors have proposed using several rounds of random dispersal for complete barrier coverage [9,18]. Another approach is to have the sensors relocate from their initial position to a new position on the barrier so as to achieve complete coverage [5,6,8,16]. Further, this relocation may be centralized (cf. [5,6]) or distributed (cf. [8]).

Closely related to our work is [13]. In this paper, n sensors were placed in the unit interval uniformly and independently at random and the cost of displacement was measured by the sum of the respective displacements of the individual sensors in the unit line segment $[0,1]$. Lets call the positions $\frac{i}{n} - \frac{1}{2n}$, for $i = 1, 2, \ldots, n$, anchor positions. The sensors have the sensing radius $r = \frac{1}{2n}$ each. Notice that the only way to attain complete coverage is for the sensors to occupy the anchoir positions. The following result was proved in [13].

Theorem 1 (cf. [13]). *Assume that, n mobile sensors are thrown uniformly and independently at random in the unit interval. The expected sum of displacements of all n sensors to move from their current location to the equidistant anchor locations $\frac{i}{n} - \frac{1}{2n}$, for $i = 1, 2, \ldots, n$, respectively, is in $\Theta(\sqrt{n})$.*

In [12], Theorem 1 was extended to when the cost of displacement is measured by the sum of the respective displacements raised to the power $a > 0$ of the respective sensors in the unit line segment $[0,1]$. The following result was proved.

Theorem 2 (cf. [12]). *Assume that n mobile sensors are thrown uniformly and independently at random in the unit interval. The expected sum of displacements to a given power a of all n sensors to move from their current location to the respective equidistant anchor positions $\frac{i}{n} - \frac{1}{2n}$, for $i = 1, 2, \ldots, n$, is in $\Theta\left(1/n^{\frac{a}{2}-1}\right)$, when a is natural number, and in $O\left(1/n^{\frac{a}{2}-1}\right)$, when $a > 0$.*

An analysis similar to the line segment was provided for the unit square in [13]. Our paper focuses on the analysis of sensor displacement for a group of sensors placed uniformly at random on the unit square. In particular, our approach is the first to generalize the results in [13] for the unit square using as cost metric the a-total movement, and also obtain sharper bounds for this case.

1.2 Preliminaries and Notation

We define the cost measure *a-total movement* as follows.

Definition 1 (a-total movement). *Let $0 < a \leq 4$ be a constant. Suppose the displacement of the i-th sensor is a distance d_i. The a-total movement is defined as the sum $M_a := \sum_{i=1}^{n} d_i^a$. (We assume that, r and n are chosen so as to allow full coverage of the square and $0 < a \leq 4$.)*

Motivation for this cost metric arises from the fact that there are obstacles that obstruct the sensor movement from their initial to their final destinations. Therefore the a-total movement is a more realistic metric than the one previously considered for $a = 1$.

To simplify our proofs we define below the concept *Square Sensing Radius* which refers to a coverage area having the shape of a square.[1]

---

[1] Recall that the generally accepted coverage area of a sensor is a circle. Our results can be easily converted to this model by describing a minimum circle outside this square.

Definition 2 (Square Sensing Radius). *Consider a sensor $Z_{(x,y)}$ located in position (x,y), where $0 \leq x, y \leq 1$. We define the range of the sensor $Z_{(x,y)}$ to be the area delimited by the square with the four vertices $(x \pm a, y \pm a)$, and call a the sensing radius of the sensor.*

In the analysis we consider the Beta distribution. We say that a random variable concentrated on the interval $[0, 1]$ has the $B(a, b)$ distribution with parameters a, b, if it has the probability density function

$$f(x) = \frac{1}{B(a,b)} x^{a-1}(1-x)^{b-1}, \tag{1}$$

where the Euler Beta function (see [17])

$$B(a,b) = \int_0^1 x^{a-1}(1-x)^{b-1} dx \tag{2}$$

is defined for all complex numbers a, b such as $\Re(a) > 0$ and $\Re(b) > 0$. Let us notice that for any integer numbers $a, b \geq 0$, we have

$$B(a,b)^{-1} = \binom{a+b-1}{a} a. \tag{3}$$

1.3   Results of the Paper

We consider n mobile sensors with identical square sensing radius r placed independently at random with the uniform distribution in the unit square. We want to have the sensors move from their current location to positions that cover the square in the sense that every point in the square is within the range of at least one sensor. When a sensor is displaced on the square a distance equal to d the cost of the displacement is d^a for some (fixed) power $0 < a \leq 4$ of the distance d traveled. We assume that r and n are chosen so as to allow full coverage of the square, i.e., every point of the region is within the range of at least one sensor.

The main contribution of the paper in Sect. 2 is to show the existence of a tradeoff between square sensing radius and a-total movement that can be summarized as follows:

1. For the case of the square sensing radius $\frac{1}{2\sqrt{n}}$ and n is the square of a natural number we present an algorithm that uses $O\left(n^{1-\frac{a}{4}}\right)$ total expected movement (see Algorithm 1 and Theorem 3).

2. If the square sensing radius is greater than $\frac{2\sqrt{3}}{\sqrt{n}}$ and n is a natural number then the expected movement is $O(n^{1-a/2}(\ln n)^{a/4})$ (see Algorithm 3 and Theorem 4).

Notice that, for $a = 2$ Algorithm 1 uses $O(\sqrt{n})$ total expected movement while Algorithm 3 uses $O(\sqrt{\ln n})$ total expected movement. Therefore this sharp decrease from $O(n^{1-a/4})$ to $O(n^{1-a/2}(\ln n)^{a/4})$ in the a-total movement of the sensors to attain complete coverage of the square indicates the presence of an interesting threshold on the square sensing radius when it increases from $\frac{1}{2\sqrt{n}}$ to $\frac{2\sqrt{3}}{\sqrt{n}}$.

In Sect. 3 we simulate our algorithms and provide the results of the simulations. Finally, Sect. 4 is the conclusion.

2   Displacement in 2D

Assume that, n mobile sensors with the same square sensing radius are thrown uniformly at random and independently in the unit square $[0,1] \times [0,1]$. Let $0 < a \leq 4$.

Our first result is an upper bound on the expected a-total movement for the case, where the square sensing radius is $\frac{1}{2\sqrt{n}}$. Observe that in this case the only way for the sensors to attain complete coverage of the square is to occupy the positions

$$\left(\frac{k}{\sqrt{n}} - \frac{1}{2\sqrt{n}}, \frac{l}{\sqrt{n}} - \frac{1}{2\sqrt{n}} \right),$$

for $1 \leq k, l \leq \sqrt{n}$. Let us also notice that the number n must be itself the square of a natural number.

We present Algorithm 1 that uses $O\left(n^{1-\frac{a}{4}}\right)$ expected a-total movement. The algorithm is in two-phases. During the first phase (see steps (1–6)) we apply a greedy strategy and move all the sensors only according to the first coordinate. Figure 1 illustrates the steps (1–6) of Algorithm 1.

As a result of the first phase we have $\sqrt{n}$ columns each with $\sqrt{n}$ random sensors. Hence the first phase reduces the sensor movement on the unit square to the sensor movement on the unit interval. During the second phase (see steps (7–9)) we move sensors only according to the second coordinate in each column to equidistant positions.

We prove the following theorem.

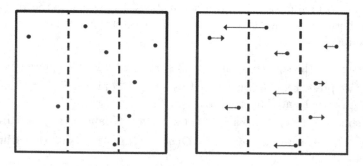

Fig. 1. Nine mobile sensors located in the interior of a cube move to new positions according to steps (1–6) of Algorithm 1.

Algorithm 1. Sensor displacement on a unit square when n is the square of the natural number.

Require: n mobile sensors with identical square sensing radius $r = \frac{1}{2\sqrt{n}}$ placed uniformly and independently at random on the square $[0,1] \times [0,1]$.

Ensure: The final positions of the sensors are at the locations $\left(\frac{k}{\sqrt{n}} - \frac{1}{2\sqrt{n}}, \frac{l}{\sqrt{n}} - \frac{1}{2\sqrt{n}}\right)$, $1 \le k, l \le \sqrt{n}$ (so as to attain coverage of the square $[0,1] \times [0,1]$.)

1: sort the initial locations of sensors according to the first coordinate; the locations after sorting $(x_1, y_1), (x_2, y_2), \ldots (x_n, y_n)$, $x_1 \le x_2 \le \cdots \le x_n$.

2: **for** $j = 1$ to $\sqrt{n}$ **do**

3:　　**for** $i = 1$ to $\sqrt{n}$ **do**

4:　　　　move the sensor $S_{i+(j-1)\sqrt{n}}$ at position $\left(\frac{j}{\sqrt{n}} - \frac{1}{2\sqrt{n}}, y_{i+(j-1)\sqrt{n}}\right)$

5:　　**end for**

6: **end for**

7: **for** $j = 1$ to $\sqrt{n}$ **do**

8:　　move sensors $S_{1+(j-1)\sqrt{n}}, S_{2+(j-1)\sqrt{n}}, \ldots, S_{\sqrt{n}+(j-1)\sqrt{n}}$ to equidistant points that are sufficient to cover the rectangle with vertices $\left(\frac{j-1}{\sqrt{n}}, 0\right)$, $\left(\frac{j}{\sqrt{n}}, 0\right)$, $\left(\frac{j-1}{\sqrt{n}}, 1\right)$, $\left(\frac{j}{\sqrt{n}}, 1\right)$

9: **end for**

Theorem 3. *Let n be the square of a natural number and let $0 < a \le 4$. Assume that n sensors of square sensing radius equal to $\frac{1}{2\sqrt{n}}$ are thrown randomly and uniformly and independently with the uniform distribution on a unit square. The total expected displacement to power a required so that the sensors cover the unit square is in $O\left(n^{1-\frac{a}{4}}\right)$.*

Proof. Let $0 < a \le 4$. We will estimate the expected a-total movement at the steps (1–6). Let X_i be the ith order statistic, i.e., the position of the ith sensor in the interval $[0,1]$ after sorting in step (1). It turns out (see [1]) that X_i obeys the Beta distribution with parameters $i, n-i+1$. We know that the density function for X_i (see Equation (1)) is

$$f_{X_i}(x) = i\binom{n}{i} x^{i-1}(1-x)^{n-i}.$$

Therefore, the expected a-total movement in steps (1–6) of the algorithm is equal to

$$E_{(1-6)}^{(a)} = \sum_{j=1}^{\sqrt{n}} \sum_{i=(j-1)\sqrt{n}+1}^{j\sqrt{n}} i\binom{n}{i} \int_0^1 \left| x - \left(\frac{j}{\sqrt{n}} - \frac{1}{2\sqrt{n}}\right) \right|^a x^{i-1}(1-x)^{n-i} dx.$$

Firstly, we compute the expected 4–total movement in steps (1–6). We apply the definition of the Beta function (see Equation (2)) with parameters $k+i, n-i+1$, as well as Equation (3) to deduce that

$$E_{(1-6)}^{(4)} = \sum_{j=1}^{\sqrt{n}} \sum_{i=(j-1)\sqrt{n}+1}^{j\sqrt{n}} \sum_{k=0}^{4} \binom{4}{k} \left(\frac{1}{2\sqrt{n}} - \frac{j}{\sqrt{n}}\right)^{4-k} \frac{\int_0^1 x^{k+i-1}(1-x)^{n-i}dx}{i^{-1}\binom{n}{i}^{-1}}$$

$$= \sum_{j=1}^{\sqrt{n}} \sum_{i=(j-1)\sqrt{n}+1}^{j\sqrt{n}} \sum_{k=0}^{4} \frac{i\binom{n}{i}}{(k+i)\binom{k+n}{k+i}} \binom{4}{k} \left(\frac{1}{2\sqrt{n}} - \frac{j}{\sqrt{n}}\right)^{4-k}.$$

Then, using the identity

$$\sum_{j=1}^{N} j^m = \frac{1}{m+1} \sum_{k=0}^{m} B_k \binom{m+1}{k} N^{m+1-k},$$

where $m \in N$, B_k are Bernoulli numbers (see [10]) with $N = \sqrt{n}$ we get

$$E_{(1-6)}^{(4)} = \frac{42 - 211n + 194n^2 + 47n^3}{240n(6 + 11n + 6n^2 + n^3)}. \tag{4}$$

Notice that, applying the Formula for $E_{(1-6)}^{(4)}$ in any mathematical software that performs symbolic calculation we get the expressions confirming Formula (4).

Therefore, the expected 4-total movement in steps (1–6) of the algorithm is equal to

$$E_{(1-6)}^{(4)} = \Theta\left(\frac{1}{n}\right). \tag{5}$$

Now we define

$$F_{(i,j)}^{(a)} = i\binom{n}{i} \int_0^1 \left| x - \left(\frac{j}{\sqrt{n}} - \frac{1}{2\sqrt{n}}\right)\right|^a x^{i-1}(1-x)^{n-i}dx.$$

Observe that,

$$E_{(1-5)}^{(a)} = \sum_{j=1}^{\sqrt{n}} \sum_{i=(j-1)\sqrt{n}+1}^{j\sqrt{n}} F_{(i,j)}^{(a)}.$$

Then, we use the discrete Hölder inequality with parameters $\frac{4}{a}$ and $\frac{4}{4-a}$ to derive

$$\sum_{j=1}^{\sqrt{n}} \sum_{i=(j-1)\sqrt{n}+1}^{j\sqrt{n}} F_{(i,j)}^{(a)} \leq \left(\sum_{j=1}^{\sqrt{n}} \sum_{i=(j-1)\sqrt{n}+1}^{j\sqrt{n}} \left(F_{(i,j)}^{(a)}\right)^{\frac{4}{a}}\right)^{\frac{a}{4}} \left(\sum_{j=1}^{\sqrt{n}} \sum_{i=(j-1)\sqrt{n}+1}^{j\sqrt{n}} 1\right)^{\frac{4-a}{4}}$$

$$= \left(\sum_{j=1}^{\sqrt{n}} \sum_{i=(j-1)\sqrt{n}+1}^{j\sqrt{n}} \left(F_{(i,j)}^{(a)}\right)^{\frac{4}{a}}\right)^{\frac{a}{4}} (n)^{\frac{4-a}{4}}. \tag{6}$$

Next, we use Hölder inequality for integrals with parameters $\frac{4}{a}$ and $\frac{4}{4-a}$ and get

$$\int_0^1 \left| x - \left(\frac{j}{\sqrt{n}} - \frac{1}{2\sqrt{n}} \right) \right|^a x^{i-1}(1-x)^{n-i} i\binom{n}{i} dx$$

$$\leq \left(\int_0^1 \left(\left| x - \left(\frac{j}{\sqrt{n}} - \frac{1}{2\sqrt{n}} \right) \right|^a \right)^{\frac{4}{a}} x^{i-1}(1-x)^{n-i} i\binom{n}{i} dx \right)^{\frac{a}{4}},$$

so $\left(F_{(i,j)}^{(a)} \right)^{\frac{4}{a}} \leq F_{(i,j)}^{(4)}$. Putting together Eqs. (5) and (6) we obtain

$$E_{(1-6)}^{(a)} = O\left(n^{1-\frac{a}{2}} \right).$$

Observe that in step (8) of the algorithm we have that $\sqrt{n}$ mobile sensors are thrown uniformly and indendently at random in the unit interval. According to Theorem 2 [cf. [12]] the expected a-total movement at the step (8) is equal $O((\sqrt{n})^{1-\frac{a}{2}})$. Hence the expected a-total movement in steps (7–9) is in $O(\sqrt{n}(\sqrt{n})^{1-\frac{a}{2}})$. Therefore, the expected cost of displacement to power a of Algorithm 1 is in $O(n^{1-\frac{a}{4}})$. This completes the proof of Theorem 3. $\qquad\square$

Now we study a lower bound on the total displacement, when the square sensing radius of the sensors is larger than $\frac{1}{2\sqrt{n}}$. First, we give a lemma which indicates how to scale the results of Theorem 3 to intervals of arbitrary length. The following lemma states that Algorithm 2 uses $O\left(x^a n^{1-\frac{a}{4}} \right)$ expected a-total movement.

Algorithm 2. Sensor displacement on a square when n is the square of the natural number.

Require: n mobile sensors with identical square sensing radius $r = \frac{x}{2\sqrt{n}}$ placed uniformly and independently at random on the square $[0, x] \times [0, x]$.

Ensure: The final positions of sensors at the location $\left(\frac{kx}{\sqrt{n}} - \frac{x}{2\sqrt{n}}, \frac{lx}{\sqrt{n}} - \frac{x}{2\sqrt{n}} \right)$, $1 \leq k, l \leq \sqrt{n}$ so as to attain coverage of the square $[0, x] \times [0, x]$.

1: sort the initial locations of sensors according to the first coordinate; the locations after sorting $(x_1, y_1), (x_2, y_2), \ldots (x_n, y_n)$, $x_1 < x_2 < \cdots \leq x_n$.
2: **for** $j = 1$ to $\sqrt{n}$ **do**
3:    **for** $i = 1$ to $\sqrt{n}$ **do**
4:      move the sensor $S_{i+(j-1)\sqrt{n}}$ at the position $\left(\frac{jx}{\sqrt{n}} - \frac{x}{2\sqrt{n}}, y_{i+(j-1)\sqrt{n}} \right)$
5:    **end for**
6: **end for**
7: **for** $j = 1$ to $\sqrt{n}$ **do**
8:    move the sensors $S_{1+(j-1)\sqrt{n}}, S_{2+(j-1)\sqrt{n}}, \ldots, S_{\sqrt{n}+(j-1)\sqrt{n}}$ to equidistant points so as to cover the rectangle with vertices $\left(\frac{(j-1)x}{\sqrt{n}}, 0 \right)$, $\left(\frac{jx}{\sqrt{n}}, 0 \right)$, $\left(\frac{(j-1)x}{\sqrt{n}}, 1 \right)$, $\left(\frac{jx}{\sqrt{n}}, 1 \right)$
9: **end for**

Lemma 1. *Let n be the square of a natural number and let $0 < a \leq 4$. Assume that, n sensors of square sensing radius equal to $\frac{x}{2\sqrt{n}}$ are thrown randomly, uniformly and independently with the uniform distribution on the square $[0, x] \times [0, x]$. The total expected displacement to the power a required for the sensors to cover the square $[0, x] \times [0, x]$ is in $O\left(x^a n^{1-\frac{a}{4}}\right)$.*

Proof. Assume that, n sensors are in the square $[0, x] \times [0, x]$. Then, multiply their coordinates by $1/x$. From Theorem 3 the expected a-total movement in the unit square is in $O\left(n^{1-\frac{a}{4}}\right)$. Now by multiplying their coordinates by x we get the result in the statement of the lemma. □

A natural question to ask is: how to exploit the proposed Algorithm 1 when the number n of nodes is not a square of natural number? Assume that, n sensors have have the square sensing radius $r = \frac{f}{2\sqrt{n}}$ and $f \geq \frac{\sqrt{n}}{\lfloor\sqrt{n}\rfloor}$. To attain coverage of the square $[0, 1] \times [0, 1]$ choose $\lfloor\sqrt{n}\rfloor^2$ sensors at random and use Algorithm 1 for the chosen sensors. Then similar arguments hold for Algorithm 2.

Notice that for $f > 4\sqrt{3}$ we can do better. The following theorem states that Algorithm 3 uses $O\left(n^{1-\frac{a}{4}}\left(\frac{\ln n}{n}\right)^{\frac{a}{4}}\right)$ expected a-total movement.

Algorithm 3. Sensor displacement to the power a when $0 < a \leq 4$ on a square; $p = \frac{9}{4}\left(2 + \frac{a}{2}\right)$, $A = \frac{3}{4}\left(2 + \frac{a}{2}\right)$.

Require: $n \geq 31$ mobile sensors with identical square sensing radius $r = \frac{f}{2\sqrt{n}}$,
  $f \geq 4\sqrt{3}$ placed uniformly and independently at random on the square $[0, 1] \times [0, 1]$.
Ensure: The final positions of sensors to attain coverage of the square $[0, 1] \times [0, 1]$.
1: Divide the unit square into subsquares of side $\frac{1}{\left\lfloor\sqrt{\frac{n}{p\ln n}}\right\rfloor}$;
2: **if** there is a subsquare with fewer than $\frac{1}{3}\frac{n}{\left\lfloor\sqrt{\frac{n}{p\ln n}}\right\rfloor^2}$ sensors **then**
3:     choose $\lfloor\sqrt{n}\rfloor^2$ sensors at random
4:     use Algorithm 1 that moves all $n := \lfloor\sqrt{n}\rfloor^2$ sensors to equidistant points that are sufficient to cover the square
5: **else**
6:     In each subsquare choose $\left\lfloor\sqrt{A\ln n}\right\rfloor^2$ sensors at random and use Algorithm 2
    with $n := \left\lfloor\sqrt{A\ln n}\right\rfloor^2$, $x := \frac{1}{\left\lfloor\sqrt{\frac{n}{p\ln n}}\right\rfloor}$ to move the chosen sensors to equidistant
    positions so as to cover the square
7: **end if**

Theorem 4. *Let n be a natural number and let $f \geq 4\sqrt{3}$, $n \geq 31$ and $0 < a \leq 4$. Assume that, n sensors of square sensing radius $r = \frac{f}{2\sqrt{n}}$ are thrown randomly and independently with uniform distribution on a unit square. Then the total expected movement to power a of sensors required to cover the unit square is $O\left(n^{1-\frac{a}{4}}\left(\frac{\ln n}{n}\right)^{\frac{a}{4}}\right)$.*

Proof. Assume that $0 < a \le 4$. Let $p = \frac{9}{4}\left(2 + \frac{a}{2}\right)$ and $A = \frac{3}{2}\left(2 + \frac{a}{2}\right)$. First of all, observe that $\frac{n}{p\ln(n)} > 1$ for $n \ge 31$.

We will prove that Algorithm 3 uses $O\left(n^{1-\frac{a}{4}}\left(\frac{\ln n}{n}\right)^{\frac{a}{4}}\right)$ expected a-total movement.

There are two cases to consider.

Case 1: There exists a subsquare with fewer than

$$\frac{1}{3}\frac{n}{\left\lfloor\sqrt{\frac{n}{p\ln n}}\right\rfloor^2}$$

sensors. In this case choose $\lfloor\sqrt{n}\rfloor^2$ sensors uniformly and randomly from n sensors. Applying the inequality $\lfloor x \rfloor > x - 1$ we deduce that

$$\frac{\lfloor\sqrt{n}\rfloor^2}{n} > 1 - \frac{2\sqrt{n} - 1}{n} \ge 1 - \frac{2\sqrt{31} - 1}{31},$$

for $n \ge 31$. Since $f \ge 4\sqrt{3}$ we have

$$\lfloor\sqrt{n}\rfloor^2 \frac{f^2}{n} > \left(1 - \frac{2\sqrt{31} - 1}{31}\right)48 \approx 32.3 > 1$$

for $n \ge 31$. Therefore, the $\lfloor\sqrt{n}\rfloor^2$ chosen sensors are enough to attain the coverage. The expected a-total movement is $O\left((\lfloor\sqrt{n}\rfloor^2)^{1-\frac{a}{4}}\right) = O\left(n^{1-\frac{a}{4}}\right)$ by Theorem 3.

Case 2: All subsquares contain at least $\frac{1}{3}\frac{n}{\left\lfloor\sqrt{\frac{n}{p\ln n}}\right\rfloor^2}$ sensors. From the inequality $\lfloor x \rfloor \le x$ we deduce that,

$$\left\lfloor\sqrt{A\ln n}\right\rfloor^2 \le \frac{1}{3}\frac{n}{\left\lfloor\sqrt{\frac{n}{p\ln n}}\right\rfloor^2}.$$

Hence it is possible to choose $\left\lfloor\sqrt{A\ln n}\right\rfloor^2$ sensors at random in each subsquare with more than $\frac{1}{3}\frac{n}{\left\lfloor\sqrt{\frac{n}{p\ln n}}\right\rfloor^2}$ sensors. Let us consider the sequence

$$a_n = \frac{4\sqrt{3}}{\sqrt{n}}\left\lfloor\sqrt{\frac{3}{2}\ln n}\right\rfloor\left\lfloor\sqrt{\frac{n}{9\ln n}}\right\rfloor$$

for $n \ge 31$. Numerical calculation for small values of n ($31 < n < 250$) confirms that $a_n > 1$. Applying inequality $\lfloor x \rfloor > x - 1$ we see that

$$a_n > 2\sqrt{2}\left(1 - \sqrt{\frac{2}{3\ln n}}\right)\left(1 - \sqrt{\frac{9\ln n}{n}}\right)$$

$$\geq 2\sqrt{2}\left(1 - \sqrt{\frac{2}{3\ln 250}}\right)\left(1 - \sqrt{\frac{9\ln 250}{250}}\right)$$

$$\approx 1.02277 > 1.$$

Hence,

$$\left\lfloor\sqrt{A\ln n}\right\rfloor^2\frac{f^2}{n}\left\lfloor\sqrt{\frac{n}{p\ln n}}\right\rfloor^2 \geq a_n^2 > 1.$$

Therefore, $\left\lfloor\sqrt{A\ln n}\right\rfloor^2$ chosen sensors are enough to attain the coverage. By the independence of the sensors positions, the $\left\lfloor\sqrt{A\ln n}\right\rfloor^2$ chosen sensors in any given subsquare are distributed randomly and independently with uniform distribution over the subsquare of side $x = \frac{1}{\left\lfloor\sqrt{\frac{n}{p\ln n}}\right\rfloor}$. By Lemma 1 the expected a-total movement inside each subsquare is

$$O\left(\left(\frac{1}{\left\lfloor\sqrt{\frac{n}{p\ln n}}\right\rfloor}\right)^a\left(\lfloor\sqrt{A\ln n}\rfloor^2\right)^{1-\frac{a}{4}}\right) = O\left(\frac{(\ln n)^{\frac{a}{4}}}{n^{\frac{a}{2}}}(\ln n)\right).$$

Since, there are $\left\lfloor\sqrt{\frac{n}{p\ln n}}\right\rfloor^2$ subsquares, the expected a-total movement over all subsquares must be in $O\left(n^{1-\frac{a}{4}}\left(\frac{\ln n}{n}\right)^{\frac{a}{4}}\right)$. It remains to consider the probability with which each of these cases occurs. The proof of the theorem will be a consequence of the following Claim.

Claim 5. Let $p = \frac{9}{4}\left(2 + \frac{a}{2}\right)$. The probability that fewer than $\frac{1}{3}\frac{n}{\left\lfloor\sqrt{\frac{n}{p\ln n}}\right\rfloor^2}$ sensors fall in any subsquare is $< \frac{\left\lfloor\sqrt{\frac{n}{p\ln n}}\right\rfloor^2}{n^{1+\frac{a}{4}}}$.

Proof. (Claim 5) First of all, from the inequality $\lfloor x\rfloor \leq x$ we get

$$\sqrt{\frac{\left(2 + \frac{a}{2}\right)\ln n}{n}}\left\lfloor\sqrt{\frac{n}{p\ln n}}\right\rfloor^2 \leq \frac{2}{3}.$$

Hence,

$$\frac{1}{3}\frac{n}{\left\lfloor\sqrt{\frac{n}{p\ln n}}\right\rfloor^2} \leq \frac{n}{\left\lfloor\sqrt{\frac{n}{p\ln n}}\right\rfloor^2} - \sqrt{\frac{\left(2 + \frac{a}{2}\right)n\ln n}{\left\lfloor\sqrt{\frac{n}{p\ln n}}\right\rfloor^2}}. \tag{7}$$

The number of sensors falling in a subsquare is a Bernoulli process with probability of success $\frac{1}{\left\lfloor\sqrt{\frac{n}{p\ln n}}\right\rfloor^2}$. By Chernoff bounds, the probability that a given subinterval has fewer than

$$\frac{n}{\left\lfloor\sqrt{\frac{n}{p\ln n}}\right\rfloor^2} - \sqrt{\frac{\left(2+\frac{a}{2}\right)n\ln n}{\left\lfloor\sqrt{\frac{n}{p\ln n}}\right\rfloor^2}}$$

sensors is less than $e^{-\left(1+\frac{a}{4}\right)\ln n} < \frac{1}{n^{1+\frac{a}{4}}}$. Specifically we use the Chernoff bound

$$\Pr[X < (1-\delta)m] < e^{-\delta^2 m/2},$$

$m = \frac{n}{\left\lfloor\sqrt{\frac{n}{p\ln n}}\right\rfloor^2}, \delta = \sqrt{\frac{\left(2+\frac{a}{2}\right)\ln n}{n}}\left\lfloor\sqrt{\frac{n}{p\ln n}}\right\rfloor$. As there are $\left\lfloor\sqrt{\frac{n}{p\ln n}}\right\rfloor^2$ subsquares, the event that one has fewer than

$$\frac{n}{\left\lfloor\sqrt{\frac{n}{p\ln n}}\right\rfloor^2} - \sqrt{\frac{\left(2+\frac{a}{2}\right)n\ln n}{\left\lfloor\sqrt{\frac{n}{p\ln n}}\right\rfloor^2}}.$$

sensors occurs with probability less than $\frac{\left\lfloor\sqrt{\frac{n}{p\ln n}}\right\rfloor^2}{n^{1+\frac{a}{4}}}$. This and Eq. (7) completes the proof of Claim 5. $\qquad\square$

Using Claim 5 we can upper bound the expected a-total movement as follows:

$$\left(1 - \frac{\left\lfloor\sqrt{\frac{n}{p\ln n}}\right\rfloor^2}{n^{1+\frac{a}{4}}}\right) O\left(n^{1-\frac{a}{4}}\left(\frac{\ln n}{n}\right)^{\frac{a}{4}}\right) + \left(\frac{\left\lfloor\sqrt{\frac{n}{p\ln n}}\right\rfloor^2}{n^{1+\frac{a}{4}}}\right) O\left(n^{1-\frac{a}{4}}\right)$$

$$= O\left(n^{1-\frac{a}{4}}\left(\frac{\ln n}{n}\right)^{\frac{a}{4}}\right),$$

$\qquad\square$

which proves Theorem 4.

3 Simulation Results

In this section we use simulation results to analyze how random placement of sensors on the square impacts the expected a-total movement.

We repeated 3 times the following experiments. Firstly, for each number of sensors $n \in \{2^2, 3^2, 4^2, \ldots, 60^2\}$ we generated 32 random placements. Then we calculated the expected a-total movement according to Algorithms 1 and 2, respectively. Let $E_{n,32}$ be the average of 32 measurements of the expected a-total movement. Then, we placed the points in the set $\{(n, E_{n,32}) : n = 2^2, 3^2, 4^2, \ldots, 60^2\}$ into the picture.

Figure 2 illustrates the described experiments for Algorithm 1 when $a = 2$ and $a = 4$. The additional line in the above pictures is the plot of the function which is the theoretical estimation. Black dots which represent numerical results are situated near the theoretical line. According to the proof of Theorem 3 the

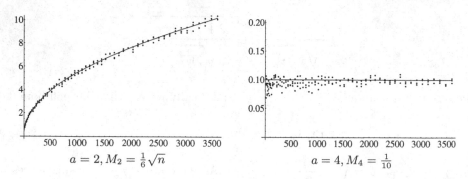

Fig. 2. The expected a-total movement of Algorithm 1.

steps (7–9) of Algorithm 1 contribute the asymptotics. Notice that, the expected a-total movement in steps (7–9) of Algorithm 1 is equal to

$$E^{(a)}_{(7-9)} = \sqrt{n} \sum_{i=1}^{\sqrt{n}} i \binom{\sqrt{n}}{i} \int_0^1 \left| x - \left(\frac{j}{\sqrt{n}} - \frac{1}{2\sqrt{n}} \right) \right|^a x^{i-1}(1-x)^{\sqrt{n}-i} dx.$$

Applying the Formulas for $E^{(2)}_{(7-9)}$ and $E^{(4)}_{(7-9)}$ in any mathematical software that performs symbolic calculation we get

$$E^{(2)}_{(7-9)} \sim \frac{1}{6}\sqrt{n} \text{ and } E^{(4)}_{(7-9)} \sim \frac{1}{10}.$$

Therefore, $M_2 = \frac{1}{6}\sqrt{n}$ and $M_4 = \frac{1}{10}$.

Figure 3 illustrates the described experiments for Algorithm 2 ($a = 2, x = \frac{1}{3}$) and ($a = 4, x = 2$). The additional line in the above pictures is the plot of the function which is the theoretical estimation. Black dots which represent numerical results are situated near the theoretical line. According to the proof of Lemma 1 we have $M_2 = \left(\frac{1}{3}\right)^2 \frac{1}{6}\sqrt{n} = \frac{1}{54}\sqrt{n}$ and $M_4 = 2^4 \frac{1}{10} = \frac{4}{5}$.

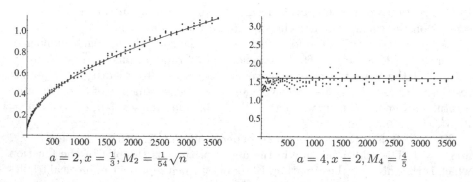

Fig. 3. The expected a-total movement of Algorithm 2.

4 Conclusion

In this paper we studied the movement of n sensors with identical square sensing radius in two dimensions when the cost of movement of sensor is proportional to some (fixed) power $0 < a \le 4$ of the distance traveled. We obtained bounds on the movement depending on the range of sensors. An interesting problem would be to study the displacement to a power a for all $a > 0$ in higher dimensions.

References

1. Arnold, B., Balakrishnan, N., Nagaraja, H.: A First Course in Order Statistics, vol. 54. SIAM, Philadelphia (1992)
2. Balister, P., Bollobas, B., Sarkar, A., Kumar, S.: Reliable density estimates for coverage and connectivity in thin strips of finite length. In: Proceedings of MobiCom 2007, pp. 75–86. ACM (2007)
3. Bhattacharya, B., Burmester, M., Hu, Y., Kranakis, E., Shi, Q., Wiese, A.: Optimal movement of mobile sensors for barrier coverage of a planar region. Theoret. Comput. Sci. **410**(52), 5515–5528 (2009)
4. Chen, D.Z., Gu, Y., Li, J., Wang, H.: Algorithms on minimizing the maximum sensor movement for barrier coverage of a linear domain. In: Fomin, F.V., Kaski, P. (eds.) SWAT 2012. LNCS, vol. 7357, pp. 177–188. Springer, Heidelberg (2012)
5. Czyzowicz, J., et al.: On minimizing the maximum sensor movement for barrier coverage of a line segment. In: Ruiz, P.M., Garcia-Luna-Aceves, J.J. (eds.) ADHOC-NOW 2009. LNCS, vol. 5793, pp. 194–212. Springer, Heidelberg (2009)
6. Czyzowicz, J., et al.: On minimizing the sum of sensor movements for barrier coverage of a line segment. In: Nikolaidis, I., Wu, K. (eds.) ADHOC-NOW 2010. LNCS, vol. 6288, pp. 29–42. Springer, Heidelberg (2010)
7. Dobrev, S., et al.: Complexity of barrier coverage with relocatable sensors in the plane. In: Spirakis, P.G., Serna, M. (eds.) CIAC 2013. LNCS, vol. 7878, pp. 170–182. Springer, Heidelberg (2013)
8. Eftekhari, M., Kranakis, E., Krizanc, D., Morales-Ponce, O., Narayanan, L., Opatrny, J., Shende, S.: Distributed local algorithms for barrier coverage using relocatable sensors. In: Proceeding of ACM PODC Symposium, pp. 383–392 (2013)
9. Eftekhari, M., Narayanan, L., Opatrny, J.: On multi-round sensor deployment for barrier coverage. In: Proceedings of 10th IEEE MASS, pp. 310–318 (2013)
10. Graham, R., Knuth, D., Patashnik, O.: Concrete Mathematics a Foundation for Computer Science. Addison-Wesley, Reading (1994)
11. Huang, C.F., Tseng, Y.C.: The coverage problem in a wireless sensor network. In: WSNA 2003: Proceedings of the 2nd ACM International Conference on Wireless Sensor Networks and Applications, pp. 115–121. ACM (2003)
12. Kapelko, R., Kranakis, E.: On the displacement for covering a unit interval with randomly placed sensors (2014, preprint)
13. Kranakis, E., Krizanc, D., Morales-Ponce, O., Narayanan, L., Opatrny, J., Shende, S.: Expected sum and maximum of displacement of random sensors for coverage of a domain. In: Proceedings of SPAA, pp. 73–82. ACM (2013)
14. Kumar, S., Lai, T.H., Arora, A.: Barrier coverage with wireless sensors. In: Proceedings of MobiCom 2005, pp. 284–298. ACM (2005)

15. Meguerdichian, S., Koushanfar, F., Potkonjak, M., Srivastava, M.B.: Coverage problems in wireless ad-hoc sensor networks. In: Proceedings of INFOCOM, vol. 3, pp. 1380–1387 (2001)
16. Mehrandish, M., Narayanan, L., Opatrny, J.: Minimizing the number of sensors moved on line barriers. In: Proceedings of IEEE WCNC 2011, pp. 1464–1469 (2011)
17. NIST Digital Library of Mathematical Functions. http://dlmf.nist.gov/8.17
18. Yan, G., Qiao, D.: Multi-round sensor deployment for guaranteed barrier coverage. In: Proceedings of IEEE INFOCOM 2010, pp. 2462–2470 (2010)

Election-Based Sensor Deployment
and Coverage Maintenance
by a Team of Robots

Qiao Li[1], Venkat Narasimhan[2], and Amiya Nayak[1(✉)]

[1] School of Electrical Engineering and Computer Science,
University of Ottawa, Ottawa, Canada
nayak@uottawa.ca
[2] Les Enterprises Norleaf Networks, Ottawa, Canada

Abstract. Wireless sensor and robot networks (WSRNs) are an integration of wireless sensor network (WSNs) and multi-robot systems. They are comprised of networked sensor and mobile robots that communicate via wireless links to perform distributed sensing and actuation tasks in a region of interest (ROI). In addition to gathering and reporting data, sensors may also report failures of neighboring sensors or lack of coverage in certain neighbourhood to a nearby mobile robot. We propose a solution, called Election-Based Deployment (EBD), for simultaneous sensor deployment and coverage maintenance in multi-robot scenario in failure-prone environment. To the best of our knowledge, it is the first carrier-based localized algorithm that is able to achieve 100 % coverage in a ROI of any shape with multiple robots in failure-prone environment since it combines both sensor deployment and coverage maintenance process. We can observe from the simulation results that EBD outperforms the existing algorithms while reducing the communication overhead to a great extent.

Keywords: Wireless sensor and actuator networks · Sensor deployment · Coverage maintenance

1 Introduction

WSRNs are an integration of wireless sensor network (WSNs) and multi-robot systems. A WSN is generally composed of a large number of distributed sensor nodes, with each one having limited and similar sensing and communication capability [1–3]. A WSN gathers information in a specific area by covering the whole region in sensors sensing range. It aims to collect accurate data which can reflect the real condition of the monitored area, which is the ROI. However, sensing holes caused by unbalanced formation or sensor failure after long-term usage will affect the accuracy of information captured by the network. Thus, the coverage is an important metric for evaluating the performance and capability of the WSN.

The WSN problem can be divided into two separate categories: sensor deployment and coverage maintenance. In the former case, robots place sensors in an empty ROI to establish a sensor network, whereas in the latter case, the task of robots is to patch sensing holes by replacing failed sensors after long usage.

© Springer International Publishing Switzerland 2015
S. Papavassiliou and S. Ruehrup (Eds.): ADHOC-NOW 2015, LNCS 9143, pp. 163–177, 2015.
DOI: 10.1007/978-3-319-19662-6_12

Carrier-based sensor deployment aims at placing static sensors in a two-dimensional empty ROI with preliminary target positions in WSRN [4]. Each robot loads a sufficient number of sensors and is able to move throughout the ROI. Robots are aware of the positions where they enter into the region and are able to detect physical obstacles and boundaries of the ROI. Although sensors are designed with low energy consumption in mind, they can survive with a very limited lifetime with current technologies [1, 3, 5–7]. An important issue in carrier-based sensor deployment is how to guide robots to achieve full coverage within the shortest time and lowest energy consumption and communication overhead. Once an event (i.e., failure) has been detected, robots coordinate with each other to make a decision about the most appropriate way to perform the action. Coverage can be established and improved in different ways in wireless sensor and robot networks. Initial random sensor placement, if applied, may be improved via robot-assisted sensor relocation or additional placement. One or more robots may carry sensors and move within the ROI; while traveling, they drop sensors at proper positions to construct desired coverage. Robots may relocate and place spare sensors according to certain energy optimality criteria.

Considering that none of the existing algorithms guarantees full coverage in failure-prone environment since they do not have a corresponding maintenance algorithm to help them support the network after initial coverage, a localized solution which can combine deployment and maintenance for multi-robot scenario is needed. The reason of choosing localized algorithm is that it is able to minimize communication consumption and deal with multiple sensor failures in multi-robot scenario [8].

We propose a novel algorithm, named Election-Based Deployment (EBD), which combines localized sensor deployment and localized network maintenance to achieve full coverage for multi-robot scenario in failure-prone environment. The proposed algorithm outperforms the existing solution in terms of energy and communication consumption and hole patching latency, while providing coverage guarantee and network support in a failure-prone scenario. The deployment phase of EBD is an extension of an existing approach to achieve a higher working efficiency by allocating different strategies to robots according to their starting positions and modifying the backtracking protocol to deal with the dead end situation. The hole fixing phase aims at minimizing the latency of hole patching by imitating the election process; it selects the most suitable robot to patch a hole first by choosing candidates and then elects a 'worker' among them to fix the sensing hole.

The remainder of the paper is organized as follows. Section 2 provides a brief review of the related work about sensor deployment and coverage maintenance approaches. In Sect. 3, the EBD approach is discussed in detail while the simulation results are presented in Sect. 4, followed by conclusion in Sect. 5.

2 Related Work

2.1 Sensor Deployment Approach

Deploying sensor by robots is an active research subject which is continuously drawing large amount of attention. There are only a few algorithms proposed to address this problem till now. Three well known approaches will be introduced in details.

Batalin *et al.* [9] proposed the least recently visited (LRV) approach to solve the sensor placement problem in a single robot environment. In LRV, the robot chooses the locally least recently visited direction when it has to make a decision about which direction to move to. However, it is unknown under what situation the approach terminates, and LRV takes extra time and redundant movement for the robot to fully cover the region.

Chang *et al.* [10] proposed the obstacle-resistant robot deployment (ORRD) approach, which is also known as Snake-like Deployment (SLD) approach. This is another algorithm which aims to address sensor deployment in a single mobile robot scenario. However, the termination condition of ORRD is also not clear, and the algorithm in fact does not guarantee full coverage in some circumstance according to the current algorithm description. When the robot meets a dead end situation made up of boundaries and obstacles, it cannot move out and leaves an empty uncovered area.

Li *et al.* [11] proposed a back-tracking deployment (BTD) approach to resolve deployment problem over the ROI for both single- and multi-robot scenario. Robots follow a predefined order when they choose moving direction in BTD algorithm and perform backtracking when they get stuck in a dead end situation. BTD approach guarantees full coverage in a failure-free environment. In failure-prone scenario, the holes which are located along the backtracking path will be detected and patched; however, sensing holes not affected by backtracking will be left untreated.

Falcon *et al.* [13] proposed a carrier-based coverage augmentation (CBCA) approach for the purpose of solving both sensor deployment and coverage maintenance problems in a failure-prone environment. The authors use mobile robots to deploy sensors to achieve focused coverage around a point of interest. However, it is not clear how CBCA outperforms other approaches since the authors mentioned there is no competing algorithm.

2.2 Coverage Maintenance Approach

Coverage maintenance approach is proposed to patch holes caused by sensor failures after long usage in a full covered area.

Mei *et al.* [12] proposed a cluster-based approach to patch sensing holes by replacing failed sensors in the region. Three protocols are proposed for distinct scenarios: centralized manager algorithm, fixed distributed manager algorithm and dynamic distributed algorithm. In the centralized approach, a central manager is in charge of receiving failure reports and then forwarding them to the responsible robot, whereas in distributed approaches each robot can serve as the manager and maintainer in the meanwhile. However, message resending and unclear subarea dividing principle can result in high communication overhead. Moreover, the relationship built among sensors is likely to be broken along with robots movement.

Wang *et al.* [14] proposed a localized carrier-based algorithm to maintain the coverage of network for both single-robot and multi-robot scenarios, called localized ant-based sensor relocation algorithm with greedy walk (LASR-G). In LASR-G, robots are designed to load a limited number of sensors so that sensor reloading is taken into consideration. Robots have to calculate pickup and dropping probability to reduce sensing holes, since there is no preset map for them to follow.

Li *et al.* [15] proposed a localized market-based sensor relocation (MSR) approach, with mobile robots and static sensors, to solve the coverage maintenance problem caused by sensing holes and redundant sensors after a random scatter in both single-robot and multi-robot scenarios. Robots following the MSR are assumed to have limited carrying capacity and have to find themselves in proper locations to minimize the area of uncovered ROI. However, the authors did not take node failure into consideration in order to simplify their analysis in simulation.

Falcon *et al.* [16] proposed a centralized one-commodity traveling salesman with selective pickup and delivery (1- TSP- SELPD) protocol to solve the carrier-based coverage problem around a point of interest by replacing failed sensors with spare ones. The capacity of robots is designed to be adaptive, which can be defined by users, leaving an impact on both picking and dropping decision and route.

3 Proposed Election-Based Deployment (EBD) Approach

We now describe the design of a *localized election-based sensor deployment and coverage maintenance approach* based on BTD algorithm, which is able to place sensors in an empty ROI and maintain the network subsequently. It has low energy consumption and communication overhead due to its localized characteristic, and guarantees full coverage as well. Furthermore, it can minimize the latency of hole patching, which improves the accuracy of data gathered by the WSN in failure-prone scenarios.

3.1 Assumptions

In EBD approach, the WSRN is composed of mobile robots and static homogeneous sensors, with both having the same communication radii. The sensing ranges of sensors r_S are of the same size, which is less than a half of the communication radii r_C. Each device is able to communicate directly with others in its communication range without transmission collision and failure. All the established message transmitting relationships can be considered bidirectional.

Each robot is aware of the starting locations of all other robots in the ROI and is able to move within the ROI. Robots are able to load sufficient number of sensors during deployment, and in the hole covering stage, the number of sensors they can pick up is unlimited as well. Furthermore, robots have enough energy to move throughout the ROI so as to achieve full coverage and maintain the network afterwards.

3.2 Sensor Deployment in EBD

Robots drop sensors on a square grid, having four geographic directions, namely *West, East, North* and *South*. As for the guidance of robots movements, four moving strategies, *WN, SW, ES* and *NE,* have been proposed. The four strategies can be described as follows:

$$\left\{ \begin{array}{l} WN : West > East > North > South \\ SW : South > North > West > East \\ ES : East > West > South > North \\ NE : North > South > East > West \end{array} \right.$$

Robots are assigned different moving strategies according to their starting locations. For example, robot with starting location closed to northwest corner of the ROI will be assigned with strategy *WN*, and so on so forth. The robot assigned strategy *WN* will check directions in the following order: *West > East > North > South* before it chooses the moving direction of its next step. Thus, the robot will keep moving west until it reaches the boundary. Now that it cannot move west anymore, it will check the east. However, the east node is already been occupied with a sensor. So the robot has to move north or south before it moves back to the east. In this way, the robot moves in a snake-like fashion following its assigned strategy.

Each sensor is dynamically assigned a color according to the conditions of its neighbors. A sensor will color itself "white" if at least one of its one hop neighbor spots has not been placed with a sensor yet, "black" if all of its four neighbors are deployed with sensors working normally, or "gray" if there is at least one failed sensor among its one hop neighbors. Besides the color, each sensor has a sequence number which represents the sensor deployment order and ID of the robot it was dropped by. Sensors share information, such as color, robot ID and sequence number with its one hop neighbors by sending "Hello" messages periodically.

Sometimes robots may get stuck when they reach a dead end situation. It happens when all the four moving directions are already occupied with sensors, so that the robot cannot move forward any more. In this scenario, a trapped robot will move to the nearest white sensor deployed by itself and resume placing sensors from there onwards. If there are more than one nearby white sensors found, the robot will randomly pick one. If the robot cannot find a white sensor deployed by itself, it will search among the sensors placed by other robots.

In a failure-prone environment, sensors may fail after long usage. Failures generate sensing holes in the ROI. Therefore, if the route of a robot is interrupted by a gray sensor, or a gray sensor is found in the nearest white sensor searching, the robot will patch the hole before it moves to the original destination.

Figure 1 indicates sensor deployment approach in a two-robot scenario. There are two robots in this figure, the blue and red circles represent sensors deployed by robot A and B respectively. The number in each circle is the sequential number, for example, the blue node 10 is the tenth sensor placed by robot A. In Fig. 1(a), robot B gets stuck at red node 33, whereas robot A just drops a sensor at blue node 33. Robot B will send searching messages to four geographical directions; the messages will spread among red nodes until node 23 is found as the nearest white sensor. After it reaches node 33, which is no longer a white sensor since robot A deployed a sensor on its one hop neighbor, robot B will keep moving towards south to the nearest white node 34, as shown in Fig. 1(b). In Fig. 1(c), the algorithm terminates since all the nodes in the ROI are deployed with sensors. The area composed of blue sensors is called subarea of robot A, while the red area is called subarea of robot B.

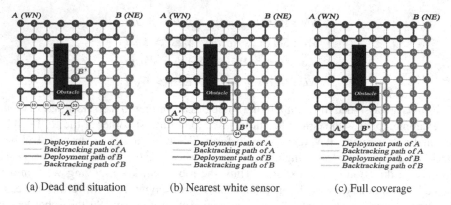

(a) Dead end situation   (b) Nearest white sensor   (c) Full coverage

Fig. 1. Sensor deployment with two robots (Color figure online).

In some cases, subareas of robots following the same moving strategy overlap when the robots encounter each other. However, the coverage maintenance algorithm requires subareas to be independent. Thus, a reallocation strategy is proposed to ensure that the subareas are regular.

Each strategy has a *capacity* which represents the number of robots following it. A strategy not assigned to any robot is an *empty strategy*, whereas the one assigned to several robots called *general strategy*. Strategies corresponding to two neighbor corners are *neighbor strategies*. For example, northwest and northeast are neighbor corners, so corresponding strategies *WN* and *NE* are neighbor strategies. Each strategy has two neighbors with the last one called *diagonal strategy*.

When a general strategy has an empty neighbor, one of the robots following the general strategy will be assigned with the empty neighbor to avoid encounter of robots with the same moving strategy. The rules for picking a robot are as follows:

1. If the empty strategy is *WN* or *ES*, the robot will be the one closest to the northernmost or southernmost border of the ROI. If several robots hold the shortest distance from these two boundaries, the one with the longest distance to the westernmost or easternmost border of its corner will be selected. If multiple strategies meet the conditions in both rounds, a strategy will be picked randomly.

2. If the empty strategy is *NE* or *SW*, the robot will be the one closest to the westernmost or easternmost border of the ROI. If several robots hold the shortest distance from these two boundaries, the one with the longest distance to the northernmost or southernmost border of its corner will be selected. In the event of a tie, a strategy will be picked randomly.

Figure 2 illustrates the *strategy reallocation*. Black nodes represent robots. Empty strategy in both figures is *WN* and the general ones are *NE* and *SW*. In Fig. 2(a), the robot closest to the northernmost or southernmost border of the ROI is robot C, which is just located on the southernmost boundary. In Fig. 2(b), both A and C hold the shortest distance from the northernmost or southernmost border. Compared the distance A holds to the easternmost boundary and C holds to the westernmost boundary, it is

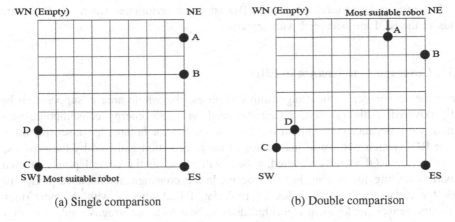

(a) Single comparison   (b) Double comparison

Fig. 2. Strategy reallocation.

obvious that A-e is larger than C-w. So, A will be selected as the one which will be assigned with *WN*.

Strategy reallocation improves the sensor deployment algorithm to a great extent. It guides robots to deploy sensors which form more regular subareas, so that the coverage maintenance algorithm can perform better. Figure 3 indicates the comparison of fully covered ROI with and without using strategy reallocation. Robot A and B start from northwest corner while robot C starts from northeast corner. Sensors deployed by robot A are marked blue, and red for B, yellow for C.

In Fig. 3(a), both robot A and B follow *WN*, and C follows *NE*. The subareas of A and B cross with each other since they are assigned the same moving strategy and encounter each other. Furthermore, the subarea of robot A is divided into two parts. In Fig. 3(b) whereas, Robot B is assigned strategy *SW*, which makes the subareas more regular and independent of one another.

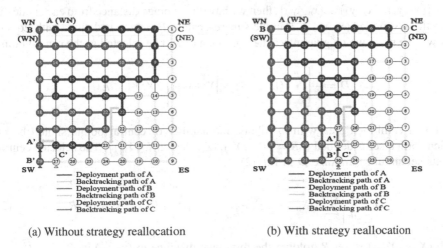

(a) Without strategy reallocation   (b) With strategy reallocation

Fig. 3. Comparison of reallocation and non-reallocation (Color figure online).

The sensor deployment phase of EBD approach terminates when all the square grids in the ROI are occupied with sensors.

3.3   Coverage Maintenance in EBD

Once the sensor deployment algorithm completes, the whole area is supposed to be fully covered. Although sensors are designed with low energy consumption, they cannot work permanently without breaking down. Sensor failures leave sensing holes in the ROI, which will cause inaccuracy of the information gathered by the network.

After the ROI is fully covered, robots will move to the central points in their subareas waiting for sensing holes to occur. In the coverage maintenance phase, the task for robots is to patch holes by replacing failed sensors with the new ones. Although sensor deployment algorithm does its best to avoid irregular subareas, they might still exist in the ROI. As we mentioned before, the coverage maintenance algorithm requires subareas to be independent and regular. Thus, before operation of the coverage maintenance algorithm, irregular subareas should be eliminated by combining the subareas to shorten moving distance.

The "combine" means treating two subareas as one large region. Combining subareas of robot A and B algorithmically means that A and B will share the information of the position of all the sensors they deploy. The subareas will be treated as a combined region. Both A and B have the responsibility of patching any hole in this new area.

Subareas of robots which are assigned the same strategy except diagonal strategy before reallocation should be combined after reallocation, since the subareas which meet the condition above overlap and be irregular inevitably. Apparently, this principle is only related to the starting points and the strategy each robot follows.

After combination, robots will move to the central points if there is no sensing hole to patch. Since a sensing hole can appear randomly in the ROI, waiting at the central points can minimize the average moving distance for the robots.

If all the coordinates of sensors in a subarea form a set E, defined as $E = \{(x_1, y_1), (x_2, y_2)...(x_n, y_n)\}$, then we have the average distance from each node (X, Y) to the rest nodes $(x_k, y_k) \in E$ in the subarea, called D, which is defined with Eq. (1) below. The node which guarantees the minimum D will be selected as the central point.

$$D = \frac{1}{n}\left[\sum_{k=1}^{n}(|X - x_k| + |Y - y_k|)\right], (X, Y) \in E \tag{1}$$

For a combined area with m robots, the number of central points should be m, which form a set R defined as $R = \{(X_1, Y_1), (X_2, Y_2)...(X_m, Y_m)\}$. The selected central points set should minimize D in Eq. (2):

$$D = \frac{1}{n}\left[\sum_{k=1}^{n}(|X_{min} - x_k| + |Y_{min} - y_k|)\right], (X_{min}, Y_{min}) \in R \tag{2}$$

with (X_{min}, Y_{min}) in set R holding the minimum distance to (x_k, y_k) in E.

After the robots move to the central points, they will inform all the sensors in the subareas with their positions. Every time a robot finishes patching a sensing hole, it will replace the coordinates of failed sensors stored in its memory with the new ones. Thus, central points will keep updating during the coverage maintenance phase.

Once a sensing hole appears, the gray sensors of the hole will send failure reports to their respective deployers. Only one robot will be selected as the *worker* to patch a hole. Before the worker is chosen, a set of robots will be selected as candidates, which have the opportunity to be elected as the worker. Candidates are those robots whose subareas contain gray sensors. For a sensing hole with gray sensors all located in one subarea, there is only one candidate which is in charge of the hole; whereas for a sensing hole located on the boundary of several subareas, there will be multiple candidates.

For multi-candidate scenario, the one holding the shortest distance to the sensing hole will be selected as the worker. The distance can be calculated directly if all the candidates are free and waiting at the central points. Sometimes, the candidates may be busy patching another sensing hole when a new one appears. In this case, the distance is comprised of three independent segments, which are:

1. The distance between current position of the candidate and the hole it is going to patch. For candidates which have already started working, the distance equals to zero.
2. The remaining size of the currently patched hole, which can be calculated with the coordinates of gray sensors. If all coordinates with the same x value are arranged in descending order as follows:

$$\begin{bmatrix} (x_1, y_{1,1}) & (x_2, y_{2,1}) & \cdots & (x_m, y_{m,1}) \\ (x_1, y_{1,2}) & (x_2, y_{2,2}) & \cdots & (x_m, y_{m,2}) \\ \vdots & \vdots & \ddots & \vdots \\ (x_1, y_{1,i}) & (x_2, y_{2,j}) & \cdots & (x_m, y_{m,k}) \end{bmatrix}$$

The remaining size of sensing hole can be calculated with the Eq. (3) below, with col representing each column, and row representing the number of sensors in each corresponding col:

$$D_{\text{Remain}} = \sum_{col=1}^{m} \left((y_{col,1} - y_{col,2} - 1) + (y_{col,2} - y_{col,3} - 1) + \cdots + (y_{col,row-1} - y_{col,row} - 1) \right)$$

$$(3)$$

3. The last one is the distance from the currently patched hole to the next one.

The distance is the sum of three independent parts in total. It should be noted that robots which finish patching a hole and travel back to central points will be considered as free since they can to move to the next hole immediately.

Figure 4 indicates how our coverage maintenance algorithm works. Figure 4(a) illustrates the full covered ROI. Since robot B and C meet the requirement of combination, their subareas are irregular and crossed, which should be combined.

In Fig. 4(b), the white area represents the combined region. Central points are marked with green circles with black border. When a sensing hole appears, all the three robots are free and selected as candidates. Since robot A holds the shortest distance to the failed sensor, it is elected as the worker.

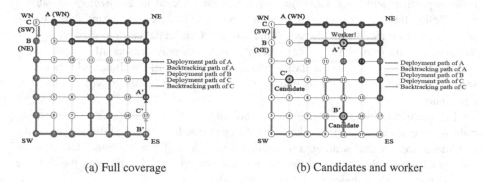

(a) Full coverage   (b) Candidates and worker

Fig. 4. Coverage maintenance algorithm (Color figure online).

The hole patching algorithm does not have a specific terminating time. Robots will not stop working as long as they are workers of any hole in the ROI. When there is no failure report, the robots will turn to a "*sleep*" mode waiting at the central points. Upon receiving a failure report, they will turn to the "*work*" mode and start working on the hole patching. The switch between "*sleep*" and "*work*" modes enables robots to avoid wasting unnecessary energy and react quickly when a hole appears.

4   Performance Evaluation

We evaluated the performance of the two phases of the proposed EBD through simulation. For the sensor deployment phase, we compared the performance of EBD with the existing BTD approach, whereas in the case of coverage maintenance phase, we use the dynamic distributed manager algorithm (DDMA) with slight modifications, which is one of the three algorithms proposed by Mei *et al.* [12] and described in Sect. 2.

4.1   Simulation Environment Setup

Our simulator is written in Java. During the simulation, we have assumed the sensing radius (r_S) and communication radius (r_C) of all the sensors to be the same. The environment is set up as shown in Fig. 5, with three obstacles. There are 192 grid points in the ROI with 176 of them can be deployed with sensors.

We placed m robots randomly in the ROI initially and each robot moves in a constant speed (i.e. 2 units in any four directions). We implemented both algorithms (BTD and EBD) with different number of robots, varying from 2 to 6. For creating/simulating sensing holes of size h, we randomly pick a sensor as the first failed sensor;

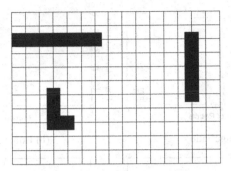

Fig. 5. Simulation environment.

then, the next one is selected among its adjacent sensors which have not failed yet, until there are h sensors in the hole. During the simulation, we randomly generated 2 to 4 sensing holes in the ROI whose size (the number of consecutive failed sensors) is a random number varies from 1 to 6.

4.2   Sensor Deployment

There are four metrics used to evaluate the performance; for each metric, we did simulation for twenty times and average the results. The four metrics are given below:

1. *Coverage Ratio (CR)*. It is the ratio of grid points covered with sensors to the total number of points.
2. *Robot Moves (MV)*. It is the number of moves of each robot during the deployment, and represents the time cost and the energy consumption for covering the ROI as well.
3. *Robot Messages (RM)*. It is the number of messages sent by robots during the deployment which reflects the total energy consumption for communication among robots and sensors.
4. *Sensor Messages (SM)*. It is the number of messages sent among sensors during the deployment which reflects the total energy consumption for communication among sensors.

In reality, the *CR* should be maximized, whereas the other four metrics should be minimized for lower energy consumption and higher working efficiency. Figure 6(a) indicates the *CR*, we can see that both algorithms can guarantee coverage higher than 90 %, we can observe an ascending trend on both two algorithms, since the more robots we have in the ROI, the higher probability for them to discover the holes. Sensor deployment phase of EBD alone cannot guarantee full coverage since sensors may fail at any time. However, with coverage maintenance phase, EBD can achieve full coverage in failure-prone environment. Figure 6(b) shows the *MV* of two algorithms, in both cases, *MV* decreases as the number of robots increases since the total number of move steps is approximately same. The more robots in the ROI, the fewer steps each

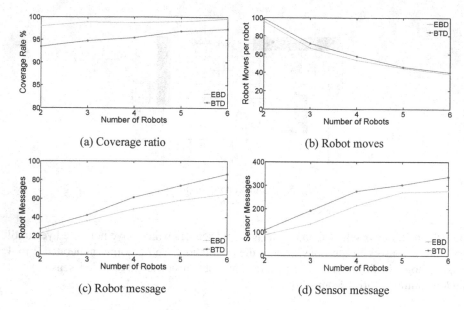

(a) Coverage ratio

(b) Robot moves

(c) Robot message

(d) Sensor message

Fig. 6. Sensor deployment algorithm performance evaluation.

robot moves. *MV* in case of EBD is slightly lower (better) than that in the case of BTD, which indicates that EBD can have lower energy and time consumption.

RM contributes to the communication overhead for robots. In EBD, all the robot messages are sent in searching for the nearest white or gray sensor when the robots get stuck; whereas in BTD, robots send messages in backtracking and load balancing. Thus, robot communication overhead increases as their number increases, since the more robots in the region, the higher probability for them to get stuck. Figure 6(c) indicates the relationship between *RM* and the number of robots *(m)*. We can notice that EBD outperforms BTD with respect to robot communication overhead. There is an ascending trend on both lines.

Figure 6(d) illustrates how *SM* varies with *m*. In EBD, *SM* increases as *m* increases, since robots are more likely to get stuck as their number grows. In BTD, the number of sensor messages is higher than that in the case of EBD; moreover, there is also an ascending trend in the number of sensor messages, which is probably due to back-tracking and load balancing, since searching for white or gray sensors along the boundaries of the ROI will require large number of sensor messages.

4.3   Coverage Maintenance

Since BTD does not maintain coverage after all the grid points in the ROI get fully occupied with sensors, it cannot be considered for comparison. The coverage maintenance capability of EBD will be tested by compared with DDMA with some adjustments. The environment is the same as that described in Sect. 4.1, which is shown in Fig. 5. The coverage maintenance algorithm will start after the completion of the sensor deployment phase which may have created sensing holes in the ROI. Robots need to patch the remaining sensing holes left from the deployment phase.

For simulation, we adjusted the DDMA slightly as follows. Each sensor selects the nearest robot as its manager. Information of robots position and the managers of sensors update every time they finish patching a hole. Once a sensing hole arises, the gray sensors will compare the distances from their current positions to their managers first, the one holding the shortest distance will send a failure report to its manager which is chosen as the worker.

During the simulation, we generated ten sensing holes whose sizes varied from 1 to 6 randomly, with each hole randomly appearing anywhere in the ROI at any time during the simulation period. The evaluation parameters are slightly different from the ones used in the sensor deployment phase. They are described below:

1. The *Robot Message (RM)* and *Sensor Messages (SM)* remain the same.
2. *Robot Moves (MV)*. It has a different meaning (i.e., average number of moves) in the sensor deployment phase, which accounts for the total number of moves by all robots during hole fixing.
3. *Coverage Time (CT)*. It is the total time for robots to patch all holes.

All the parameters should be minimized since they are proportional to energy consumption, communication overhead, and time to completion.

Figure 7(a) indicates how *MV* behaves as the number of robots (m) increases. For both cases, we notice slight decrease in *MV* as m increases. This is due to the distance between a hole to its worker or maintainer will decrease with the increase in the number of robots in the ROI.

Robots in EBD send messages in central points updating, which happens when they change their subareas. However, in DDMA, robots send messages to four geographical directions to update subareas once they finish patching a hole. Since there are ten sensing holes in total, the robot messages will remain the same as forty. Figure 7(b) illustrates that how *RM* varies with m. For DDMA, *RM* does not vary as m. While for EBD, we can notice there is an ascending trend in *RM* as m increases. More robots there are in the ROI, the smaller each subarea is. As we know, smaller regions are less stable than larger ones, whose central point is more likely to change due to the adjustment of area, resulting in higher number of robot messages.

Figure 7(c) shows the relationship between *SM* and m. In this case, we notice a descending trend for DDMA and a nearly flat trend for EBD. For DDMA, the number of *SM* is high since the robots send messages to inform all the sensors in the ROI with their new position once they cover a hole. Compared with DDMA, EBD has much lower sensor message overhead.

Figure 7(d) demonstrates the relationship between *CT* and m. We can observe decreasing trends in both approaches, since the distance from robots to sensing holes decreases as m increases. In this case, EBD outperforms DDMA by approximately 20 time units.

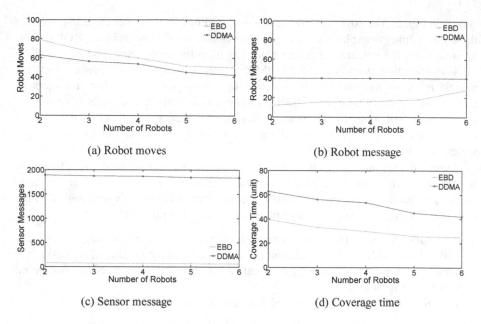

Fig. 7. Sensor deployment algorithm performance evaluation.

5   Conclusion

This paper provided a solution for sensor deployment and coverage maintenance problem in multi-robot scenario in failure-prone environment based on the BTD approach. EBD is the first carrier-based localized algorithm that is able to achieve full coverage with multiple robots in failure-prone environment since it combines both sensor deployment and coverage maintenance process. We are now improving EBD to consider reloading of robots and incorporating a new white sensor searching method to further reduce energy consumption and communication overhead consumption.

References

1. Ma, M., Yang, Y.: Adaptive triangular deployment algorithm for unattended mobile sensor networks. IEEE Trans. Comput. **56**, 946–958 (2007)
2. Akyildiz, I.F., Su, W., Sankarasubramaniam, Y., Cayirci, E.: A survey on sensor networks. IEEE Commun. Mag. **40**, 102–114 (2002)
3. Asada, G., Dong, M., Lin, T.S., Newberg, F., Pottie, G., Kaiser, W.J., Marcy, H.O.: Wireless integrated network sensors: low power systems on a chip wireless integrated network sensor. In: Proceedings of the 24th European Solid-State Circuits Conference, pp. 9–16 (1998)
4. Nayak, A., Stojmenovic, I.: Wireless Sensor and Actuator Networks: Algorithms and Protocols for Scalable Coordination and Data Communication. Wiley, Hoboken (2010)
5. Pottie, G.J., Kaiser, W.J.: Wireless Integrated Networks Sensors (WINS) Principles and Practice. Sensoria Corporation, Los Angeles (2001)

6. Min, R., Bhardwaj, M., Ickes, N., Wang, A., Chandrakasan, A.: The hardware and the network: total-system strategies for power-aware wireless microsensors. In: Proceedings of the IEEE CAS Workshop Wireless Communications and Networking (2002)
7. Enz, C.C., El-Hoiydi, A., Decotignie, J.-D., Peiris, V.: WiseNet: an ultralow-power wireless sensor network solution. IEEE Comput. **37**, 62–70 (2004)
8. Fletcher, G.: Sensor placement and relocation by a robot team. M.Sc. thesis, School of Electrical Engineering & Computer Science. University of Ottawa, Ottawa (2010)
9. Batalin, M., Sukhatme, G.: The analysis of an efficient algorithm for robot coverage and exploration based on sensor network deployment. In: Proceedings of the IEEE International Conference on Robotics and Automation, pp. 3478–3485 (2005)
10. Chang, C., Chen, Y., Chang, H.: Obstacle-resistant deployment algorithms for wireless sensor networks. IEEE Trans. Veh. Technol. **58**, 2925–2941 (2009)
11. Li, X., Fletcher, G., Nayak, A., Stojmenovic, I.: Placing sensors for area coverage in a complex environment by a team of robots. ACM Trans. Sens. Networks **11**, 3–11 (2014)
12. Mei, Y., Xian, C., Das, S., Hu, Y.C., Lu, Y.H.: Sensor replacement using mobile robots. Comput. Commun. **30**, 2615–2626 (2007)
13. Falcon, R., Li, X., Nayak, A.: Carrier-based coverage augmentation in wireless sensor and robot networks. In: IEEE 30th International Conference on Distributed Computing System Workshops, Genoa, Italy, pp. 234–239 (2010)
14. Wang, Y., Barnawi, A., Mello, R.F.D., Stojmenovic, I.: Localized ant colony of robots for redeployment in wireless sensor networks. J. Multi-Valued Logic & Soft Comput. **23**, 35–51 (2014)
15. Li, H.T., Barnawi, A., Stojmenovic, I., Wang, C.: Market-based sensor relocation by robot team in wireless sensor network. Ad Hoc Sens. Wireless Netw. **22**, 259–280 (2014)
16. Falcon, R., Li, X., Nayak, A., Stojmenovic, I.: The one-commodity traveling salesman problem with selective pickup and delivery: an ant colony approach. In: IEEE Congress on Evolutionary Computation, pp. 1–8 (2010)

Distributed Computing
with Mobile Agents

Wireless Autonomous Robot Evacuation from Equilateral Triangles and Squares

J. Czyzowicz[1], E. Kranakis[2($\boxtimes$)], D. Krizanc[3], L. Narayanan[4],
J. Opatrny[4], and S. Shende[5]

[1] Département d'informatique, Université du Québec en Outaouais,
Gatineau, QC, Canada
[2] School of Computer Science, Carleton University, Ottawa, ON, Canada
kranakis@scs.carleton.ca
[3] Department of Mathematics and Computer Science,
Wesleyan University, Middletown, CT, USA
[4] Department of Computer Science and Software Engineering,
Concordia University, Montreal, Canada
[5] Department of Computer Science, Rutgers University, Camden, NJ, USA

Abstract. Consider an equilateral triangle or square with sides of length 1. A number of robots starting at the same location on the perimeter or in the interior of the triangle or square are required to evacuate from an exit which is located at an unknown location on its perimeter. At any time the robots can move at identical speed equal to 1, and they can cooperate by communicating with each other wirelessly. Thus, if a robot finds the exit it can broadcast "exit found" to the remaining robots which then move in a straight line segment towards the exit to evacuate. Our task is to design robot trajectories that minimize the evacuation time of the robots, i.e., the time the last robot evacuates from the exit. Designing such optimal algorithms turns out to be a very demanding problem and even the case of equilateral triangles turns out to be challenging.

We design optimal evacuation trajectories (algorithms) for two robots in the case of equilateral triangles for any starting position and for squares for starting positions on the perimeter. It is shown that for an equilateral triangle, three or more robots starting on the perimeter cannot achieve better evacuation time than two robots, while there exist interior starting points from which three robots evacuate faster than two robots. For the square, three or more robots starting at one of the corners cannot achieve better evacuation time than two robots, but there exist points on the perimeter of the square such that three robots starting from such a point evacuate faster than two robots starting from this same point. In addition, in either the equilateral triangle or the square it can be shown that a simple algorithm is asymptotically optimal (in the number k of robots, as $k \to \infty$), provided that the robots start at the centre of the corresponding domain.

J. Czyzowicz, E. Kranakis, L. Narayanan, and J. Opatrny — Research supported in part by NSERC grant.

S. Papavassiliou and S. Ruehrup (Eds.): ADHOC-NOW 2015, LNCS 9143, pp. 181–194, 2015.
DOI:10.1007/978-3-319-19662-6_13

Keywords: Evacuation · Mobile robots · Square · Triangle · Wireless communication

1   Introduction

Searching is an important computational task and many different variants have been studied in the mathematical and computer science literature. Numerous fascinating search "excursions" can be found in several books, e.g., [1] (search problems and group testing), [4] (search and game theoretic applications), [6] (cops and robbers problems), [16] (classic pursuit and evasion problems), as well as the seminal treatise [17], to mention a few.

The problem studied here differs from well-known search studies involving multiple robots in that we are not minimizing the time the first robot finds the object of search, but rather the time the last robot reaches the search object. This reminds us of the situation when several people are searching for an exit from a closed area in order to evacuate it and it is important to minimize the time the last person leaves. We call this type of a search an *evacuation problem*, where the object of search is the *exit* and the time the last robot reaches the search object is the *evacuation time*.

In this paper, robot search within a pre-specified geometric domain (in our case, an equilateral triangle or square) is combined with the use of wireless communication between robots in order to minimize the evacuation time. It is clear that finding optimum length evacuation trajectories requires some form of co-operation and co-ordination between the robots (e.g., taking advantage of the possibility of wireless communication between the robots) so that at the time when one of the robots finds the exit, the rest of the robots are as close to it as possible so as to minimize the remaining time to achieve evacuation.

1.1   Model

Three important aspects of the model are required to describe the evacuation problem: (a) the search domain, (b) participating robots and their capabilities, and (c) evacuation algorithms. We elaborate on each of these three aspects in the sequel.

Search Domain. In general, it includes the interior of a non-intersecting, closed curve (e.g., cycle, polygon, or any Jordan curve) in the plane together with the curve itself forming the boundary. The exit is a point on the closed curve at a location unknown to the robots but which they can recognized when they go through it. In a typical setting of interest (which also justifies the term *evacuation*) the robots are enclosed by the boundary of the search domain and can leave it only thorough the exit. The search domains considered in this paper are restricted to equilateral triangles and squares with unit sides.

Robots and their Capabilities. The robots are labeled and may move at maximum speed 1. Although the movement of the robot is unrestricted inside the boundary of the domain, the robot can recognize the boundary when it reaches it and traverse it in its search for the exit. All robots will be assumed to start at the same location. We assume that the robots are aware of their initial position and the search area (for example they can trace the perimeter) and they can change direction when needed at no additional cost. They are also endowed with compass to orient themselves, but they do not know the position of the exit though they know it is located in the boundary. Further, the robots have communication capabilities in that they can communicate with each other wirelessly.

Evacuation Algorithms. An evacuation algorithm is the specification of a trajectory for each robot. A trajectory is a sequence of geometric instructions to be followed by a robot. Such instructions may include move along a specified curve, or move in a particular direction on that curve, or move in a straight line towards the exit, etc. Different robots may follow different trajectories. The closed curve on which the exit is located, as well as the complete evacuation algorithm will be known to each robot. From this knowledge each robot can determine the location of other robots at any time. A robot that finds the exit broadcasts a message "Exit found" and its label to the rest of the robots. Each robot then calculates the location of the exit and moves to it in the shortest possible time (usually along a straight line).

1.2 Related Work

Several previously studied problems can be seen as related to our work, even though they differ in some aspects.

In traditional search problems in an unknown environment, the goal is to construct a map of the environment. In [11], one is seeking for algorithms that achieve a bounded ratio of the worst-case distance traversed in order to see all visible points of the environment, [2] is concerned with exploration where the robot has to construct a complete map of an unknown environment, modelled by a directed, strongly connected graph, and in [3] the robot has to explore either n rectangles or a grid graph with obstacles in a piecemeal fashion. The pursuit-evasion problem [13] is also related to our work; surveys of recent results and autonomous search as related to applications in mobile robotics is given in [8] and additional material with interesting discussions can also be found in [16].

There is extensive literature on robotic exploration in simple polygons. Here we only mention [14] which presents an on-line strategy that enables a mobile robot with vision to explore an unknown simple polygon, and [15] which considers on-line search and exploration problems from the perspective of robot navigation in a simple polygon, with a robot that does not have access to the map of the polygon. Frontier-based exploration which directs mobile robots to regions on the boundary between unexplored space and space that is known to be open is considered in [18]. The problem of creating fast evacuation plans for buildings that are modelled as grid polygons and containing many cells was investigated in [12].

The evacuation problem on equilateral triangles and squares as considered in our paper has, as far as we know, never been studied, but the key motivation of the current study comes from the recent work of [9]. In that paper the authors introduce the evacuation problem as a new search paradigm for autonomous robots, whose initial location is the center of a unit cycle, and an unknown exit on the perimeter of the cycle. The authors obtain tight bounds for evacuation time in the wireless and non-wireless (also known as face to face) communication models depending on the number of robots. Also related is the recent work of [10] which considers evacuation for two robots from a cycle in the face to face communication model and [7], a related evacuation study on a line.

1.3   Outline and Results of the Paper

Here is an outline of our main results. In Sect. 2 we study evacuation in an equilateral triangle with unit side. We design optimal evacuation algorithms for two robots in any initial position on the perimeter or inside the equilateral triangle. We then show that three or more robots cannot improve the evacuation time for any starting position on the perimeter, but three robots can improve the search for the exit for some initial position inside the triangle, in particular, for the centre of the triangle. Section 3 considers a unit side square. Results include: (1) we design optimal evacuation algorithms for two robots starting on the perimeter of the square, (2) we show that three or more robots cannot improve the evacuation time when starting at a corner of the square, but (3) three robots can improve the evacuation time for some points on its perimeter, in particular the midpoint of one of its sides. As a consequence of our analysis, the shortest evacuation time is obtained when the starting position of the robots on the perimeter is (a) at the midpoint of any edge for an equilateral triangle, and (b) at a point at distance $1/2 - 1/\sqrt{12}$ from a vertex for a square. In addition, in either the equilateral triangle or the square it can be shown that a simple algorithm is asymptotically optimal (in the number k of robots, as $k \to \infty$), provided that the robots start at the corresponding centre of the domain. Details of missing proofs can be found in the full paper.

2   Equilateral Triangle

For the equilateral triangle with unit sides we first prove tight upper and lower bounds on the evacuation time for two robots. As stated above, the robots are initially located at the same location. To make the proofs easier to understand, despite some repetition, two cases are considered separately depending on whether or not the robots start on the perimeter or in the interior of the triangle.

2.1   Two Robots Starting on the Perimeter of the Triangle

Theorem 1. *Assume that two robots are initially located on the perimeter of an equilateral triangle at distance x from the closest midpoint of an edge of the triangle. Then $x + \frac{3}{2}$ is a tight bound for evacuating these two robots.*

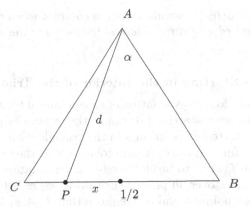

Fig. 1. Evacuating from an equilateral triangle with the two robots starting at a point P at distance x from the midpoint of edge BC.

Proof. Let the robots start at a point P on the perimeter of triangle ABC and at distance x from the midpoint of edge BC (see Fig. 1). To prove the upper bound the robots execute the following algorithm:

1. The robots walk together to the midpoint of edge BC.
2. From the midpoint they move in opposite directions along the perimeter, i.e. Robot 1 towards vertex A via vertex B, and the other Robot 2 towards vertex A via vertex C.
3. When a robot finds the exit it broadcasts "Exit found" to the other robot which immediately goes in a straight line segment to the exit.

The analysis of this algorithm is as follows. The time required for evacuation is $T = T_1 + T_2$, where T_1 is the time from the departure until one of the two robots broadcasts that the exit was found, and T_2 is the time from the broadcast "Exit found" of one robot until the other robot reaches the exit, i.e., the distance of the other robot from the exit. Since the triangle is equilateral, it is easily seen that $T_1 + T_2 \leq x + \frac{3}{2}$, which proves the upper bound.

The value $x + \frac{3}{2}$ is also a lower bound. Without loss of generality assume the starting position of the robots is at point P on edge BC and at distance x from its midpoint, as depicted in Fig. 1. Observe that one of the two vertices A, B is visited first by a robot. We distinguish two cases.

Case 1: A is visited first.

Let $\alpha := \angle PAB$. Since $\alpha < \pi/3$ is the smallest angle of the triangle $\triangle PAB$, it is easy to see that $d := |PA| \geq x + \frac{1}{2}$. Now the adversary puts the exit at B and the robot will need to travel an extra distance 1 to the vertex B; therefore $d + 1$ is a lower bound. However $d + 1 \geq x + \frac{3}{2}$.

Case 2: B is visited first.

Clearly, it takes time at least $x + \frac{1}{2}$ to visit point B from the starting point P. Now, if the adversary puts the exit at A it will take an additional 1 time unit to evacuate. This proves the theorem. ∎

Corollary 1. *The shortest evacuation time is obtained when the two robots start at the midpoint of an edge of the equilateral triangle and the resulting trajectory has length $\frac{3}{2}$.* ∎

2.2 Two Robots Starting in the Interior of the Triangle

Let $\mathcal{A}$ be an algorithm for the evacuation of an equilateral triangle by two robots initially placed at the *same* starting position in the interior or in the perimeter of the triangle. When started at position s in the triangle with exit position e, we say $\mathcal{A}$ terminates in time $T(\mathcal{A}, s, e)$ if both robots are at the exit position in time $T(\mathcal{A}, s, e)$. We define $T(\mathcal{A}, s)$ to be the worst-case termination time of algorithm $\mathcal{A}$ for starting position s over all possible exit positions e. Let $d(a, b)$ denote the distance between two points a and b. We prove that $T(\mathcal{A}, s) \geq 3/2 + x$ where x is the minimum distance from s to any of the midpoints of edges.

We start with a series of simple observations and lemmas. Assume without loss of generality that the triangle is $\triangle ABC$, with A, B, C its three vertices, the starting position s is closest to midpoint D and corner A as depicted in Fig. 2. Thus s is located inside $\triangle AFD$.

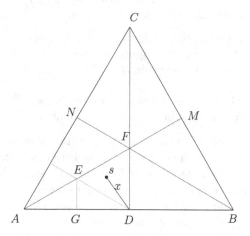

Fig. 2. The triangle to be evacuated. The starting position of the two robots is in $\triangle AFD$. N, M, D are the midpoints of the corresponding sides.

Observation 1 (Distances).

1. $d(s, B) \geq 1/2$, $d(s, C) \geq 1/2$.
2. If $s \in \triangle AEG$, then $d(s, B) \geq 3/4$.
3. If $s \in \triangle AED$, then $d(s, C) \geq 3/4$.

Observation 2. *If s is in the quadrilateral $EFGD$, then $3/2 + x \leq 3/2 + d(D, E) = 3/2 + \frac{\sqrt{3}}{6} < 1.79$.*

Observation 3. *If a robot visits a midpoint and then a corner and at this time, there remains an unvisited corner, then $T(\mathcal{A}, s) \geq 3/2 + x$.*

Proof. Place the exit at the as-yet unvisited corner.

Observation 4. *If a robot visits the corner B and then either (a) visits a point P on the edge AB with $d(B, P) \geq 1/2$ before the corner C is visited by either robot, or (b) visits a point Q on the edge BC with $d(B, Q) \geq 1/2$ before the corner A is visited by either robot, then $T(\mathcal{A}, s) \geq 3/2 + x$.*

Proof. If $s \in \triangle AEG$, then the robot takes at least time $3/4$ to visit B, and at least $1/2$ to visit P. At this point since C is not visited by either robot, the adversary can place the exit at C, forcing the robot to take at least $\sqrt{3}/2$ time to reach C. Therefore $T(\mathcal{A}, s) \geq 3/4 + 1/2 + \sqrt{3}/2 > 2$. If instead s is the quadrilateral $EFGD$, the above trajectory takes time at least $1/2 + 1/2 + \sqrt{3}/2 > 1.86 > 1/2 + x$ by Observation 2 for this case.

Observation 5. *Similar to Observation 4 about corner C.*

Lemma 1. *If a robot visits corner B and then visits corner C before the entire edge BA has been explored (by a combination of both robots), then $T(\mathcal{A}, s) > 2$. Also, if it visits corner B and then visits corner A before the entire edge BC has been explored, then $T(\mathcal{A}, s) > 2$.*

We are now ready to show that for two robots the evacuation time when starting inside the triangle follows the same rule as when starting on the perimeter.

Lemma 2. *Similar statement to that of Lemma 1 for a robot that first visits corner C.*

Theorem 2. *Assume that two robots are initially located at point s inside the equilateral triangle, and let $x = \min\{d(s, m) \mid m \text{ is a mid point of an edge}\}$. Then $x + \frac{3}{2}$ is a tight bound for evacuating these two robots.*

Proof. We give a proof of the lower bound by contradiction. Assume $\mathcal{A}$ is a two-robot evacuation algorithm such that $T(\mathcal{A}, s) < 3/2 + x$. Consider the first two distinct corners to be visited by either of the robots. If they were both visited by the same robot, then we place the exit at the remaining corner and thus $T(\mathcal{A}, s) \geq 2$. Therefore, we assume that robot 1 visited corner X and robot 2 visited corner Y where $X, Y \in \{A, B, C\}$ and $X \neq Y$. We consider the three cases separately.

First assume that $X = B$ and $Y = C$. Then Lemmas 1 and 2 imply that the entire edge BC must have been explored before either robot can visit the corner A. However, if either robot visits the midpoint M before visiting the corner A, then by Observation 4, we have $T(\mathcal{A}, s) \geq 3/2 + x$, a contradiction.

Next, suppose $X = A$ and $Y = B$. We claim that if Robot 1 (having visited corner A) visits corner C before exploring the entire segment AD, then $T(\mathcal{A}, s) \geq 2$. Define Q such that $[AQ)$ is the longest contiguous segment visited by Robot 1 before visiting corner C, and suppose for the purpose of contradiction that $y = d(A, Q) \leq 1/2$ (see Fig. 3). By Observation 4, Robot 2 cannot have visited the point Q and neither has Robot 1. The adversary now

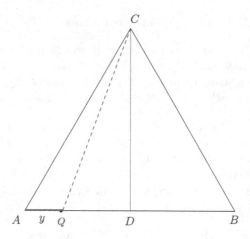

Fig. 3. Q is the leftmost point that is unvisited by A before going to C.

places the exit at point Q. The time taken by Robot 1 to reach Q is at least $d(A,Q)+2d(Q,C) = y+2\sqrt{3/4+(1/2-y)^2} \geq y+2(1-y/2) = 2$. We conclude that Robot 1 must explore the entire segment AD before visiting the corner C. Now, suppose P is a point on the edge AC such that Robot 1 has explored the contiguous segment $[AP)$ before arriving at D, with $d(A,P) = z$. Depending on which of the points P and C is visited first (by either robot), the adversary places the exit at the other point.

First we claim that if $d(A,P) \geq 1/2$, then $T(\mathcal{A},s) \geq 3/2+x$. Recall that A is visited before D (by Observation 3). So for robot 1 to explore the segment AP as well as AD takes at least time $d(s,P)+d(P,A)+d(A,D) \geq x+1$. If the exit is placed at C, it takes an additional $\sqrt{3}/2$ time for robot 1 to reach the exit. We conclude that $z < 1/2$, as illustrated in Fig. 4.

Suppose of the two points P and C, the first to be visited is C. If it is robot 1 that arrives first at C, the exit is placed at P, and the time it takes for robot 1 to reach the exit is at least $z + 1/2 + \sqrt{3/2} + 1 - z > 2$. If instead it is robot 2 that arrives first at C, the time it takes to reach the exit at P is at least $1/2 + 1 + 1 - z > 2$ since $z < 1/2$.

Suppose instead that the first of the two points P and C to be visited by either robot is P. The adversary now places the exit at C. If it was robot 1 that arrived first at P, then the time it needs to reach the exit is at least $d(s,P) + d(P,A) + d(A,D) + d(D,P) + d(P,C) = d(s,P) + z + 1/2 + d(D,P) + 1 - z = d(s,P) + d(D,P) + 3/2 > d(s,D) + 3/2 = 3/2 + x$. If instead it was robot 2 that arrived first at P, we consider two cases. Either robot 2 visited P before it visited B or after. If it visited P before it visited B, observe that the closest point to P in the triangle ADF (where s is by assumption) is found by dropping a perpendicular to the line segment AF, say to point L. Now in the right triangle APL, the angle PAL is 30 degrees, and $d(A,P) = z$, therefore, $d(A,L) = z/2$.

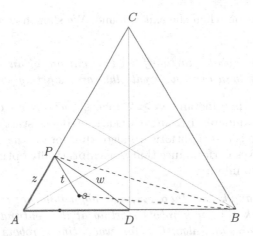

Fig. 4. Robot 1 has visited the segments AP and AD before going to C.

Therefore, the time robot 2 needs to reach C is $d(s, P) + d(P, B) + d(B, C) \geq z/2 + \sqrt{3/4 + (1/2 - z)^2} + 1 > z/2 + 1 - z/2 + 1 = 2$.

Otherwise, it visited B and then P, and then the time it needs is at least $d(s, B) + d(B, P) + d(P, C) \geq d(s, B) + \sqrt{3}/2 + 1/2$ (see Fig. 4). Now, by Observation 1, if $s \in \triangle ADG$, this time is at least $3/4 + \sqrt{3}/2 + 1/2 > 2$, and if instead $s \in EFGD$, this time is at least $1 + \sqrt{3}/2 > 1.86 > 3/2 + x$ by Observation 2. Thus if robot 1 visits A and robot 2 visits B, then we are done.

The remaining case $X = A$ and $Y = C$ is very similar to the previous case and details are left to the reader.

Clearly, the evacuation algorithm stated in Theorem 1 gives evacuation time of at most $x + 3/2$, thus proving a matching upper bound. ∎

2.3 More Than Two Robots

The following rather surprising result, states that more robots don't help for starting points on the perimeter.

Theorem 3. *Assume that $k \geq 3$ robots are initially located at point s on the perimeter of an equilateral triangle and let $x = \min\{d(s, m) : m$ is a mid point of an edge $\}$. Then $x + \frac{3}{2}$ is a tight bound for evacuating these k robots.*

Proof. An examination of the proof of the lower bound of Theorem 1 shows that it does not depend upon the number of robots starting on the perimeter. ∎

The above result shows that starting on the boundary of the triangle limits the nature of the search for the exit. But if the robots were to start at an interior point, then they have more options, including travelling through the interior to different points on the perimeter. Thus during an active search, they can not only search more cooperatively but also remain somewhat close to each other to allow faster evacuation when the exit is found. We show how this extra freedom can help:

Theorem 4. *Three robots starting from the centroid of an equilateral triangle can evacuate faster than two robots with the same starting position.*

Above we showed that for any $k \geq 2$ robots there is an optimal algorithm for evacuating an equilateral triangle when the robots start on the perimeter. Here we show there is a straightforward algorithm for evacuation starting at the centroid that achieves performance that is asymptotically optimal as the number of robots, k, goes to infinity.

Theorem 5. *For any $k \geq 2$, there exists an algorithm that achieves evacuation in time $\sqrt{3}/3 + 3/k + 1$ for k robots starting at the centroid of the equilateral triangle. Furthermore, any algorithm for evacuating k robots from the centroid of the equilateral triangle requires at least $\sqrt{3}/3 + 1$ time.*

Remark 1. Consider the case where the robots start at a point along a line from the centroid to one of the corners of the triangle. Let x be the distance from the starting point to the two other corners. It is easy to see, using essentially the same arguments, there is an algorithm that achieves evacuation with k robots in time $1 + x + 3/k$ and that any algorithm will take at least $1 + x$ time. I.e., there is an asymptotically optimal algorithm starting at any of these points as well.

3   Square

In this section we consider the square with unit sides. We first analyze the case of two robots starting on the perimeter searching for the exit.

3.1   Two Robots Starting on the Perimeter of the Square

The surprising fact (to be proved below) is that the starting point for the robots that minimizes the evacuation time is at a point at distance $\frac{1}{\sqrt{12}}$ from the midpoint of an edge. This differs from the equilateral triangle where the shortest evacuation time is obtained when starting from the midpoint of an edge.

Theorem 6. *Assume that two robots are initially located on the perimeter of a square at distance x from the closest vertex of the square. Then $1 + x + \sqrt{1 + (1 - 2x)^2}$ is a tight bound for evacuating these two robots.*

Proof. Consider evacuation of two robots starting from a point on the perimeter of a square $ABCD$ at distance $x \leq 0.5$ from the vertex A (see Fig. 5). To prove the upper bound consider the following algorithm for the robots:

1. Robots move from s along the perimeter in opposite directions. One robot moves on the perimeter towards vertex D via vertex A. The other robot moves on the perimeter towards C via vertex B.

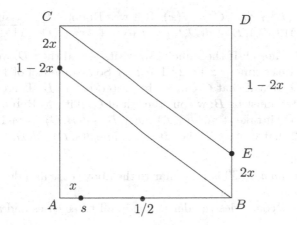

Fig. 5. Square with side 1, $0 \leq x \leq 1/2$. Observe that $d(E,C) = \sqrt{1 + (1 - 2x)^2}$, $d(s,C) = \sqrt{1 + x^2}$, $d(s,D) \geq \sqrt{5/4}$, and $d(A,E) = \sqrt{1 + 4x^2}$

2. If a robot finds the exit it broadcasts "Exit found" to the other robot which moves directly to the exit in a straight line segment, i.e., without necessarily moving on the perimeter.

Next we determine the running time until evacuation. It is easy to see that when the robots are starting at distance x from vertex A, the worst case running time of this algorithm is obtained when the exit is located at C or E. Let $d = d(E,C) = \sqrt{1 + (1 - 2x)^2}$. Thus the worst case running time of this algorithm is $f(x) := 1 + x + d = 1 + x + \sqrt{1 + (1 - 2x)^2}$.

Next we show the lower bound. We consider the visits to the vertices by either of the robots. Suppose at the time C is visited by a robot, the vertex B is as yet unvisited. Then we place the exit at B. The time taken by the robot who visits C to go to the exit is $\geq dist(s,C) + dist(C,B) = \sqrt{1 + x^2} + \sqrt{2} \geq f(x)$ for all $0 \leq x \leq 1/2$. Similarly, suppose D is visited before A is visited by either robot. Then we place the exit at A. The time taken by the robot who visits D to get to A is $\geq dist(s,D) + dist(D,A) = \sqrt{1 + (1 - x)^2} + \sqrt{2} \geq \sqrt{1 + x^2} + \sqrt{2} \geq f(x)$.

By the observations above, the following cases are exhaustive for the first two points to be visited:

A and B are the first two vertices to be visited: If the same robot visits both, we place the exit at the vertex diagonally opposite the second visited vertex, causing the robot to take at least $1 + x + \sqrt{2} \geq f(x)$. So we assume A was visited by Robot 1 and B by Robot 2. We observe now which is visited next: C or E as shown in Fig. 5, and put the exit at the other one. Suppose C is visited first. If Robot 1 visited C, to get to E, it takes time $d(s,A) + d(A,C) + d(C,E) = f(x)$. If Robot 2 visited C, to get to E it takes time $d(s,B) + d(B,C) + d(C,E) > f(x)$. Suppose instead E was visited first. If Robot 2 visited E first, to get to C, it takes time

$\geq d(s, B) + d(B, E) + d(E, C) = f(x)$. If it was Robot 1 that visited E first, it takes time $d(s, A) + d(A, E) + d(D, C) = x + \sqrt{1 + 4x^2} + \sqrt{1 + (1 - 2x)^2} \geq f(x)$.

A is visited first, then D: If the same robot visits A and then D, we put the exit at C. The robot takes time $\geq x + \sqrt{2} + 1 \geq f(x)$. So assume Robot 1 visits A, then Robot 2 visits D. Recall that C cannot be visited before B. Therefore, the next vertex to be visited must be B: we put the exit at C. If it was Robot 1 that visited B, then to get to C it takes time $d(s, A) + d(A, B) + d(B, C) = x + 1 + \sqrt{2} \geq f(x)$. If it was Robot 2 that visited B, then it takes time $d(s, D) + d(D, B) + d(B, C) > 2 + \sqrt{2} > f(x)$.

B is visited first, then C: This is similar to the above case and details are left to the reader.

This completes concludes the derivation in all three cases and also the proof of the theorem. ∎

Corollary 2. *Among all possible starting positions of the two robots at distance x from a vertex of the unit square the optimal length of a trajectory is obtained when $x = \frac{1}{2} - \frac{1}{\sqrt{12}}$. Moreover, the length of the resulting trajectory for evacuating two robots starting together on the perimeter of a unit square is $\frac{3}{2} + \frac{\sqrt{3}}{2}$.*

Proof. To find the initial position of the robots that minimizes the evacuation time, we differentiate the evacuation time from Theorem 6 with respect to x and we obtain that the maximum is obtained for $x = \frac{1}{2} - \frac{1}{\sqrt{12}}$. Substituting this value of x into the evacuation time function shows that the maximum distance traversed in this case is at most

$$f\left(\frac{1}{2} - \frac{1}{\sqrt{12}}\right) = \frac{3}{2} + \frac{\sqrt{3}}{2},$$

thus proving the corollary. ∎

3.2   More Than Two Robots

In this section we show that for a square, using more that two robots to search for the exit can improve the search time for some starting point on the perimeter, but not for all. Thus, it differs from the results for an equilateral triangle. First we show that having more robots doesn't help if the starting point is one of the corners.

Theorem 7. *If $k \geq 2$ robots start evacuation from a corner of the square, then it takes them at least $1 + \sqrt{2}$ time before the evacuation is completed.*

As opposed to the case of the triangle, there exist points on the perimeter where even one extra robot helps.

Theorem 8. *Let p be the midpoint of one of the sides of the square. Three robots starting from p can solve the evacuation problem faster than two robots starting from p.*

For the square, our results for $k > 2$ robots are optimal only if the robots start at one of the corners of the square. Again we can show an asymptotically optimal (in k) result holds for the case where they start at the center. We have:

Theorem 9. *For any $k \geq 2$, there exists an algorithm that achieves evacuation in time $3\sqrt{2}/2 + 4/k$ for k robots starting at the center of the square. Furthermore, any algorithm for evacuating k robots from the center of the square requires at least time $3\sqrt{2}/2$.*

Proof. For the upper bound, starting at one of the corners of the square, mark off points at distance $4/k$ in clockwise order around the perimeter. Send one robot to each marked point and have them search in a clockwise manner the segment between their point and the next marked point. The robot that finds the exit broadcasts "Exit found" (as well as its label) and the other robots move directly to the exit (the position of which they can calculate from the time and the label). The last robot to search its segment finishes at time $\sqrt{2}/2 + 4/k$ and all robots are at most distance $\sqrt{2}$ from the exit at that time so that the all the robots are evacuated by time $3\sqrt{2}/2 + 4/k$.

For the lower bound, we will inform the robots that the exit is either at vertex A or C (opposing corners of the square). Let t be the time when the first robot reaches A or C. If this robot first reaches A the exit will be placed at C and vice versa. Clearly t is greater or equal to $\sqrt{2}/2$ so that at least this robot must take time greater or equal to $\sqrt{2}/2 + \sqrt{2} = 3\sqrt{2}/2$ to exit. ∎

4   Conclusion and Open Problems

In this paper, we investigated the problem of evacuating $k \geq 2$ robots from equilateral triangles and squares. Our results assume that the robots start from the same initial point: in the interior or on the perimeter for triangles, and on the perimeter or centre for squares. We design robot trajectories that optimize the evacuation time, and also show that in some specific cases, three robots can evacuate faster than two robots, e.g. when they start at the centroid of the triangle, or start at the midpoint of a side of the square. However, in other cases, having more than two robots makes no difference, e.g. when the robots start anywhere on the perimeter of the triangle or start at a corner of the square.

In addition to problems that remain unsolved from our investigations, there are numerous interesting and surprisingly not easy to solve open problems. Very little is known when the robots do not start at the same point be that on the perimeter or in the interior of the polygon. In this regard, the techniques of parametrized complexity as employed in [5] should be useful. Throughout our investigations we considered only the wireless communication model. However, nothing is known for the face-to-face communication model whereby robots can exchange information only if and when they meet at the same location. Further, as in [9], it would be interesting to investigate the evacuation time for k robots asymptotically in k and compare the wireless and non-wireless models. Evacuation in the more general case of arbitrary simple polygons is entirely open.

References

1. Ahlswede, R., Wegener, I.: Search Problems. Wiley, New York (1987)
2. Albers, S., Henzinger, M.: Exploring unknown environments. SIAM J. Comput. **29**(4), 1164–1188 (2000)
3. Albers, S., Kursawe, K., Schuierer, S.: Exploring unknown environments with obstacles. Algorithmica **32**(1), 123–143 (2002)
4. Alpern, S., Gal, S.: The Theory of Search Games and Rendezvous. International Series in Operations Research and Management Science, Vol. 55. Springer, Heidelberg (2003)
5. Bhattacharya, B., Burmester, M., Hu, Y., Kranakis, E., Shi, Q., Wiese, A.: Optimal movement of mobile sensors for barrier coverage of a planar region. Theoret. Comput. Sci. **410**(52), 5515–5528 (2009)
6. Bonato, A., Nowakowski, R.: The Game of Cops and Robbers on Graphs. American Mathematical Society, Providence (2011)
7. Chrobak, M., Gąsieniec, L., Gorry, T., Martin, R.: Group search on the line. In: Italiano, G.F., Margaria-Steffen, T., Pokorný, J., Quisquater, J.-J., Wattenhofer, R. (eds.) SOFSEM 2015-Testing. LNCS, vol. 8939, pp. 164–176. Springer, Heidelberg (2015)
8. Chung, T., Hollinger, G., Isler, V.: Search and pursuit-evasion in mobile robotics. Auton. Robots **31**(4), 299–316 (2011)
9. Czyzowicz, J., Gąsieniec, L., Gorry, T., Kranakis, E., Martin, R., Pajak, D.: Evacuating robots via unknown exit in a disk. In: Kuhn, F. (ed.) DISC 2014. LNCS, vol. 8784, pp. 122–136. Springer, Heidelberg (2014)
10. Czyzowicz, J., Georgiou, K., Kranakis, E., Narayanan, L., Opatrny, J., Vogtenhuber, B.: Evacuating robots from a disk using face-to-face communication (extended abstract). In: Paschos, V.T., Widmayer, P. (eds.) CIAC 2015. LNCS, vol. 9079, pp. 140–152. Springer, Heidelberg (2015)
11. Deng, X., Kameda, T., Papadimitriou, C.: How to learn an unknown environment. In: Proceedings of the 32nd Annual Symposium on Foundations of Computer Science (FOCS 1991), pp. 298–303. IEEE (1991)
12. Fekete, S., Gray, C., Kröller, A.: Evacuation of rectilinear polygons. In: Wu, W., Daescu, O. (eds.) COCOA 2010, Part I. LNCS, vol. 6508, pp. 21–30. Springer, Heidelberg (2010)
13. Gluss, B.: An alternative solution to the lost at sea problem. Nav. Res. Logistics Q. **8**(1), 117–122 (1961)
14. Hoffmann, F., Icking, C., Klein, R., Kriegel, K.: The polygon exploration problem. SIAM J. Comput. **31**(2), 577–600 (2001)
15. Kleinberg, J.: On-line search in a simple polygon. In: Proceedings of the Fifth Annual ACM-SIAM Symposium on Discrete Algorithms, pp. 8–15. Society for Industrial and Applied Mathematics (1994)
16. Nahin, P.: Chases and Escapes: The Mathematics of Pursuit and Evasion. Princeton University Press, Princeton (2012)
17. Stone, L.: Theory of Optimal Search. Academic Press, New York (1975)
18. Yamauchi, B.: Frontier-based exploration using multiple robots. In: Proceedings of the Second International Conference on Autonomous Agents, pp. 47–53. ACM (1998)

Rendezvous of Many Agents with Different Speeds in a Cycle

Evan Huus and Evangelos Kranakis[✉]

School of Computer Science, Carleton University Ottawa, Ottawa, ON, Canada
kranakis@scs.carleton.ca

Abstract. Rendezvous is concerned with enabling $k \geq 2$ mobile agents to move within an underlying domain so that they meet, i.e., rendezvous, in the minimum amount of time. In this paper we study a generalization from 2 to k agents of a deterministic rendezvous model first proposed by [5] which is based on agents endowed with different speeds. Let the domain be a continuous (as opposed to discrete) ring (cycle) of length n and assume that the k agents have respective speeds $s_1, \ldots, s_k$ normalized such that $\min\{s_1, \ldots, s_k\} = 1$ and $\max\{s_1, \ldots, s_k\} = c$. We give rendezvous algorithms and analyze and compare the rendezvous time in four models corresponding to the type of distribution of agents' speeds, namely *Not-All-Identical, One-Unique, Max-Unique, All-Unique*. We propose and analyze the *Herding Algorithm* for rendezvous of $k \geq 2$ agents in the *Max-Unique* and *All-Unique* models and prove that it achieves rendezvous in time at most $\frac{1}{2}\left(\frac{c+1}{c-1}\right)n$, and that this rendezvous is *strong* in the *All-Unique* model. Further, we prove that, asymptotically in k, no algorithm can do better than time $\frac{2}{c+3}\left(\frac{c+1}{c-1}\right)n$ in either model. We also discuss and analyze additional efficient algorithms using different knowledge based on either n, k, c as well as when the mobile agents employ pedometers.

Keywords: Cycle · Mobile agents · Rendezvous · Speeds

1 Introduction

Rendezvous of mobile agents (or robots) is a well-known paradigm in distributed computing, networking and robotics; it can be stated as follows: Given an underlying domain have the agents move along the domain so that they meet in the minimum possible amount of time. The rendezvous problem has been widely studied with several researchers emphasizing either of the following: domain of exploration, agent capabilities, deterministic, randomized rendezvous algorithms, etc.

Evan Huus—A preliminary version of this work was part of the author's undergraduate honours project [9].
Evangelos Kranakis—Research supported in part by NSERC Discovery grant.

© Springer International Publishing Switzerland 2015
S. Papavassiliou and S. Ruehrup (Eds.): ADHOC-NOW 2015, LNCS 9143, pp. 195–209, 2015.
DOI:10.1007/978-3-319-19662-6_14

The present paper is concerned with a deterministic rendezvous model recently introduced in [5] which is making use of the agents' speeds in order to show that speed differences between two agents can be useful for accomplishing rendezvous in optimal time. They consider two agents in a ring which have different speeds but are otherwise identical. Clearly, because they have different speeds, if the two agents choose any arbitrary starting positions (on the ring) and any arbitrary starting directions (clockwise or counter-clockwise) of movement they will renderzvous if they just stay the course. What is not trivial is that the authors in [5] can use this asymmetry to achieve guaranteed rendezvous with tight time bounds even though the agents do not know their own speed. They also note that this technique has much broader potential applications, both in rendezvous and in other symmetry-breaking problems. We note that the work of [5] has recently been extended to randomized rendezvous in [10].

1.1 Related Work

There are numerous variants of the rendezvous problem. The domain of exploration may be continuous [3], or discrete [15]. It may be a specific graph, such as the ring [12], a general graph [4] or a two dimensional grid [2,11]. Agent capabilities can have a considerable effect on the rendezvous problem leading to a variety of studies. E.g., [8] explored the scenario where agents can exchange arbitrary information with any agents in their "line of sight", [13] considered a weaker model where the agents were restricted to dropping indistinguishable tokens to mark their current location, while [6] studied the case when tokens may fail. Of particular interest are cases where the agents must all run the same algorithm, which is generally known as the *symmetric rendezvous problem* [1]. If agents can run different algorithms, generally known as the *asymmetric rendezvous problem*, then the problem is typically much easier, though not always trivial.

In many of these variants, rendezvous cannot be achieved without symmetry breaking. For example, in the simple case of two deterministic agents on a ring graph, rendezvous cannot be achieved if the agents are perfectly symmetric; they will always remain exactly the same distance apart. Even in cases where rendezvous can be achieved without it, breaking symmetry often leads to much more efficient algorithms than would otherwise be possible. A frequently used method for breaking symmetry is to use random bits to make probabilistic choices. Although such algorithms may provide reasonable expected time to rendezvous, they typically have poor worst-case behaviour. If no good source of random bits is available various tricks must be used in order to exploit other sources of asymmetry. One such trick is to let the agents drop tokens and count inter-token distances. This method successfully achieves rendezvous as long as the agents do not start in antipodal positions on the ring [13]. Breaking symmetry has applications to other distributed computing problems as well. For example, [14] developed a new symmetry-breaking technique which gives an exponential speed-up in certain distributed graph colouring problems.

A source of asymmetry that has recently received more attention is the difference in speed between two otherwise identical agents. While agent speeds have been considered as a problem parameter before, the traditional assumption has been that all agents move at a common fixed speed. Even when agent speed has been allowed to vary, as in [3], agents have typically had a uniform range of possible speeds and have been required to choose their speed deterministically, maintaining the symmetry of the problem.

1.2 Rendezvous Models

Throughout this paper we will consider rendezvous across several different variants of a single core model, similar but not identical to the one used by [5]. We have $k \geq 2$ mobile agents placed on a continuous ring of length n. We denote the agents by $A_1, \ldots, A_k$, but emphasize that both the agents and the ring are anonymous. The agents are identical except possibly for their speeds. Each agent A_i has its own maximum speed, denoted s_i. In [5] agents could be either stopped or travelling at s_i; we relax this requirement slightly to allow agents to travel at any speed (among the given speeds) in the interval $[0, s_i]$. Without loss of generality, we normalize to $\min\{s_1, \ldots, s_k\} = 1$ and denote $\max\{s_1, \ldots, s_k\} = c$, the ratio between the speeds of the fastest and slowest agents.

Agents can travel in either direction on the ring, and can turn without cost. However, the ring is not oriented. Agents can detect when they encounter another agent, as well as whether that agent is travelling in the same or opposite direction. Co-located agents may exchange arbitrary amounts of information. We wish to explicitly note that this is a deterministic model, and as such agents must have a reason to turn or change speeds, whether that reason is an encounter with another agent, a certain amount of time elapsed, or some other event.

When discussing running-time and optimal bounds, we take an adversarial model where the adversary can choose the initial position and orientation of each agent. The adversary may also choose the maximum speed of each agent, within the restrictions of the speed distributions discussed below.

The above conditions define the portions of our model that remain constant throughout the paper. We generate variants from this base across two dimensions: agent knowledge, and the distribution of agent speeds. In the weakest case of agent knowledge, the agents know nothing except their own speed s_i, and they do not have pedometers or timers. In stronger cases they may know the values of n, k, or c; they may also have pedometers or timers.

The distribution of agent speeds is a more interesting source of differences. We consider the following four possible distributions, sorted in increasing order of apparent power:

Not-All-Identical. This is the weakest possible model. Its only restriction on the adversary's choice of speeds is that there is some pair of agents, A_i and A_j such that $s_i \neq s_j$. No agent is guaranteed to have a unique speed.

One-Unique. In this slightly stronger model, the adversary is required to give at least one agent a unique speed, though that speed is not required to be maximal or minimal among the agents.

Max-Unique. In this model, the adversary is required to give at least one agent a unique speed, and that speed is required to be maximal (or, symmetrically, minimal) among the agents.

All-Unique. In this strongest model, the adversary is required to give each agent a unique speed. Note that since agents are aware of their own speeds, this effectively provides each agent with a unique label, which simplifies a number of problems [7].

The last component of our model is the very concept of rendezvous. The obvious, simple definition is that rendezvous occurs when all k agents meet at the same location, but for certain real-world problems this is not quite enough. For example, if some quorum of agents is necessary to execute an algorithm, actually achieving rendezvous is not sufficient if the agents are not aware of that fact. Specifically, if k is not known to the agents then they may not realize that they have achieved rendezvous even when all k agents are present.

The authors in [4] address a more general version of this problem. In their model, all agents have a unique label, and the problem is solved when all agents know all k labels and are aware of that fact. They call this problem Strong Global Learning. Following this terminology, we define the concept of *strong rendezvous* for k agents as the situation where all k agents meet at the same location, and are aware of that fact. Note that in the event of strong rendezvous, all agents are therefore trivially able to calculate k by counting the agents present.

1.3 Outline and Results of the Paper

In Sect. 2 we present the *Herding Algorithm* for rendezvous of $k \geq 2$ agents in the `Max-Unique` and `All-Unique` models with no additional knowledge. We prove that it achieves rendezvous in time at most $\frac{1}{2}\left(\frac{c+1}{c-1}\right)n$, and that this rendezvous is *strong* in the `All-Unique` model. We also prove that, asymptotically in k, no algorithm can do better than time $\frac{2}{c+3}\left(\frac{c+1}{c-1}\right)n$ in either model. In Sect. 3 we consider variants of the problem with more agent knowledge. We show that knowledge of n is insufficient on its own to make a difference; the algorithm from Sect. 2 does not receive any advantage. We also show that knowledge of k is somewhat more interesting; while it does not materially affect the existing algorithm, it does permit the construction of a variant that is capable of achieving rendezvous even in the weaker `One-Unique` model. This variant is presented and a proof of correctness is given. The cases where the agents have pedometers or know the value of c are also discussed. As with knowledge of n, these additions are insufficient on their own to provide any obvious advantage. However, we show that both of these elements can be used by the agent to gain useful knowledge. Finally, we consider the combined case where the agents have a pedometer and know both c and n. We use this knowledge to construct a variant of the herding algorithm which takes $n/(c^2 - 1)$ less time than the original. Finally, we present our conclusion in Sect. 4 and consider possible areas for future work. All missing proofs can be found in the full paper.

2   The Herding Algorithm

In this section we present the basic herding algorithm, which achieves rendezvous in the `Max-Unique` model and strong rendezvous in the `All-Unique` model while requiring no additional agent knowledge. As this section deals strictly with those two stronger models, we introduce one extra notational convenience: we define A_{max} to be the agent with the unique maximal speed. Further, in this section the agents cannot measure time or distance in the model considered.

We prove the correctness of the presented algorithm, analyze its running-time, and prove a lower bound on the running-time of the optimal algorithm. While our algorithm was developed independently, it turns out that it shares some similarities with a randomized algorithm proposed by Steve Alpern in Sect. 7.4.1 of his paper [1]. That algorithm belongs in the same family as the one presented here, but to the best of our knowledge was never published in full. Due to its use of random bits, it provides only a probabilistic running-time, as opposed to the guaranteed running-time of our algorithm.

2.1   Presentation of the Algorithm

Intuitively the herding algorithm is quite simple. Agents begin to move until they encounter another agent. When that occurs, the two agents attempt to circumnavigate the ring in opposite directions, in order to "herd" all the other agents to rendezvous. When two herding agents meet, the agent with the faster remembered speed dominates, meaning that eventually A_{max} and some partner will herd all other agents to a single rendezvous point.

Formally, the algorithm requires each agent to maintain in memory a simple state machine consisting of four states. We denote the current state of A_i as σ_i. The algorithm also requires each agent to remember a single agent speed (not necessarily its own), which we denote m_i.

So far we have said nothing bounding the size of agent speeds; since c represents the largest ratio, not the largest absolute speed, the intuitive $\lceil \log c \rceil$ has no actual relation to the amount of memory agents require to store m_i. However, the algorithm only needs enough precision to do comparisons on the speeds. Therefore we can take $\lceil \log k \rceil$ as a bound, since there cannot be more than k distinct speeds in any model. As the state machine requires only a constant number of bits to store one of the five states, this gives us a total memory requirement of $O(\log k)$ bits per agent (Note that during the execution of the algorithm the robots will travel only with any one of the k speeds $s_1, s_2, \ldots, s_k$).

Since there are only four states, we provide an intuitive name and description for each one:

Searching. Agents begin in this state, and while in it they simply move in their current direction at their maximum speed until they encounter another agent. At this point they transition to some other state and continue the algorithm. Once an agent has left this state, it never re-enters it.

Herding. Agents enter this state when they have reason to believe that they are the fastest agent, A_{max}. This occurs when they have encountered at least one other agent, and all agents encountered so far have been slower than themselves. While in this state, agents attempt to herd all other agents to rendezvous by circumnavigating the ring. They travel at their maximum speed or, if other agents are travelling with them, the maximum speed achievable by the group.

Herded. Agents enter this state when they become aware of an agent faster than themselves, and therefore know that they are not the fastest agent. While in this state, agents are moving towards rendezvous with the fastest agent they are currently aware of. As in **Herding** they travel at their maximum speed or, if other agents are travelling with them, the maximum speed achievable by the group.

Penned. Agents stop and enter this state when they have reached what they believe to be the rendezvous location. Agents in this state are not moving. They may transition back to **Herded** if they are reached by an even faster agent (thus invalidating their current rendezvous location) or they may terminate if they learn conclusively that their current location is correct.

Initially, each agent sets σ_i to **Searching**, and m_i to s_i. When encountering another agent, the two agents transmit their values for s_i, m_i and σ_i. The agents perform a state transition if necessary, and store the largest of all communicated speeds as their new value of m_i. If an encounter involves more than two agents, every distinct pair of agents present performs this process in arbitrary order.

The rules for state transitions are relatively straightforward, though not trivial. For convenience, we denote the two agents in an encounter as A and B, and without loss of generality we assume $m_A > m_B$, meaning that A's stored speed is initially greater than B's stored speed. The transitions for agent A are as follows:

- If A is in state **Searching** then it transitions to state **Herding**. Agent A turns and begins moving at its maximum speed in the opposite direction.
- Otherwise, agent A does not change its state or direction. If A is in either state **Herding** or state **Herded**, and agent B is slower than any other agent present, then agent A reduces its current speed to match.

The transitions for agent B are slightly more complex:

- If A is in state **Searching** then B transitions to state **Herded**. Agent B begins moving at maximum speed in the direction opposite to A.
- If A is in state **Penned** then B transitions to state **Penned** and stops moving.
- Otherwise, agent B transitions to state **Herded**. It begins moving in the same direction as A, at the smallest speed present in the group.

The case where $m_B > m_A$ is perfectly symmetric, which leaves only the case where $m_A = m_B$ (the agents have the same stored speed). In this case, both agents stop and enter state **Penned**. If an agent enters state **Penned** in the **All-Unique** model, it knows that rendezvous has been achieved and that it may terminate.

2.2 Proof of Correctness

Theorem 1. *When running the herding algorithm in the All-Unique or Max-Unique models, all agents eventually stop at the same location on the ring, achieving rendezvous.*

Proof. To see this, we must follow the state transitions of the fastest agent, A_{max}. Like all agents, it starts in state **Searching** and begins moving at maximum speed. Since it is the fastest agent and has a unique speed, by moving at maximum speed it is guaranteed to encounter another agent eventually. At this encounter, following the state transition rules, it transitions to state **Herding** and begins moving at its maximum speed in the opposite direction. The agent it encountered transitions to state **Herded** and does the same. At this point, both agents have s_{max} stored in memory.

Here we know several important facts. First, the two agents are moving towards each other around the entire ring from their initial point of encounter. This means that all other agents must be between them in their direction of movement. Second, we know that no other agent has s_{max} in memory. Agent A_{max} has a unique speed, and has only encountered a single agent, thus only those two agents have s_{max} in memory.

Now consider the segment of the ring that lies between the two agents. As the two agents move, the segment gets smaller and each other agent will encounter either A_{max} or its partner. At each such encounter, the agent will be dominated (since its stored speed must be less than s_{max}) and so it transitions to state **Herded** and joins in the circumnavigation of the ring.

Eventually, A_{max} and its partner will meet again. At this point all agents stop and transition to state **Penned**. Rendezvous has been achieved.   □

Theorem 2. *In the All-Unique model, the herding algorithm results in strong rendezvous.*

Proof. Due to Theorem 1 we know that all agents must enter state **Penned** when rendezvous is reached. Now we simply show that no agent can enter that state prior to rendezvous being reached. Agents can then know that rendezvous has been reached when they transition to state **Penned**, making the rendezvous strong. Consider some agent A_i. For A_i to enter state **Penned** it must encounter some other agent A_j such that either $m_i = m_j$ (case 1), or $m_i < m_j$ and $\sigma_j =$ **Penned** (case 2).

Since all agents speeds are unique in this model, the first case can only occur when the two agents are connected by some chain of previous encounters. This implies that their stored speed is maximal among all agents in the chain, which means that they are both participating in the same circumnavigation of the ring (headed by the agent A_x such that $s_x = m_i = m_j$). A_x must have encountered some other agent at this point, so must be in state **Herding**.

Since an agent that is participating in a circumnavigation joins the other agents in that process and travels in lockstep with them, A_i and A_j cannot both be travelling in the same "prong" of the herd. This implies that the circumnavigation instigated by A_x must be complete when A_i encounters A_j, meaning that rendezvous has been achieved and that $A_x = A_{max}$.

The second possible case ($m_i < m_j$ and $\sigma_j = \mathtt{Penned}$) is handled by induction; since there is no agent faster than A_{max}, that agent must only enter state $\mathtt{Penned}$ when rendezvous is really achieved. Given this, it must be true for the second-fastest agent as well, and so on for the third-fastest, etc.   □

Theorem 3. *In the* $\mathtt{Max\text{-}Unique}$ *model, the herding algorithm cannot result in strong rendezvous.*

Proof. To show that an agent may enter state $\mathtt{Penned}$ prematurely in the $\mathtt{Max\text{-}Unique}$ model, consider some agent A_i that is not A_{max}. As in Theorem 2, for A_i to enter this state it must encounter another agent A_j such that $m_i = m_j$, or $m_i < m_j$ and $\sigma_j = \mathtt{Penned}$. But unlike in Theorem 2, the first of these cases can happen in any number of trivial scenarios where rendezvous hasn't been reached. Since an arbitrary number of agents can have the same speed as long as that speed isn't maximal, an adversary can simply place two agents facing each other with the same speed to start.

If an agent enters state $\mathtt{Penned}$ prematurely in this model, then both agents will transition back into state $\mathtt{Herded}$ when an agent knowing some larger speed comes along. As such, no agent can ever know that some agent with an even larger speed won't eventually arrive and wake it up again.   □

2.3   Time Analysis

Theorem 4. *The herding algorithm finishes in time at most* $\frac{1}{2}\left(\frac{c+1}{c-1}\right)n$.

Proof. The algorithm can be analyzed in two parts. The first part consists of the time before A_{max} encounters any other agent. Since A_{max} is moving at its full speed this devolves in the worst case into the "Distributed Race" discussed by [5]. This part takes at most $n/(c-1)$ time.

The second part is the time for A_{max} and its partner to herd all other agents to rendezvous after their initial meeting. The two agents travel the entire ring (size n) in opposite directions, and each trivially moves with a speed of at least 1. Therefore this step takes time at most $n/2$. Putting the two together we get:

$$\frac{n}{c-1} + \frac{n}{2} = \frac{2n + n(c-1)}{2(c-1)} = \frac{n(2 + (c-1))}{2(c-1)} = \frac{n(c+1)}{2(c-1)} = \frac{1}{2}\left(\frac{c+1}{c-1}\right)n,$$

which completes the proof.   □

Remark 1. This bound implies that if the agents have knowledge of n, knowledge of c, and some ability to count time, then strong rendezvous is possible even in the $\mathtt{Max\text{-}Unique}$ model simply by achieving normal rendezvous and waiting for $\frac{1}{2}\left(\frac{c+1}{c-1}\right)n$ time to elapse, thus guaranteeing that there are no other agents left.

Remark 2. The algorithm actually achieves a slightly faster running-time than proved here, because the longer the first part (the distributed race) takes, the shorter the second part (the herding) must take. Assume the first part takes time t. Then any agent with speed 1 must be at least distance $t(c-1)$ "behind" A_{max} at the point of its first encounter. Therefore when it turns, it must spend at least time $t\frac{c-1}{c+1}$ travelling faster than speed 1, so the second part of the algorithm (the herding) must take less than $n/2$ time.

Remark 3. When $k = 2$ we can give a slightly tighter bound because when circumnavigating the ring there are no other agents to encounter. This means agent A_{max} has no reason to slow down, and ends up travelling at speed c the entire time. The resulting circumnavigation takes time exactly $n/(c+1)$, leading to a bound of

$$\frac{n}{c-1} + \frac{n}{c+1} = \frac{n(c+1) + n(c-1)}{(c+1)(c-1)} = \frac{n(c+1+c-1)}{c^2 - 1} = \frac{2cn}{c^2 - 1}.$$

2.4 Optimality

In order to prove a useful bound on the optimal algorithm, we first need a few intermediate results whose utility will eventually become clear.

Lemma 1. *An agent may not turn or change its speed except when encountering another agent.*

We now consider a particular initial layout of agents that is available to the adversary. In this layout, the adversary places the agents clumped together on the ring facing the same direction. The clump is tight (the two "outside" agents are only a tiny distance, ϵ, apart) and the agents are ordered by their speed such that the fastest agent, A_{max} is at the "head" of the clump and the slowest agent is at the tail. The speed of each agent A_i is chosen as $s_i = 1 + \frac{(i-1)(c-1)}{k-1}$. For convenience we shall name this layout "alpha".

Lemma 2. *When starting from layout alpha, no algorithm can achieve the first agent encounter faster than $(n - \epsilon)/(c - 1)$.*

Proof. The proof of this lemma follows trivially from Lemma 1. No agent can turn or change its speed prior to the first encounter, so the fastest way of achieving rendezvous is for all agents to travel at maximum speed, taking time $(n - \epsilon)/(c - 1)$. This is effectively the Distributed Race algorithm from [5]. $\square$

Note that since the first encounter *is* rendezvous when $k = 2$, this is trivially the optimal algorithm. We therefore only consider the case $k > 2$ going forward.

Lemma 3. *When the agents are started in layout alpha, then at the time of first rendezvous the agents are spread equidistant at $k-1$ locations along the ring (we call this layout "beta").*

We ignore the ϵ term going forward as it complicates the equations without materially affecting the result.

Lemma 4. *In position beta, every agent A_i has some other agent A_j at distance at least $\frac{n}{2}\left(\frac{k-2}{k-1}\right)$ from it. Without loss of generality, $i < j$ and $j = i + \lfloor(k-1)/2\rfloor$.*

Proof. Since the agents are equidistant in position beta, we have two cases. When $k - 1$ is even, each agent A_i $(1 \le i < (k-1)/2)$ has another agent A_j exactly opposite it on the ring, where $j = i + \frac{k-1}{2}$. This agent is at distance $n/2$ from A_i, and A_i is at distance $n/2$ from it. The lemma holds since $n/2 > \frac{n}{2}\left(\frac{k-2}{k-1}\right)$.

If $k - 1$ is odd then the point opposite A_i $(1 \le i < k/2)$ is equidistant between agent A_{j-1} and agent A_j where $j = i + k/2$. These agents are neighbours and are thus at distance $\frac{n}{k-1}$ from each other. Both of these agents are therefore at distance $(n - \frac{n}{k-1})/2 = \frac{n}{2}\left(\frac{k-2}{k-1}\right)$ from A_i. □

Lemma 5. *Asymptotically in k, no algorithm can achieve rendezvous in time better than $\frac{n}{c+3}$ from position beta.*

Theorem 5. *Asymptotically in k, no algorithm can achieve rendezvous in time better than $\frac{2}{c+3}\left(\frac{c+1}{c-1}\right)n$.*

Proof. This result follows simply from Lemmas 1, 2 and 5. Given starting position alpha, any algorithm must take at least time $n/(c-1)$ to reach position beta. No algorithm can do better than time $n/(c+3)$ to achieve rendezvous from position beta. Together they sum:

$$\frac{n}{c-1} + \frac{n}{c+3} = \frac{n(c+3) + n(c-1)}{(c-1)(c+3)} = \frac{cn + 3n + cn - n}{(c-1)(c+3)}$$

$$= \frac{2cn + 2n}{(c-1)(c+3)} = \frac{2n(c+1)}{(c-1)(c+3)} = \frac{2}{c+3}\left(\frac{c+1}{c-1}\right)n,$$

which proves the theorem. □

Note that the ratio of the rendezvous time of Theorem 5 divided by the rendezvous time of Theorem 4 tends to 1 as $c \to 1$, and diverges as c gets bigger.

3 Discussion of Stronger Models

In this section we consider variants of the problem with more agent knowledge. We consider the cases where the agents know the values of n, k, or c. We also consider the case where the agents have pedometers, and the case where agents have pedometers as well as knowledge.

We construct two variants of the herding algorithm presented in Sect. 2. One depends on knowledge of k but is capable of achieving rendezvous even in the weaker One-Unique model. The other uses a result of [5] to reduce the running-time of the original algorithm by $\frac{n}{c^2-1}$, but requires the agents have pedometers as well as knowledge of n and c.

3.1   Knowledge of n

Giving the agents knowledge of n seems to be relatively useless on its own. The agents do not have pedometers or timers, so they still cannot tell when they've traversed any given fraction of the ring. As the value of n is the same for each agent, it does not provide any asymmetry in its own right, and combining it in some way with an agent's speed provides no more asymmetry than the speed would alone. Of particular interest is that Lemma 1 still holds (with effectively the same proof) when the agents know n. The rest of Sect. 2.4 follows unchanged, meaning that Theorem 5 also holds in this model.

3.2   Knowledge of k

Unlike knowledge of n, knowledge of k is obviously useful. It trivially permits the herding algorithm to achieve strong rendezvous in the Max-Unique model, while Theorem 3 showed that this was impossible for the Herding algorithm. In fact, knowledge of k makes strong and normal rendezvous equivalent; if normal rendezvous is achieved in any circumstance, the agents can simply count how many other agents are present to make the rendezvous strong.

In [9], it was conjectured that rendezvous could not be achieved in the One-Unique model without some additional agent knowledge. Here we present a variation of the herding algorithm which achieves rendezvous in this model, given that the agents know k and have sufficient memory. We call this the "non-maximal variant".

As in Sect. 2 we introduce one additional notational convenience. Since there may be multiple agents with the fastest speed in the One-Unique model, the label A_{max} is no longer useful. Instead, there is at least one agent who's speed is unique even if it is not necessarily maximal. We therefore denote this agent A_U. We emphasize that, like A_{max}, the agents are not aware of this label.

The Non-Maximal Herding Algorithm. In the non-maximal variant of the herding algorithm, each agent is required to store the set of all speeds it has encountered so far, as well as whether it believes each of those speeds is unique. As there are k agents potentially each with a unique speed, this requires at least $k(1 + \lceil \log k \rceil) = O(k \log k)$ bits of memory per agent.

Each agent requires an additional 3 bits to store one of five states. Four of these states should be familiar from Sect. 2; Searching, Herding, Herded, and Penned are all nearly the same. The new state we call Wandering; when an agent knows it is not A_U but does not have any other information, it begins Wandering simply by moving at its maximal speed in some direction; this is necessary to prevent the algorithm from getting stuck.

Agents initialize their memory in the obvious way; their set of speeds contains only their own speed, and that speed is currently believed to be unique. They start in state Searching, and behave as expected in that state; each agent simply begins moving at its maximal speed. Since A_U's speed is unique, it is guaranteed to encounter another agent at some point.

When two agents meet (call them A_i and A_j) they first count the number of agents present and check against k; if rendezvous has been achieved then the agents terminate. Otherwise they exchange their set of known speeds and their state. If A and B have the same speed ($s_A = s_B$) then clearly that speed is not unique and is marked as such by both agents. If one of the agents already has the other's speed in memory, then that speed is also marked as not unique by both. If either agent has seen a speed that the other has not, then that speed and its corresponding uniqueness flag is copied.

At this point the two agents have the same set of speeds in memory. If the agents do not know of any unique speeds (for example they have the same speed, and have only encountered each other) then the agents begin to Wander. Otherwise what happens next depends on the agents' states. Assume without loss of generality that prior to the encounter, the speed that both agents now believe to be maximally unique was stored in the memory of A_i (either because it was A_i's speed or because A_i had already encountered an agent knowing that speed).

The transitions here are very similar to the normal algorithm. If $\sigma_i \neq$ Herding and the maximal unique speed is s_i, then A_i and A_j both transition to state Herding and begin moving at their respective maximum speeds in opposite directions. Otherwise A_j transitions to state Herded and beings moving in tandem with A_i. Agent A_i does nothing except potentially slow down to permit A_j to keep up.

Theorem 6. *The non-maximal variant of the herding algorithm achieves strong rendezvous in the* One-Unique *model.*

3.3 Knowledge of c

When we discuss knowledge of c we specifically do not mean knowledge of the actual fastest speed; as we have already seen, c is defined as the normalized ratio between the fastest and slowest speed. If the agents instead know the actual fastest speed then this leads to a trivial algorithm: all agents know whether or not they are the fastest or not, so the fastest stays put and all other agents move to join. This takes time at most n and requires no memory.

Knowledge of c (the ratio) can potentially be useful in its own right however. Specifically, an agent A_i can use just its knowledge of its own speed s_i combined with its knowledge of c to determine the range of valid speeds for all other agents: no agent can possibly have a speed slower than s_i/c, and no agent can possibly have a speed faster than $s_i c$. When two agents meet, they can exchange bounds and narrow the window of valid speeds. If a fastest agent (with normalized speed c) meets a slowest agent (with normalized speed 1) then they both immediately know that they are the fastest/slowest agents, respectively.

While this knowledge does not obviously lead to a better general-case algorithm in any model, it does potentially lend itself to improvements in specific cases.

3.4   Using Pedometers

Like knowledge of n, giving agents pedometers on their own is relatively useless without something to compare against. However, pedometers do give us one small advantage: since agents know their own speed, the two can be used in combination to construct a timer. As with knowledge of c, this does not obviously lead to any improvements.

3.5   Pedometers and Knowledge

While knowledge of n (Sect. 3.1), knowledge of c (Sect. 3.3) and pedometers (Sect. 3.4) have not proved particularly fruitful lines of enquiry on their own, in combination they produce a much more interesting result. [5] proved in their paper that the optimal two-agent rendezvous in this case required time $\frac{cn}{c^2-1}$. Using this as a component, we can construct a second variant of the original herding algorithm, which we will call the "fast herding algorithm".

The Fast Herding Algorithm. The fast herding algorithm is in almost all respects identical to the herding algorithm presented in Sect. 2. The only difference is in the initial step. Instead of using the plain "distributed race" of [5] we have each agent execute the optimal two-agent algorithm given as part of Theorem 1 in that paper. As the two agent case is the degenerate one, this guarantees that A_{max} will reach its first encounter in time at most $\frac{cn}{c^2-1}$. The proof of correctness given for Theorem 1 can be trivially adapted to show that this fast variant also achieves rendezvous. More interesting is the running-time.

Theorem 7. *In the* Max-Unique *and* All-Unique *models, the fast herding algorithm achieves rendezvous in time that is* $\frac{n}{c^2-1}$ *faster than the normal herding algorithm, specifically in time* $\frac{1}{2}\left(\frac{c+1}{c-1}\right)n - \frac{n}{c^2-1}$.

Remark 4. Since the agents do have knowledge of n, c and a pedometer (which we mentioned in Sect. 3.4 could be used to construct a timer) this implies that even in the Max-Unique model, strong rendezvous is achievable by the fast herding algorithm given enough memory. Agents simply run the algorithm and then wait for their timer to expire.

Remark 5. As with the original algorithm (see Remark 3) we can give a slightly tighter bound when $k = 2$.

$$\frac{cn}{c^2-1} + \frac{n}{c+1} = \frac{cn}{(c+1)(c-1)} + \frac{n}{c+1} = \frac{cn+(c-1)n}{c^2-1} = \frac{2cn}{c^2-1} - \frac{n}{c^2-1}.$$

4   Conclusion

We have studied the rendezvous problem for $k \geq 2$ agents on the ring, using differences in agent speed to break symmetry. We extended the $k = 2$ case to $k \geq 2$

for four different models, and presented the herding algorithm which achieves rendezvous in the two stronger of these models. We also proved a bound on the optimal algorithm in those models. We further studied the cases where agents had knowledge or capabilities above and beyond those available in the base model. We presented two variants on the herding algorithm, one of which makes use of additional knowledge to achieve a better running-time, and one of which is capable of achieving rendezvous even when guarantees about agent speed distribution are weakened. Many interesting open problems remain especially in tightening our bounds and improving our algorithms. Studying the possibility of using these models in other domains (more general than the ring) as well as probabilistic versions similar to the work of [10] remains entirely open.

References

1. Alpern, S.: Rendezvous search: a personal perspective. Oper. Res. **50**(5), 772–795 (2002)
2. Bampas, E., Czyzowicz, J., Gasieniec, L., Ilcinkas, D., Labourel, A.: Almost optimal asynchronous rendezvous in infinite multidimensional grids. In: Lynch, N.A., Shvartsman, A.A. (eds.) DISC 2010. LNCS, vol. 6343, pp. 297–311. Springer, Heidelberg (2010)
3. Czyzowicz, J., Ilcinkas, D., Labourel, A., Pelc, A.: Asynchronous deterministic rendezvous in bounded terrains. TCS **412**(50), 6926–6937 (2011)
4. Dieudonné, Y., Pelc, A., Villain, V.: How to meet asynchronously at polynomial cost. In: Proceedings of ACM PODC, pp. 92–99. ACM (2013)
5. Feinerman, O., Korman, A., Kutten, S., Rodeh, Y.: Fast rendezvous on a cycle by agents with different speeds. In: Chatterjee, M., Cao, J., Kothapalli, K., Rajsbaum, S. (eds.) ICDCN 2014. LNCS, vol. 8314, pp. 1–13. Springer, Heidelberg (2014)
6. Flocchini, P., An, H.-C., Krizanc, D., Luccio, F.L., Santoro, N., Sawchuk, C.: Mobile agents rendezvous when tokens fail. In: Kralovic, R., Sýkora, O. (eds.) SIROCCO 2004. LNCS, vol. 3104, pp. 161–172. Springer, Heidelberg (2004)
7. Flocchini, P., An, H.-C., Krizanc, D., Santoro, N., Sawchuk, C.: Multiple mobile agent rendezvous in a ring. In: Farach-Colton, M. (ed.) LATIN 2004. LNCS, vol. 2976, pp. 599–608. Springer, Heidelberg (2004)
8. Hegarty, P., Martinsson, A., Zhelezov, D.: A variant of the multi-agent rendezvous problem. CoRR, abs/1306.5166 (2013)
9. Huus, E.: Knowledge in rendezvous of agents of different speeds, Honours project, Carleton University, School of Computer Science, Spring 2014
10. Kranakis, E., Krizanc, D., MacQuarrie, F., Shende, S.: Randomized rendezvous on a ring for agents with different speeds. In: Proceedings of ICDCN 2015 (2015)
11. An, H.-C., Krizanc, D., Markou, E.: Mobile agent rendezvous in a synchronous torus. In: Correa, J.R., Hevia, A., Kiwi, M. (eds.) LATIN 2006. LNCS, vol. 3887, pp. 653–664. Springer, Heidelberg (2006)
12. Kranakis, E., Krizanc, D., Markou, E.: The Mobile Agent Rendezvous Problem in the Ring: An Introduction. Synthesis Lectures on Distributed Computing Theory Series. Morgan & Claypool Publishers, San Rafael (2010)
13. Sawchuk, C.: Mobile agent rendezvous in the ring. Ph.D. thesis. Carleton University (2004)

14. Schneider, J., Wattenhofer, R.: A new technique for distributed symmetry breaking. In: Proceedings of ACM PODC, pp. 257–266. ACM, New York (2010)
15. Yu, X., Yung, M.: Agent rendezvous: a dynamic symmetry-breaking problem. In: Meyer auf der Heide, F., Monien, B. (eds.) ICALP 1996. LNCS, vol. 1099. Springer, Heidelberg (1996)

The Random Bit Complexity
of Mobile Robots Scattering

Quentin Bramas[1,2](✉) and Sébastien Tixeuil[1,2,3]

[1] Sorbonne Universités, UPMC University Paris 06, UMR 7606, 75005 Paris, France
quentin.bramas@lip6.fr
[2] CNRS, UMR 7606, LIP6, 75005 Paris, France
[3] Institut Universitaire de France, Paris, France

Abstract. We consider the problem of scattering n robots in a two dimensional continuous space. As this problem is impossible to solve in a deterministic manner [6], all solutions must be probabilistic. We investigate the amount of randomness (that is, the number of random bits used by the robots) that is required to achieve scattering.

We first prove that $n \log n$ random bits are necessary to scatter n robots in any setting. Also, we give a sufficient condition for a scattering algorithm to be random bit optimal. As it turns out that previous solutions for scattering satisfy our condition, they are hence proved random bit optimal for the scattering problem.

Then, we investigate the time complexity of scattering when strong multiplicity detection is not available. We prove that such algorithms cannot converge in constant time in the general case and in $o(\log \log n)$ rounds for random bits optimal scattering algorithms. However, we present a family of scattering algorithms that converge as fast as needed without using multiplicity detection. Also, we put forward a specific protocol of this family that is random bit optimal ($n \log n$ random bits are used) and time optimal ($\log \log n$ rounds are used). This improves the time complexity of previous results in the same setting by a $\log n$ factor.

Aside from characterizing the random bit complexity of mobile robot scattering, our study also closes its time complexity gap with and without strong multiplicity detection (that is, $O(1)$ time complexity is only achievable when strong multiplicity detection is available, and it is possible to approach it as needed otherwise).

1 Introduction

We consider distributed systems consisting of multiple autonomous robots [7,10] that can move freely on a 2-dimensional plane, observe their surroundings and perform computations. The robots do not communicate explicitly with other robots and there is no central authority that communicates with the robots.

This work was performed within the Labex SMART supported by French state funds managed by the ANR within the Investissements d'Avenir programme under reference ANR-11-IDEX-0004-02. It has also been partially supported by the LINCS.

© Springer International Publishing Switzerland 2015
S. Papavassiliou and S. Ruehrup (Eds.): ADHOC-NOW 2015, LNCS 9143, pp. 210–224, 2015.
DOI:10.1007/978-3-319-19662-6_15

Such teams of robots can be deployed in areas inaccessible to humans, to perform collaborative tasks such as search and rescue operations, data collections, environmental monitoring and even extra-terrestrial exploration. From the theoretical point of view, the interest lies in determining which tasks can be performed by such robot teams and under what conditions.

One line of research is to determine the minimum capabilities required by the robots to achieve any given task [7]. A particularly weak model of robots is assumed and additional capabilities are added whenever it is necessary to solve the problem. In our model, the robots are assumed to be anonymous (*i.e.* indistinguishable from one another), oblivious (*i.e.* no persistent memory of the past is available) and disoriented (*i.e.* they do not agree on a common coordinate system nor a common chirality). The robots operate in Look-Compute-Move (LCM) cycles, where in each cycle a robot *Looks* at its surroundings and obtains a snapshot containing the locations of all robots as points on the plane with respect to its own location and ego-centered coordinate system; Based on this visual information, the robot *Computes* a destination location and then *Moves* towards the computed location. Since the robots are identical, they all follow the same algorithm. The algorithm is oblivious if the computed destination in each cycle depends only on the snapshot obtained in the current cycle (and *not* on the past history of execution). The snapshots obtained by the robots are not consistently oriented in any manner.

When processing a snapshot, a robot can distinguish whether a point is empty (*i.e.*, not occupied by any robot). However, since robots are viewed as points, the question arises of how robots occupying the same position at the same time are perceived in a snapshot. The answer to this question is formulated in terms of the capacity of the robots to detect multiplicity of robots in a point. The robots are said to be capable of *multiplicity detection* if they can distinguish if a point is occupied by one or more than one robot.

One important task useful in multi-robot coordination is *gathering* the robots at a single location, not known beforehand. The dual problem of gathering is the *scattering* problem. Scattering requires that, starting from an arbitrary configuration, eventually no two robots share the same location. It turns out that neither deterministic gathering [10] nor scattering [6] are possible without additional assumptions. Most of the work done so far in order to circumvent the impossibility of gathering focuses on required minimal additional assumptions with respect to the coordinate system or multiplicity detection [7,10] to make the problem solvable. However, the scattering problem cannot allow deterministic solutions [6].

Related Work. The first probabilistic algorithms to solve mobile robot scattering without multiplicity detection were given by Dieudonné and Petit [5,6]. The algorithms are based on the following simple scheme: after the Look phase, a robot computes the Voronoi diagram [2] of the observed positions, and then tosses a coin ($\frac{1}{4}$ [5] or $\frac{1}{2}$ [6]) to either remain in position, or move toward an arbitrary position in its Voronoi cell. The fact that a robot may only move within its Voronoi cell preserves the fact that initially distinct robots (that is robots occupying distinct positions) remain distinct thereafter. This invariant and the positive probability that two robots on the same point separate implies the eventual scattering of all robots. A later study [4] shows that the scattering algorithm [6] converges in expected $O(\log n \log \log n)$ rounds. In the same paper, a new probabilistic algorithm was presented, with the assumption that robots are aware of the total number of robots. This protocol is optimal in time as it scatters any n-robots configuration in expected $O(1)$ rounds. If the total number of robots n is known, then robots are able to choose uniformly at random a position within their Voronoi cell among $2n^2$ possibilities, inducing an expected $O(1)$ rounds scattering time. In the limited visibility setting [8] (the visibility capability of each robot has a constant radius, and visual connectivity has to be maintained throughout scattering), the time lower bound grows to expected n rounds for scattering n robots. None of the aforementioned works investigated the number of random bits used in the scattering process. The scattering problem when robots know the total number of robots (*ie*, when robots are capable of strong multiplicity detection) is closely related to the self-stabilizing *Unique Naming Problem* [1]. Indeed, scattering two robots by choosing two different destinations is equivalent to choosing two different names. In fact, Lemma 1 can be deduced from this related topic (and even from other studies on probabilistic processes). Also, in the remainder of the paper, unique naming problem can give insight about the expected results. But, except in Lemma 1, proofs are strongly dependent on the oblivious and autonomous mobile robots model that contains assumptions, such as different multiplicity detection and obliviousness, so that our results cannot be directly deduced from the unique naming problem. That is why we also choose to prove Lemma 1 using our model to be more consistent with the remaining of the paper.

Our Contribution. We investigate the amount of randomness (that is, the number of random bits used by the robots) that is necessary to achieve mobile robots scattering. In more details, we first define a canonical scattering algorithm, that encompasses all previous solutions, and is tantamount to selecting the number of possible locations that are selected uniformly at random by the robots.

Then, we prove that $n \log n$ random bits are necessary to scatter n robots in any setting for all scattering algorithms (not only canonical algorithm). Also, we give a sufficient condition for a canonical scattering algorithm to be random bit optimal (namely, the number of possible locations must be polynomial in the number of observed positions). As it turns out that previous solutions for scattering [4–6] satisfy our condition, they are hence proved random bit optimal for the scattering problem.

Finally, we investigate the time complexity of scattering algorithms, when strong multiplicity detection is not available. We prove that such algorithms cannot converge in constant time in the general case and in $o(\log \log n)$ rounds in the case of random bits optimal algorithms (in this last setting, the best known upper bound was $\log n \log \log n$ [5,6]). On the positive side, we provide a family of scattering algorithms that converge as fast (but not $O(1)$) as needed, without using multiplicity detection. Also, we give a particular protocol among this family that is random bit optimal ($n \log n$ random bits are used) and time optimal ($\log \log n$ rounds are used). This improves the time complexity of previous results in the same setting by an expected $\log n$ factor.

Due to space constraints, proofs are omitted from the main text but can be found in the technical report [3].

2 Model and Preliminaries

Robot Networks. There are n robots modeled as points on a geometric plane. A robot can observe its environment and determine the location of other robots in the plane, relative to its own location and coordinate system. All robots are identical (and thus indistinguishable) and they follow the same algorithm. Moreover, each robot has its own local coordinates system, which may be distinct from that of other robots. In this paper, robots are said to have unlimited visibility, in the sense that they are always able to sense the position of all other robots, regardless of their proximity.

Multiplicity Detection. When several robots share the same location, this location is called a point of multiplicity. Robots are capable of *strong* multiplicity detection when they are aware of the number of robots located at each point of multiplicity. In contrast, when robots are capable of weak multiplicity detection, they know which points are points of multiplicity, but are unable to count how many robots are located there. The multiplicity detection of a robot is said to be *local* if the multiplicity detection concern only the point where robot lies. If robots detect the multiplicity of each observed point, the multiplicity detection is *global*. Robots are not aware of the actual number n of robots unless they are capable of global strong multiplicity detection.

If robots are not able to detect multiplicity, they never know if the configuration is scattered and thus never stop moving. Hence, algorithms that do not use multiplicity detection cannot terminate. With local weak multiplicity detection robots are aware of the situation at their position, *e.g.* they can stop executing the algorithm if they sense they are alone at their location. However, they may not know if the global configuration is scattered (yet, if the configuration is indeed scattered, all robots are stopped and the algorithm (implicitly) terminates). With global weak multiplicity detection, algorithm can explicitly terminates when every observed position is not a multiplicity point.

System Model. Three different scheduling assumptions have been considered in previous work. The strongest model is the fully synchronous (FSYNC) model where each phase of each cycle is performed simultaneously by all robots. On the other hand, the weakest model, called asynchronous (ASYNC) allows arbitrary delays between the Look, Compute, and Move phases and the movement itself may take an arbitrary amount of time [7]. The semi-synchronous (SSYNC) model [7,10] lies somewhere between the two extreme models. In the SSYNC model, time is discretized and at each considered step an arbitrary subset of the robots are active. The robots that are active, perform exactly one *atomic* Look-Compute-Move cycle. It is assumed that a hypothetical scheduler (seen as an adversary) chooses which robots should be active at any particular time and the only restriction of the scheduler is that it must activate each robot infinitely often in any infinite execution (that is, the scheduler is *fair*).

In this paper, for the analysis, we use the FSYNC model. Lower bounds naturally extend to SSYNC and ASYNC models and upper bounds (that is, algorithms) are also valid in SSYNC. Indeed, as in [4], since all the algorithms in this paper ensure that two robots moving at different times necessarily have different destinations, the worst case scenario is when robots are activated simultaneously. In FSYNC model, robots perform simultaneously an *atomic computational cycle* composed of the following three phases: Look, Compute, and Move.

- *Look.* An observation returns a snapshot of the positions of all robots. All robots observe the exact same environment (according to their respective coordinate systems).
- *Compute.* Using the observed environment, a robot executes its algorithm to compute a destination.
- *Move.* The robot moves towards its destination (by a non-zero distance but without always reaching it).

Moreover, robots are assumed to be oblivious (*i.e.*, stateless), in the sense that a robot does not keep any information between two different computational cycles. We evaluate the time complexity of algorithms using the number of asynchronous rounds required to scatter all robots. An asynchronous round is defined as the shortest fragment of an execution where each robot executes its cycle at least once.

Notations. In the sequel, C denotes a n-robots configuration, that is, a multi-set containing the position of all robots in the plane. Removing multiplicity information (that is, multiple entries for the same position) from C yields the corresponding set $U(C)$. For a multi-set C, $|C|$ denotes its cardinality. For a particular point $P \in \mathbb{R}^2$, $|P|$ denotes the multiplicity of P. We denote by $\mathcal{C}(k,n)$ the set of 2-tuples (C,P) where C is a n-robots configuration that contains a point P of multiplicity k.

Random Bits Complexity. The number of random bits needed by a robot to choose randomly a destination among k possible locations is at least $\log_2(k)$, regardless of the distribution, as long as each destination has a non-zero probability to be chosen. We denote by $\log = \log_2$ the logarithm with respect to base 2 obtained from the natural logarithm as $\log(x) = \frac{\ln(x)}{\ln(2)}$. Of course, since there is a probabilistic process involved, starting from the same initial configuration, the exact number of random bits may not be the same for two particular executions of a protocol. So, in the sequel, we consider the expected number of random bits used for scattering.

Since we are concerned about the scattering problem, we do not take into account random bits used by robots that are not located at a point of multiplicity (*i.e*, robots that are already scattered). Of, course, all our lower bound results remain valid without this assumption, but upper bounds we provide do make use of this hypothesis when robots are not capable of weak local multiplicity (as termination cannot be insured in this case). We also assume that robots cannot use an infinite number of random bits in a single execution.

For the study of the random bits complexity, we define $Z_{C,P}$, the random variable that represents the number of random bits used by an algorithm to scatter the robots in P starting from the configuration C. Formally, for an algorithm $\mathcal{A}$, $Z_{C,P}$ is defined over all the possible executions of $\mathcal{A}$ (starting with the configuration C that contains the point P). For an execution, $Z_{C,P}$ equals b if and only if the number of random bits used to scatter all robots that are initially in P is b (ignoring the robots in C that are not initially located at P).

For a point P of multiplicity n, we can represent the way robots at P are divided over k possible destinations with a multi-index $\alpha \in \mathbb{N}^k$ such that $|\alpha| = \sum_{i=1}^k \alpha_i = n$. The resulting maximum multiplicity is denoted by $\|\alpha\|_\infty = \max_i \alpha_i$. Consider the random variable X that equals $\alpha \in \mathbb{N}^k$ if and only if the robots in P are divided in k points of multiplicity $\alpha_1, \alpha_2, \ldots$ and α_k.

It is known that:

$$\mathbb{E}(Z_{C,P}) = \sum_{\alpha \in \mathbb{N}^k, \, |\alpha|=n} \mathbb{P}(X = \alpha)\mathbb{E}(Z_{C,P}|X = \alpha) \tag{1}$$

Then, $\mathbb{E}(Z_{C,P}|X = \alpha)$ equals the number of random bits used during the first round $(n\log(k))$ plus the expected number of random bits used to scatter the k points $p_1, p_2, \ldots$ and p_k, coming from P of multiplicity $\alpha_1, \alpha_2, \ldots$ and α_k. Of course the rest of the configuration may have changed too. But since we want to bound the expectation, we can have an upper or a lower bound by taking the worst or the best resulting configuration. For all $N \in \mathbb{N}$, $n \leq N$, let

$$B(n, N) = \min_{(C,P)\in\mathcal{C}(n,N)} (\mathbb{E}(Z_{C,P})) \quad \text{and} \quad W(n, N) = \max_{(C,P)\in\mathcal{C}(n,N)} (\mathbb{E}(Z_{C,P})) \tag{2}$$

The existence of such min and max comes from the fact that for $N \in \mathbb{N}^*$, $n < N$, the set $\mathcal{C}(n, N)$, is finite. This is due to the fact that there exists an initial configuration from which some (deterministically computed but randomly chosen) paths have been followed by the robots.

Moreover, if Algorithm $\mathcal{A}$ makes sure that two robots at distinct locations in a given configuration remain at distinct locations thereafter, then for two distinct points P and P', $Z_{C,P}$ and $Z_{C,P'}$ are independent and their sum is exactly the number of random bits used to scatter P and P'. Then,

$$B(n, N) \geq n\log(k) + \sum_{\alpha \in \mathbb{N}^k, \, |\alpha|=n} \mathbb{P}(X = \alpha)\sum_{i=1}^k B(\alpha_i, N) \tag{3}$$

$$W(n, N) \leq n\log(k) + \sum_{\alpha \in \mathbb{N}^k, \, |\alpha|=n} \mathbb{P}(X = \alpha)\sum_{i=1}^k W(\alpha_i, N) \tag{4}$$

The recursive inequality (3) is used in Lemma 1 to find the lower bound, and the recursive inequality (3) in Theorem 3 to find the upper bound.

A Canonical Scattering Algorithm. Let $\mathcal{A}$ be a scattering algorithm. As $\mathcal{A}$ can't be deterministic [6], the computation of the location to go to must result from a probabilistic choice (more practically, a robot must randomly choose a destination among a previously computed set of possible destinations). We note $k_{\mathcal{A}}(C, P)$ the function that returns the number of possible destinations depending on the current observed (global) configuration C, and the current observed (local) point P (that is, the point where the robot executing the algorithm lies). Robots located at P may not be aware of P's multiplicity, but they base their computation of the possible destinations set on the same observation. We assume an adversarial setting where symmetry is preserved unless probabilistic choices are made, so we expect the local coordinate systems of all robots occupying the same position P to be identical. Thus, the set of possible destinations is the same for all robots at P. We now define a canonical scattering algorithm that generalizes previously known scattering algorithms.

Definition 1. *An algorithm $\mathcal{A}$ is a* canonical *scattering algorithm if it has the following form:*

Algorithm 1: *Canonical scattering algorithm, executed by a robot r*

---
1 $C \leftarrow$ *Observed current configuration.*
2 $P \leftarrow$ *Observed current position of r.*
3 *Compute a set of $k_{\mathcal{A}}(C, P)$ possible destinations Pos such that every point in Pos may not be chosen by a robot not currently in P.*
4 *Move toward a point in Pos chosen uniformly at random.*

The function $k_{\mathcal{A}}$ that gives the number of possible destinations depending on the current configuration and position is called the destination function *of the algorithm $\mathcal{A}$.*

Line 3 of Algorithm 1 implies that a canonical algorithm must ensure that the multiplicity of any given point never increases (*i.e.* robots located at different locations remain at different locations thereafter). Previous algorithms [4,6] are canonical in the SSYNC and FSYNC models. Both of them use Voronoi Diagrams [2] to ensure monotonicity of multiplicity points. The algorithm given in [6] is a canonical scattering algorithm with line 3 replaced by: Compute a set *Pos* of 2 points in the Voronoi cell of r. The algorithm given in [4] is a canonical scattering algorithm with line 3 replaced by: Compute a set *Pos* of $2|C|^2$ points in the Voronoi cell of r. This property holds only if Voronoi cells computations occur at the same time (that is, in the FSYNC and the SSYNC models). In the ASYNC model, two robots at different positions, activated at different times, may move towards the same destination[1].

---
[1] To our knowledge, no algorithm exists for scattering mobile robots in the ASYNC model.

An algorithm that computes a set Pos of k points but does not choose uniformly the destination from Pos can be seen as a canonical scattering algorithm if points in Pos can have multiplicity greater than 1, *i.e.* if Pos is a multi-set. For example, if an algorithm computes a set $Pos = \{x_1, x_2\}$ and chooses x_1 with probability $\frac{3}{4}$, this is equivalent to choosing uniformly at random from the multi-set$\{x_1, x_1, x_1, x_2\}$. Of course this scheme cannot be extended to irrational probability distributions, yet those distributions induce an infinite number of random bits, which is not allowed in our model. We can now state our first Theorem:

Theorem 1. *If A is an algorithm that ensures that the multiplicity of any point never increases, then A is a canonical scattering algorithm (with Pos possibly a multiset).*

Observe that any deterministic protocol for mobile robot networks (that ensure monotonicity of multiplicity points) can be seen as a canonical scattering algorithm whose destination function is identically 1. Also, if an algorithm computes a multiset Pos with duplicate positions, it uses more random bits to select its destination at any given stage of the computation. As we focus on efficient algorithms (that is, we try to minimize the number of random bits), we suppose from now on that Pos is a set (*i.e.* it has no duplicate positions). Indeed the uniform distribution is the probability distribution that has the largest entropy.

3   The Random Bit Complexity of Scattering

In this section we demonstrate that any mobile robots scattering algorithm must use at least $n \log(n)$ random bits, whether or not it uses multiplicity detection. Then we prove a sufficient condition for a canonical scattering algorithm to effectively use $O(n \log n)$ random bits. As this condition is satisfied by previously known canonical scattering algorithms [4–6], a direct consequence of our result is that those algorithms are random bit optimal.

3.1   Lower Bound

In this section we prove that the expected number of random bits used by any scattering algorithm is greater than $n \log(n)$. Actually, we implicitly prove that any execution of a scattering algorithm that scatters n robots initially located at the same position uses more than $n \log(n)$ random bits. The proof first considers *canonical* scattering algorithms, and later expands to *arbitrary* scattering algorithms (that is, algorithms that may not insure that the multiplicity of any point never increases, see Theorem 1).

Lemma 1. *Let A be a canonical scattering algorithm (Algorithm 1). The expected number of random bits needed to scatter n robots is at least $n \log(n)$.*

The sketch of the proof is as follows. We prove that any execution of the algorithm uses at least $n \log(n)$ random bits by mathematical induction on the number of robots located at a particular point. For the base case, we observe that 2 robots located at the same point and executed simultaneously must both use at least 1 random bit. So, 2 robots are scattered with more than $2 = 2 \log 2$ random bits.

To prove the induction step, we observe that the most favorable scenario is when, at each round, robots are uniformly distributed over all possible destinations. If we assume that points with multiplicity $m < n$ need more than $m \log(m)$ random bits to be scattered, then if n robots are split among two points of multiplicity m_1 and m_2, the number of random bits used to scatter those two points (which is greater than $m_1 \log m_1 + m_2 \log m_2$) is greater than $n \log(n/2)$. This result comes from the convexity of function $x \mapsto x \log x$.

Theorem 2. *Let $\mathcal{A}$ be a scattering algorithm (not necessarily canonical). The expected number of random bits needed to scatter n robots is greater that $n \log(n)$*

Corollary 1. *Algorithms defined in [4, 6] (see Sect. 2) are random bit optimal.*

3.2   Upper Bound

If an algorithm computes a set of distinct points, and if each robot chooses randomly a destination in this set, then robots must scatter. Moreover if the cardinality of the chosen set is bounded, then the expected number of random bits may be bounded too. We now prove that if the destination set cardinality is bounded by a polynomial in $|C|$, the random bit complexity of the algorithm is $O(n \log(n))$ (that is, optimal).

We start with three technical lemmas. Lemma 3 helps us to bound the maximum multiplicity obtained after a round, when n robots are randomly distributed among k possible destinations. Let Ω be the universe of the experiment of randomly and uniformly distributing n robots among k possible destinations. These lemmas are related to previous studies called *balls into bins* [9]. Some of them, Lemma 2 for instance, can be directly deduced from previous work but others differ by the approach. Indeed, the number of destinations (bins) depends on several parameters and in the Sect. 4, the resulting probability needs to be bound by non-standard parameters (namely, the number $|U(C)|$ of observed robots). More precisely, for Lemma 6, the bound we derive is with high probability on $|U(C)|$ (contrary to the number on bins in previous results [9]). That is why we choose not to deduces those lemma from previous results.

Lemma 2. *Let $X : \Omega \mapsto \mathbb{N}^k$ be the random variable that gives the distribution of n robots among k destinations. If $k \geq 2n^2$, then:*

$$\mathbb{P}(\|X\|_\infty > 1) \leq \frac{1}{2}$$

Lemma 3. *Let $X : \Omega \mapsto \mathbb{N}^k$ be the random variable that gives the distribution of n robots among k destinations (uniformly at random). If $k \leq An^K$ with $A, K \in \mathbb{N}$, then there exists $N_{A,K} \in \mathbb{N}$ (that depends only on A and K), such that, for all $n \geq N_{A,K}$:*

$$\mathbb{P}\left(\|X\|_\infty > \frac{n}{k}(1 + k^\xi)\right) \leq \frac{1}{2} \quad with \ \xi = 1 - \frac{1}{K+1}$$

From now on we use, for a canonical scattering algorithm $\mathcal{A}$, the notation $W(n, N)$ for the largest expected number of random bits used by $\mathcal{A}$ to scatter n robots gathered in a point P in a configuration of N robots (see Equation (2)).

Lemma 4. *Let $\mathcal{A}$ be a canonical scattering algorithm with a destination function that satisfies: $\exists K \in \mathbb{N}, k_\mathcal{A}(C, P) = O(|C|^K)$. Then, for all $\mathcal{N} \in \mathbb{N}^*$, there exists $R \in \mathbb{R}$ such that:*

$$\forall n \leq \mathcal{N}, \ \forall N \geq n, \qquad W(n, N) \leq R \log(N)$$

Theorem 3. *Let $\mathcal{A}$ be a canonical scattering algorithm with a destination function $k_\mathcal{A}$. Then $\mathcal{A}$ is optimal in terms of random bits, i.e. the expected number of random bits needed to scatter n robots is $O(n \log(n))$ if and only if $k_\mathcal{A}$ satisfies:*

$$\exists K \in \mathbb{N}, \quad k_\mathcal{A}(C, P) = O(|C|^K) \tag{5}$$

We observe that if $k_\mathcal{A}$ does not satisfy equation (5) then the number of random bits cannot be bounded by $O(n \log(n))$. Conversely, if $k_\mathcal{A}$ satisfies equation (5), the sketch of the proof is as follows. We use mathematical induction over the global number N of robots ($N = |C|$) and the local number n of robots located at position P. We show that there exist R and R' such that:

$$W(n, N) \leq nR \log(n) + nR' \log(N) \tag{6}$$

Then for a configuration where all n robots are gathered, we have $W(n, n) \leq n(R + R') \log(n) = O(n \log(n))$. Lemma 4 is used for the base case. Indeed, for all $\mathcal{N} \in \mathbb{N}$, we can assign a value to R' in order to make Equation (6) true for all $n \leq \mathcal{N}$ and for all $N \geq n$. Then, Lemma 3 is used in the inductive step. When n robots are randomly and uniformly distributed among k possible destinations, there is a high probability that the distribution is almost fair, and a low probability that a large number of robots moves toward the same destination. Lemma 3 indicates how fair the distribution can be with probability greater than $1/2$. Overall, two rounds (in expectation) are sufficient to have this "almost fair" distribution. We then use the inductive hypothesis with the new points.

4 Time Complexity Without Strong Multiplicity Detection

In this section, we investigate the time complexity (that is, the expected scattering time) of scattering algorithms that do not use strong multiplicity detection. We already know that global strong multiplicity detection (that permit to compute the number n of robots) enables $O(1)$ expected scattering time (see the algorithm of Clement *et al.* in the previous section). That bound obviously still holds if only *local* strong multiplicity detection is available (their scattering algorithm is canonical, so different multiplicity points are independent). There remains the case of weaker forms of multiplicity detection (that is, local and global weak multiplicity, or no multiplicity detection whatsoever). We essentially show that with respect to time complexity, weak multiplicity detection does not help. Without strong multiplicity detection, we show that: for any algorithm, the optimal expected $O(1)$ cannot be achieved; for random bit optimal algorithms, at least $\Omega(\log\log n)$ expected rounds are necessary. On the positive side, we present a family of scattering algorithms that do *not* use multiplicity detection yet can achieve arbitrarily fast (yet not constant) expected time. Of particular interest in this family is a scattering algorithm that is both random bit optimal scattering protocol and scatters n robots in $O(\log\log n)$ expected rounds.

Theorem 4. *There exists no scattering algorithm with $O(1)$ expected rounds complexity that uses only global weak multiplicity detection.*

Proof. Suppose that there exists $E \in \mathbb{N}$, such that for every $n \in \mathbb{N}$, the expected number of rounds needed by $\mathcal{A}$ to scatter n robots is less than E. Let $u \in \mathbb{N}$, and $\mathscr{P}$ be a set of u points. Consider the equivalence relation $\backsim$ over the set of configurations C that satisfy $U(C) \subset \mathscr{P}$ such that $C \backsim C'$, if C and C' cannot be distinguished with only the weak multiplicity detection. There is a finite number of equivalence classes, so the image of $k_{\mathcal{A}}$ is finite. So, after E rounds there is a maximum number of points where robots can lie, and if n is greater than that number, no n-robots configuration can be scattered in E rounds. A contradiction.   □

The following lemmas are used in Algorithm 2 and in Theorem 5.

Lemma 5. *Let X be the random variable that gives the distribution of m robots among k destinations (uniformly at random). There exists $\mathcal{N}$, such that for all $m > \mathcal{N}$ and $k \leq 8m^3$:*

$$\mathbb{P}(\|X\|_\infty > \frac{m}{k}(1 + k^{\frac{3}{4}})) \leq \frac{1}{2k}$$

Lemma 6. *Let P a point where lie m robots. For all $u \in \mathbb{N}$, $x \in \mathbb{N}$ we have: after a random and uniform distribution of robots at P among $k = \max(16x^4; u^3; 8\mathcal{N}^3)$ possible destinations, robots are divided into points of multiplicity 1 or less than m/x with probability at least $1 - \frac{1}{2u}$.*

Lemma 7. *Let C be a configuration with n robots organized in u points of multiplicity at most m (i.e., $U(C) = \{P_1, P_2, \ldots, P_u\}$). Let $x \in \mathbb{N}$. If there exists u disjoint sets of $k = \max(16x^4; u^3; 8\mathcal{N}^3)$ points $D_1, D_2, \ldots, D_u$, such that all robots in P_i are randomly distributed among points in D_i. Then the maximum multiplicity of the resulting configuration is 1 or less than m/x with probability at least $\frac{1}{2}$.*

Proof. Let $U(C) = \{P_1, P_2, \ldots, P_u\}$. We define the indicator random variable Z_i as follows: $Z_i = 1$ if all robots located at the same point P_i are located after one round on points of multiplicity either 1 or less than $\dfrac{m}{x}$. $Z_i = 0$ otherwise. Notice that $\{Z_1, Z_2, \ldots, Z_u\}$ are mutually independent because the destinations sets D_i are disjoint i.e. no two robots from different points ever reach the same position.

Since for all i, all robots at P_i are randomly distributed among k possible destinations, by Lemma 6 we have:

$$\mathbb{P}(Z_i = 1) \geq 1 - \frac{1}{2u}$$

$$\mathbb{P}(\bigwedge_{i=1}^{u} Z_i = 1) \geq \left(1 - \frac{1}{2u}\right)^u \geq 1 - \frac{u}{2u} = \frac{1}{2}$$

$\square$

Let $\mathcal{F} = \{f : \mathbb{N} \mapsto \mathbb{N} \mid f \text{ is increasing and surjective}\}$. Let $f \in \mathcal{F}$. We define f^{-1} as the maximum of the inverse function: *i.e.* $f^{-1}(y) = \max\{x \; ; \; f(x) = y\}$. Since $f \in \mathcal{F}$, f is not bounded and $f^{-1} : \mathbb{N} \mapsto \mathbb{N}$ is well defined, increasing and diverging. Moreover we have $f^{-1}(1) > 0$.

Given a function $f \in \mathcal{F}$, we now define Algorithm $\mathcal{SA}_f$ (see Algorithm 4) that converges in $O(f(n))$ rounds in expectation (see Theorem 5).

Algorithm 2: $\mathcal{SA}_f$: Scattering algorithm executed by robot r. No multiplicity detection

---

1 Compute the Voronoï diagram of the observed configuration
2 Let $u = |U(C)|$ and $x = f^{-1}(f(u) + 1)$
3 Let $k = \max(8\mathcal{N}^3, 16x^4, u^3)$ with $\mathcal{N}$ given by Lemma 5
4 Let *Pos* be a set of k distinct positions in the Voronoï cell where r is located
5 Move toward a position in *Pos* chosen uniformly at random.

$\mathcal{SA}_f$ is a canonical scattering algorithm under the FSYNC and SSYNC models. To construct the set of possible destinations, it executes the procedure given by the previous lemma with $x = f^{-1}(f(u) + 1)$ where $u = |U(C)|$. Thus, if m is the maximum multiplicity of a given configuration, then after one execution of $\mathcal{SA}_f$, the maximum multiplicity is $max(1, m/f^{-1}(f(u) + 1))$ with probability at least $1/2$.

Theorem 5. *Let $f \in \mathcal{F}$. $\mathcal{SA}_f$ is an canonical scattering algorithm, which scatters n robots in $O(f(n))$ rounds in expectation.*

The sketch of the proof is as follows. We first show that after $2i$ rounds in expectation, the maximum multiplicity of every point is less than $n/f^{-1}(i)$. Indeed we use Lemma 7 with $x = f^{-1}(f(n) + 1)$. So that the expected number of rounds of an execution is less than $2f(n)$.

Theorem 6. *Let $\mathcal{A}$ be a canonical scattering algorithm such that at each activation, the number of possible destinations computed by each robot is the same i.e. for every configuration C that contains two points P and P', $k_{\mathcal{A}}(C, P) = k_{\mathcal{A}}(C, P')$. Let $B(n)$ be the maximum number of random bits used by a robot among all n-robots configuration. Then the random bit complexity of $\mathcal{A}$ is $\Theta(n \log(n) + nB(n))$.*

A direct consequence of Theorem 6 is the following:

Theorem 7. *Let $f \in \mathcal{F}$. $\mathcal{SA}_f$ uses $\Theta(n \log(f^{-1}(f(n) + 1)))$ random bits in expectation.*

Note that the hypothesis that $f \in \mathcal{F}$ is not very restrictive. Indeed, if we want our algorithm to converge in $O(g(n))$ with g a function that may not be increasing nor surjective but such that $\lim_{n \to +\infty} g(n) = +\infty$. We can define f by : $f(0) = 0$ and $\forall x > 0$, $f(x) = \min\left(\max\left(g(x), f(x - 1)\right), f(x - 1) + 1\right)$. So that $f \in \mathcal{F}$ and $O(f(n)) \subset O(g(n))$.

Now, our algorithm converges as fast as we want. We can try it with some convenient functions. For example, with $f = \log^*$, the algorithm $\mathcal{SA}_{\log^*}$ converges in $O(\log^*(n))$ rounds in expectation. Moreover, since[2] $\log(f^{-1}(f(n) + 1)) = \log\left(^{(\log^*(n)+1)}2\right) =^{\log^* n} 2 = n$, the resulting algorithm uses $O(n^2)$ random bits in expectation. A faster algorithm can be obtained using the inverse Ackermann function A^{-1} such that the time complexity of $\mathcal{SA}_{A^{-1}}$ is in $O(A^{-1}(n)) = o(log^*log^*log^*log^*n)$.

A Random Bit Optimal Algorithm. If we want our algorithm $\mathcal{SA}_f$ to be random bit optimal, f must satisfy: $n \log(f^{-1}(f(n) + 1)) = O(n \log(n))$.

With $f = \log \circ \log$, we have:

$$n \log(f^{-1}(f(n) + 1)) = n \log 2^{2^{\log \log(n)+1}} = 2n \log(n) = O(n \log(n))$$

So that $\mathcal{SA}_{\log \circ \log}$ is random bits optimal and converge in $O(\log \log n)$ rounds in expectation. Also, the following theorem makes $\mathcal{SA}_{\log \circ \log}$ optimal in time.

Theorem 8. *There exists no random bit optimal scattering algorithm with $o(\log(\log(n)))$ expected rounds complexity that uses only global weak multiplicity detection.*

Corollary 2. *$\mathcal{SA}_{\log \circ \log}$ is optimal for both time and random bit complexity.*

The following table summarizes the dependency between time complexity and multiplicity detection.

---

[2] we use the tetration notation : $^{n}a = \underbrace{a^{a^{\cdot^{\cdot^{\cdot^{a}}}}}}_{n}$ i.e. a exponentiated by itself, n times.

Multiplicity detection	Optimal time complexity	Optimal time complexity for random bit optimal algorithm
Strong global or local	$O(1)$	$O(1)$
Weak global or local	$\forall f,\ O(f(n))$	$O(\log\log(n))$
No multiplicity detection	$\forall f,\ O(f(n))$	$O(\log\log(n))$

5   Concluding Remarks

We investigated the random bit complexity of mobile robot scattering and gave necessary and sufficient conditions for both (expected) random bit complexity and time complexity. It turns out that multiplicity detection plays an important role in the expected time complexity ($O(1)$ expected time can be achieved with strong multiplicity detection, while $\Theta(\log\log n)$ expected time complexity is optimal in the case of weak or no multiplicity detection) for the class of random bit optimal algorithms.

We also found out that without strong multiplicity detection, even if the time complexity $O(1)$ is not reachable, there exist scattering algorithms that converge as fast as needed (yet not in expected constant time). Indeed our algorithms can have a time complexity of $O(f(n))$, for every increasing and surjective f, and expect $\Theta(n\log(f^{-1}(f(n)+1)))$ random bits in return.

An interesting remaining open question would be to prove whether our algorithms can be extended to the ASYNC model.

References

1. Anagnostou, E., El-Yaniv, R.: More on the power of random walks: uniform self-stabilizing randomized algorithms. In: Toueg, S., Spirakis, P., Kirousis, L. (eds.) Distributed Algorithms. LNCS, vol. 579, pp. 31–51. Springer, Heidelberg (1992). http://dx.doi.org/10.1007/BFb0022436
2. Aurenhammer, F.: Voronoi diagrams—a survey of a fundamental geometric data structure. ACM Comput. Surv. (CSUR) **23**(3), 345–405 (1991)
3. Bramas, Q., Tixeuil, S.: The random bit complexity of mobile robots scattering. CoRR abs/1309.6603 (2015). http://arxiv.org/abs/1309.6603
4. Clément, J., Défago, X., Potop-Butucaru, M.G., Izumi, T., Messika, S.: The cost of probabilistic agreement in oblivious robot networks. Inf. Process. Lett. **110**(11), 431–438 (2010)
5. Dieudonné, Y., Petit, F.: Robots and demons (the code of the origins). In: Crescenzi, P., Prencipe, G., Pucci, G. (eds.) FUN 2007. LNCS, vol. 4475, pp. 108–119. Springer, Heidelberg (2007)
6. Dieudonné, Y., Petit, F.: Scatter of robots. Parallel Process. Lett. **19**(1), 175–184 (2009)
7. Flocchini, P., Prencipe, G., Santoro, N.: Distributed Computing by Oblivious Mobile Robots. Synthesis Lectures on Distributed Computing Theory. Morgan & Claypool Publishers (2012)

8. Izumi, T., Potop-Butucaru, M.G., Tixeuil, S.: Connectivity-preserving scattering of mobile robots with limited visibility. In: Dolev, S., Cobb, J., Fischer, M., Yung, M. (eds.) SSS 2010. LNCS, vol. 6366, pp. 319–331. Springer, Heidelberg (2010)
9. Raab, M., Steger, A.: "Balls into bins" a simple and tight analysis. In: Luby, M., Rolim, J., Serna, M. (eds.) Randomization and Approximation Techniques in Computer Science. LNCS, vol. 1518, pp. 159–170. Springer, Heidelberg (1998). http://dx.doi.org/10.1007/3-540-49543-6_13
10. Suzuki, I., Yamashita, M.: Distributed anonymous mobile robots: formation of geometric patterns. SIAM J. Comput. **28**(4), 1347–1363 (1999)

On the Relations Between SINR Diagrams
and Voronoi Diagrams

Merav Parter[(✉)] and David Peleg

The Weizmann Institute of Science, Rehovot, Israel
{merav.parter,david.peleg}@weizmann.ac.il

Abstract. In this review, we illustrate the relations between wireless communication and computational geometry. As a concrete example, we consider a fundamental geometric object from each field: SINR diagrams and Voronoi diagrams. We discuss the relations between these representations, which appear in several distinct settings of wireless communication, as well as some algorithmic applications.

1 Introduction

Wireless networks are embedded in our daily lives, with an ever-growing use of cellular, satellite and sensor networks. Subsequently, the capacity of wireless networks, i.e., the maximum *achievable rate* by which stations can communicate reliably, has received an increasing attention in recent years [2,7,14–16,18,20]. The great advantage of wireless communication, namely, the broadcast nature of the medium, also creates its biggest obstacle – interference. When a receiver has to decode a message (i.e., a signal) sent from a transmitter, it must cope with all other (legitimate) simultaneous transmissions by neighboring stations.

While the physical properties of channels have been thoroughly studied [13,22], less is known about the topology and geometry of the wireless network structure and their influence on performance. This review concerns a novel approach recently proposed to describe the behavior of multi-station networks, which is based on building a *reception map* according to the *signal-to-interference & noise ratio (SINR)* model. By now, the SINR model is the most commonly studied abstract physical model for wireless communication networks, widely used by both the Electrical Engineering community and the algorithmic computer science community. This physical model aims at gauging the quality of signal reception at the receivers while faithfully representing phenomena such as attenuation and interference. Specifically, in this model, the signal decays as it travels and a transmission is successful if its strength at the receiver exceeds the accumulated signal strength of interfering transmissions by a sufficient (technology determined) factor.

Supported in part by the Israel Science Foundation (grant 894/09) and the Israel Ministry of Science and Technology (infrastructures grant).

S. Papavassiliou and S. Ruehrup (Eds.): ADHOC-NOW 2015, LNCS 9143, pp. 225–237, 2015.
DOI:10.1007/978-3-319-19662-6_16

The SINR model gives rise to a natural geometric object, the *SINR* diagram, which partitions the plane into a reception region $H(s_i)$ per station $s_i \in S$ and the remaining area $H(\emptyset)$ where none of the stations are heard. Each of these $n+1$ regions may possibly be composed of several disconnected regions.

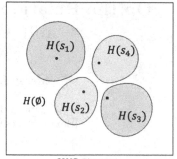

SINR Diagram

SINR diagrams have been recently studied from topological and geometric standpoints [5,6,17], and they appear to provide improved understanding on the behavior of wireless networks. Specifically, these diagram have been shown to play a central role in the development of several approximation and online algorithms (e.g., point location tasks, map drawing). Such a role is analogous perhaps to the role played by *Voronoi diagrams* in the study of proximity queries and other algorithms in computational geometry.

The ordinary Voronoi diagram on a given set of points S tessellates the space in such a way that every location is assigned to the closest point in S, thus partitioning the space into *regions* $\mathrm{VOR}(s_i)$, each consisting of the set of locations closest to one point in S (referred to as the region's *generator*).

In this review, we focus on the analogy between these two space partitions, the SINR diagrams and the Voronoi diagrams. At first glance, the connection is not immediate. SINR diagrams are based upon the physics of the wireless communication, and geometry is only one aspect of it. In contrast, Voronoi diagrams are mainly based on the geometry of the point arrangement. In addition, whereas Voronoi diagrams are mostly based on the pairwise relations between the generators (i.e., the points of S, referred to as stations in the wireless terminology), in an SINR

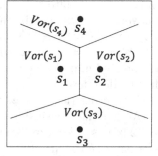

diagram the reception region of s_i is determined by a complex relation to all other stations and cannot be represented in general as a collection of pairwise relations.

Despite these distinctions, the connections between these diagrams appear to be persistent and re-occur in several *distinct* settings of wireless communications (in fact, in any setting that has been studied so-far). We also exemplify several algorithmic and theoretical applications of these relations.

Let us remark that this review restricts attention to the *static* setting (i.e., where the locations of the network stations are fixed). Turning to the stochastic setting, the relations between stochastic SINR diagram (formed by modeling the SINR as a marked point process) and classical stochastic geometry models such as Poisson–Voronoi tessellations, have been studied extensively, and are out of the scope of this review. See [8] for a detailed analysis, results and applications of this approach.

The structure of this review is as follows. In Sect. 2 we provide a brief overview of SINR diagrams. We then describe the connections between SINR diagrams and Voronoi diagrams in three main settings. Section 3 considers the *uniform power* setting, when all stations have the same transmission power. In Sect. 4, we consider the general setting of *non-uniform power* (i.e., when the transmission powers are arbitrary). Finally, in Sect. 5, we describe a setting in which receivers are allowed to employ a decoding technique known as *interference cancellation* to improve their reception quality. In each of these settings, the resulting SINR diagrams turn out to correspond to some variant of Voronoi diagrams, as elaborated on in what follows.

2   Wireless Networks and SINR

We consider a wireless network $\mathcal{A} = \langle d, S, \psi, N, \beta, \alpha \rangle$, where $d \in \mathbb{Z}_{\geq 1}$ is the dimension, $S = \{s_1, s_2, \ldots, s_n\}$ is a set of transmitting *radio stations* embedded in the d-dimensional space, ψ is an assignment of a positive real *transmitting power* ψ_i to each station s_i, $N \geq 0$ is the *background noise*, $\beta \geq 1$ is a constant that serves as the *reception threshold*, and $\alpha > 0$ is the *path-loss parameter*. The *signal to interference & noise ratio (SINR)* of s_i at point p is defined as

$$\text{SINR}_{\mathcal{A}}(s_i, p) = \frac{\psi_i \cdot \text{dist}(s_i, p)^{-\alpha}}{\sum_{j \neq i} \psi_j \cdot \text{dist}(s_j, p)^{-\alpha} + N} \tag{1}$$

The fundamental rule of the SINR model is that the transmission of station s_i is received correctly at point $p \notin S$ if and only if its SINR at p reaches or exceeds the reception threshold of the network, i.e., $\text{SINR}_{\mathcal{A}}(s_i, p) \geq \beta$. When this happens, we say that s_i is *heard* at p.

We refer to the set of points that hear station s_i as the *reception region* of s_i, defined as

$$H(s_i, \mathcal{A}) = \{p \in \mathbb{R}^d - S \mid \text{SINR}_{\mathcal{A}}(s_i, p) \geq \beta\} \cup \{s_i\} . \tag{2}$$

(Note that $\text{SINR}(s_i, \cdot)$ is undefined at points in S and in particular at s_i itself.) Analogously, the set of points that hear no station $s_i \in S$ (due to the background noise and interference), the *null region*, is defined as

$$H(\emptyset, \mathcal{A}) = \{p \in \mathbb{R}^d - S \mid \text{SINR}(s_i, p) < \beta, \ \forall s_i \in S\}.$$

An SINR diagram

$$\mathcal{H}(\mathcal{A}) = \left(\bigcup_{s_i \in S} H(s_i, \mathcal{A}) \right) \cup H(\emptyset, \mathcal{A})$$

is a "reception map" characterizing the reception regions of the stations. When the network $\mathcal{A}$ is clear from the context, we may omit it, and simply write $\text{SINR}(s_i, p), H(s_i)$ and $H(\emptyset)$.

3 Uniform SINR Diagram and Voronoi Diagram

The study of SINR diagram has been initiated by Avin et al. in [6] for the relatively simple case where all stations use the same transmission power, a.k.a., *uniform power* (i.e., $\psi_i = 1$ for every station s_i in Eq. (1)). It has been shown that under this setting, the SINR diagram assumes a rather convenient form. In particular, for SINR threshold $\beta \geq 1$, it holds that every reception region $H(s_i)$ is convex and fat (see Fig. 1(a) for schematic illustration of these notions).

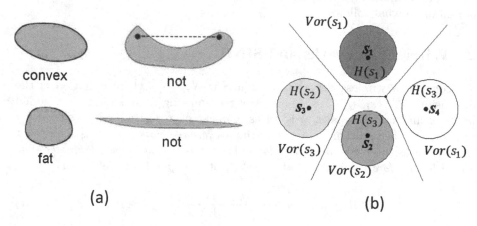

(a)   (b)

Fig. 1. (a) Uniform SINR regions are "nice": convex and fat. (b) Illustration of the relations between uniform SINR diagram and Voronoi diagram. The reception region $H(s_i)$ is fully contained in the corresponding Voronoi region $\text{VOR}(s_i)$.

Let $\text{VOR}(s_i)$ be the Voronoi region of station s_i, defined by

$$\text{VOR}(s_i) = \{p \in \mathbb{R}^d \mid \text{dist}(s_i, p) \leq \text{dist}(s_j, p), \text{ for any } j \neq i\}. \quad (3)$$

Since all transmission powers are the same, for a receiver p that successfully receives the transmission of s_i, it must hold that s_i is the closest station to p among all other network stations, i.e., that $p \in \text{VOR}(s_i)$. This condition is not sufficient for successful reception, and hence uniform SINR diagrams can be considered as a *refinement* of Voronoi diagrams. The refinement stems from that fact that the SINR model (even in the uniform case) takes into consideration not only the geometry but also other physical parameters such as attenuation and fading of signals.

The following lemma from [6] formalizes this intuition by claiming that the reception region $\mathcal{H}(s_i, \mathcal{A})$ is strictly contained in the corresponding Voronoi region $\text{VOR}(s_i)$.

Lemma 1 (Uniform SINR Diagram and Voronoi Diagram, [6]). *For uniform network $\mathcal{A} = \langle d, S, \bar{1}, N, \beta \geq 1, \alpha \rangle$, it holds that $H(s_i, \mathcal{A}) \subseteq \text{VOR}(s_i)$ for every $s_i \in S$.*

See Fig. 1(b). In fact, this analogy between the two diagrams becomes stronger when the path-loss parameter α tends to infinity and there is no ambient noise $N = 0$. The case of noisy SINR diagram with $\alpha \to \infty$ and $N > 0$ is more involved and can be shown to converge to alpha shapes [11].

Note that when $\beta < 1$, the inclusion between the SINR diagram and the Voronoi diagram may no longer hold. That is, it is possible to identify points $p \in \mathbb{R}^d$, for which $\text{SINR}(s_i, p) \geq \beta$ while $p \notin \text{VOR}(s_i)$.

Algorithmic application: Point Location. In the *Point Location* task, one is given an n-station wireless network $\mathcal{A}$ and a query point p referred to as a point-location query. The goal is to identify which of the stations is heard at p, if any. As it is assumed that the reception threshold satisfies $\beta > 1$, if there is noise $N \geq 0$, then at most one station can be heard at p. The trivial procedure for answering the query is to evaluate the SINR function $\text{SINR}(s_i, p)$ at each of the stations $s_i \in S$, which takes linear time in the number of stations n. To facilitate multiple queries, one may want to build a data structure that can guarantee faster response.

By exploiting the fatness and convexity of the reception regions as well as the relation to Voronoi diagrams, [6] proposed an *approximate* point location scheme that answers point location queries in $O(\log n)$ time. This scheme consists of a preprocessing step in which the Voronoi diagram and "approximated" reception SINR regions $\widetilde{H}(s_i)$ for every $s_i \in S$ are constructed. Answering a point-location query p then involves two main steps. First, the sole candidate station s_p that may be heard at p is identified by finding the Voronoi cell to which p belongs (this can be done in logarithmic time). The second step then uses the approximated region $\widetilde{H}(s_p)$ (constructed in the preprocessing step) to decide whether $\text{SINR}(s_p, p) \geq \beta$. For an efficient batched point location schemes, see the recent [3].

4   Non-uniform SINR Diagram and Weighted Voronoi Diagram

In many actual wireless communication systems, wireless communication devices can modify their transmission power. Moreover, it has been demonstrated convincingly that allowing transmitters to use different power levels increases the efficiency of various communication patterns. It is therefore important to study the topology of SINR diagrams with non-uniform power (i.e., when the transmission powers are arbitrary). This general setting has been considered in [17]. The first observation of [17] is the unfortunate fact that non-uniform diagrams are more complicated than in the uniform case. In particular, the reception regions are no longer convex or even connected. That is, a reception region may consist of several disconnected "reception islands".

For example, the reception region of s_1 in the figure consists of two disconnected components. The loss of the nice properties established for the uniform setting motivated the definitions of alternative notions of weaker convexity.

Another important research direction involves bounding the maximum number of connected components, that an n-station wireless network may assume.

Several important properties of SINR diagrams were established in [17]. One of the key results demonstrates that the reception regions in $\mathbb{R}^{d+1}$ (i.e., drawing the SINR diagram in one dimension higher than that in which the stations are embedded) are *hyperbolically convex*; see Fig. 2. Hence, although the d-dimensional map might be highly fractured, drawing the map in one dimension higher "heals" the regions, which become (hyperbolically) connected.

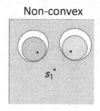

Non-convex

Disconnected

In the context of Voronoi diagrams, in contrast to the uniform case, the reception regions of a non-uniform SINR diagram are not necessarily contained in the corresponding Voronoi regions. Clearly, the reception region of a very strong station (i.e., with sufficiently large transmission power) may exceed its corresponding Voronoi region. The asymmetry that arises by non-uniform power assignments calls for a *weighted* variant of Voronoi diagram. Specifically, [17] showed that SINR diagrams with *non-uniform* powers are related to *multiplicatively-weighted Voronoi* diagrams (see [4]).

In the (multiplicatively) weighted version of Voronoi diagram [4], every generator (i.e., station) s_i is given a weight w_i that expresses the capability of s_i to influence its neighborhood. Formally, the weighted system $V = \langle S, W \rangle$ con-

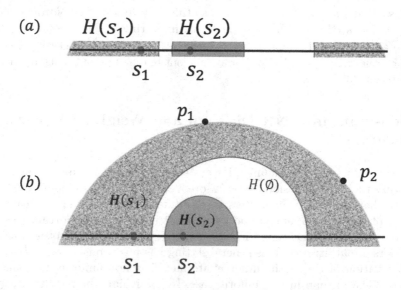

Fig. 2. Reception regions are hyperbolic-convex in $\mathbb{R}^{d+1}$. The given network consists of two stations s_1 and s_2 aligned on a line (i.e., one-dimensional network). (a) The reception region $H(s_1)$ of s_1 is not connected in $\mathbb{R}^1$. (b) The reception region of s_1 is connected in $\mathbb{R}^2$. The hyperbolic line connecting two reception points p_1 and p_2 is fully contained in the 2-dimensional reception region $H(s_1)$.

sists of $S = \{s_1, ..., s_n\}$, which represents a set of n points or generators in d-dimensional Euclidean space, and $W = \{w_1, ..., w_n\}$, which is an assignment of weights $w_i \in \mathbb{R}_{>0}$ to each point $s_i \in S$. The *weighted voronoi diagram* of $V = \langle S, W \rangle$ partitions the space into n regions, where

$$\text{WVOR}(s_i) = \left\{ p \in \mathbb{R}^d \mid \frac{w_i}{\text{dist}(s_i, p)} \geq \frac{w_j}{\text{dist}(s_j, p)}, \text{ for any } j \neq i \right\} \quad (4)$$

denotes the region of influence of the generator s_i in S, for every $i \in \{1, \ldots, n\}$. Note that when all the weights are the same, Eq. (4) is equivalent to Eq. (3), i.e., the weighted Voronoi diagram is the same as the ordinary Voronoi diagram. Unlike the ordinary Voronoi diagram, the weighted Voronoi region $\text{WVOR}(s_i)$ is *not necessarily connected* and the diagram may consist of $\Omega(n^2)$ components.

Given a non-uniform wireless network $\mathcal{A}$ with transmission powers $\psi_1, \ldots, \psi_n$, the corresponding weighted Voronoi system is given by setting the weight of each station $s_i \in S$ to

$$w_i = \psi_i^{1/\alpha}, \quad (5)$$

where α is the path-loss parameter of the wireless network. This weight adjustment yields the next lemma, the analogue of Lemma 2 for the non-uniform case.

Lemma 2 (non-uniform SINR Diagram and Weighted Voronoi Diagram, [17]). *Let $\mathcal{A}$ be a non-uniform network. Then $H(s_i, \mathcal{A}) \subseteq \text{WVOR}(s_i)$ for every $s_i \in S$, where the weights of the weighted Voronoi diagram are set according to Eq. (5) (see Fig. 3).*

Note that the weight assignment of Eq. (5) tends to 1 as the path-loss parameter α tends to infinity. Hence, the non-uniform SINR diagram converges to the ordinary (non-weighted) Voronoi diagram as α tends to infinity. (Intuitively, as α gets larger, the distances dominate the effect of the transmission powers.)

Similarly to the uniform case, the relation to weighted Voronoi diagram is found to be useful for solving point location queries efficiently. We conclude this section by providing an example in which weighted Voronoi diagrams are *not* helpful, due to some fundamental differences between the two models.

Example illustrating the gap between SINR and Voronoi diagram. Establishing a tight universal bound for the number of connected components in the SINR diagram of an n-station network is one of the main open challenges in the study of non-uniform SINR networks. A seemingly promising approach to studying this question is considering it on the corresponding weighted Voronoi diagrams. Since any non-uniform SINR region is fully contained in a corresponding weighted Voronoi region, it seems plausible that the number of weighted Voronoi cells (bounded by $O(n^2)$ [4]) might upper bound the number of connected components in the corresponding SINR diagram. Unfortunately, this does not hold in general, since it might be the case that a single weighted Voronoi component contains a disconnected SINR region. This scenario is illustrated in the figure below and emphasizes a fundamental gap between the SINR and Voronoi diagrams.

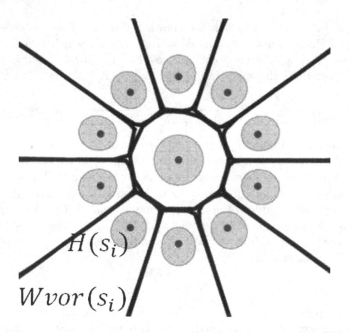

Fig. 3. The reception regions of the non-uniform SINR diagram are strictly contained in the corresponding weighted Voronoi regions upon setting the weights according to Eq. (5).

The first figure depicts a non-uniform network $\mathcal{A}$ with stations $S = \{s_1, \ldots, s_n\}$ and transmission powers $\psi = \{\psi_1, \ldots, \psi_n\}$. The powers are set such that the reception region of s_1 is not connected (the two SINR reception islands of s_1 are represented by the two white circles with dashed boundaries). However, it is connected when restricting it to a given connected component of the weighted Voronoi region of s_1. The network $\mathcal{A}$ is then transformed into another network $\mathcal{A}'$ (shown in the bottom figure) by replacing each interfering station $s_i \in S \backslash \{s_1\}$ by sufficiently many copies of interfering stations $S_i = \{s_i^1, \ldots, s_i^m\}$ co-located at the location of s_i, each transmitting with power $\psi_i' = \psi_i/m$. Hence, by Eq. (5), all the interfering stations have a

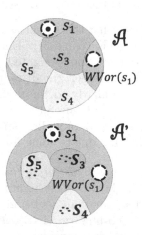

very small weight in the corresponding weighted Voronoi diagram. Since the total interference experienced by any point $p \in H(s_1, \mathcal{A})$ is the same in both networks, the reception region of s_1 is preserved (i.e., $H(s_1, \mathcal{A}) = H(s_1, \mathcal{A}')$). By taking m to be sufficiently large, the influence of s_1 is increased tremendously in the corresponding weighted Voronoi diagram for network $\mathcal{A}'$. This results in a very large connected weighted Voronoi region for s_1, which contains the two disconnected SINR reception components of s_1.

5 SINR Diagram with Interference Cancellation and High Order Voronoi Diagram

Finally, we turn to consider the third setting for studying wireless networks in which the reception points are allowed to employ a decoding technique, known as *Interference cancellation* (IC). IC is a relatively recent and promising method for efficient decoding [1]. The basic idea of interference cancellation, and in particular *successive interference cancellation (SIC)*, is quite simple. First, the strongest interfering signal is detected and decoded. Once decoded, this signal can then be subtracted ("cancelled") from the original signal. Subsequently, the next strongest interfering signal can be detected and decoded from the now "cleaner" signal, and so on. Optimally, this process continues until all interferences are cancelled and we are left with the desired transmitted signal, which can now be decoded. It should be noted that without using IC, every station can decode at most one transmission (i.e., the strongest signal it receives). In contrast, with IC, every station can decode more transmissions, or expressed dually, every transmitter can reach more receivers. This clearly increases the utilization of the network. Interference cancellation is fairly well-studied from an information-theoretic point of view [9,12,21,23].

Recently, [5] studied the reception regions of a wireless network in the SINR model with receivers that employ SIC. We next formally define the SIC-SINR diagrams, then define a generalization of Voronoi diagram known as, *high-order Voronoi diagram*, and finally describe the connection between these diagrams, established in [5].

SIC-SINR Diagrams in Uniform Power Networks. SIC changes the basic criterion for a successful reception, and hence it calls for new definitions of the reception regions that form the *SIC-SINR Diagrams*.

Let $\mathcal{A} = \langle d, S, \psi = \bar{1}, N, \beta > 1, \alpha \rangle$ be an n-station uniform power wireless network. For a subset $S' \subseteq S$, let $\mathcal{A}(S') = \langle d, S', \psi = \bar{1}, N, \beta > 1, \alpha \rangle$ be the network induced on a subset of stations S'. Let $\overrightarrow{S_i} = \{s_{i_1}, \ldots, s_{i_k}\} \subseteq S$ be an ordering of k stations.

We begin by defining the reception area $H(\overrightarrow{S_i})$ of all points that receive s_{i_k} correctly after successive cancellation of $s_{i_1}, \ldots, s_{i_{k-1}}$. For every $j \in \{2, \ldots, k\}$, let $S_{i,j} = S \setminus \{s_{i_1}, \ldots, s_{i_{j-1}}\}$ be the subset of stations excluding the first $j - 1$ stations in the ordering $\overrightarrow{S_i}$ and let $S_{1,j} = S$. The reception region $H(\overrightarrow{S_i})$ is defined by

$$H(\overrightarrow{S_i}) = \bigcap_{j=1}^{k} H(s_{i_j}, \mathcal{A}(S_{i,j})) \, , \tag{6}$$

where $H(s_{i_j}, \mathcal{A}(S_{i,j}))$ is given by Eq. (2). The reception region $H^{SIC}(s_1)$ consists of all points that can receive the transmission of s_1 by employing interference cancellation. Hence, it is the union of all $H(\overrightarrow{S_i})$ regions for which that last station

$\mathrm{Last}(\vec{S_i})$ is s_1.

$$H^{SIC}(s_1) = \bigcup_{\vec{S_i} \mid \mathrm{Last}(\vec{S_i})=s_1} H(\vec{S_i}) . \tag{7}$$

The fundamental result of [5] states that although potentially there are exponentially many possible cancellation orderings $\vec{S_i}$ with $\mathrm{Last}(\vec{S_i}) = 1$, and as a result, disconnected reception components in $H^{SIC}(s_1)$, in fact there are only $O(n^{2d})$ orderings $\vec{S_i}$, $\mathrm{Last}(\vec{S_i}) = s_1$ with nonempty reception regions $H(\vec{S_i})$. The final SIC-SINR diagram consists of n reception regions $H^{SIC}(s_1), \ldots, H^{SIC}(s_n)$ and the complementary region, the null region $H^{SIC}(\emptyset)$ in which none of the stations can be received, despite the ability to employ SIC.

SIC-SINR diagrams are related to higher order Voronoi diagrams, a natural extension of the ordinary Voronoi, briefly defined next.

Higher Order Voronoi Diagrams. In higher order Voronoi diagrams the cells are generated by more than one generator. Such diagrams provide tessellations where each region consists of the collection of points having the same k (ordered or unordered) closest points in S, for some given integer k. The are two variants of high order Voronoi diagrams: ordered (in which the order of the generator set matters) and non-ordered. SIC-SINR diagrams are related to the former.

Ordered Order-k Voronoi Diagram. Let $\vec{S_i} \subseteq S$ be an ordered set of k elements from S. When the k generators are ordered, the diagram becomes the *ordered order-k Voronoi diagram* $\mathcal{V}^{\langle k \rangle}(S)$ [19], defined as

$$\mathcal{V}^{\langle k \rangle}(S) = \{\mathrm{VOR}(\vec{S_i})\},$$

where the ordered order-k Voronoi region $\mathrm{VOR}(\vec{S_i})$, $|\vec{S_i}| = k$, is defined as

$$\mathrm{VOR}(\vec{S_i}) = \{p \in \mathbb{R}^d \mid \mathrm{dist}(p, s_{i_1}) \leq \mathrm{dist}(p, s_{i_2}) \leq \ldots$$
$$\leq \mathrm{dist}(p, s_{i_k}) \leq \min(\mathrm{dist}(p, S \setminus S_i))\}.$$

Note that each $\mathrm{VOR}(\vec{S_i})$ is an intersection of k convex shapes and hence it is convex as well. Schematic illustration for $k = 2$ is provided in Fig. 4. For example, the region $\mathrm{VOR}(s_1, s_2)$ consists of all points whose nearest neighbor is s_1 and whose second nearest neighbor is s_2. Similarly, the region $\mathrm{VOR}(s_2, s_1)$ consists of all points whose nearest neighbor is s_2 and whose second nearest neighbor is s_1.

SIC-SINR Diagrams and Ordered Order-k Voronoi Diagrams. The relation between a nonempty reception region $H(\vec{S_i})$, $\vec{S_i} = \{s_{i_1}, \ldots, s_{i_k}\}$ and an nonempty ordered order-k polygon $\mathrm{VOR}(\vec{S_i})$ is given in the next lemma.

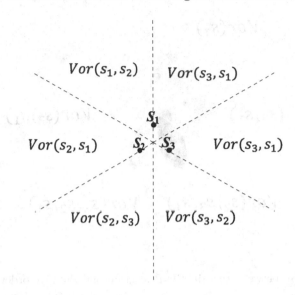

Fig. 4. A system of 3 points s_1, s_2, s_3 and their ordered order-2 Voronoi diagram.

Lemma 3 ([5]). $H(\vec{S_i}) \subseteq \text{VOR}(\vec{S_i})$, for $\beta \geq 1$.

This relation was used by [5] to provide the following characterization of SIC-SINR reception regions: every region $H^{SIC}(s_i)$ is composed of a collection of convex regions, each of which is contained in corresponding cell of the higher-order Voronoi diagram. Figure 5 illustrates this relation, and shows the reception region $H^{SIC}(s_1)$. The light grey region $H(s_1)$ corresponds to the ordinary reception region with no cancellation. The black region $H(s_2, s_3, s_1)$ consists of all points that first decoded s_2, cancelled it, then decoded s_3 and cancelled it, and finally were able to decode s_1. Each of these connected reception regions is fully contained in their corresponding high order Voronoi regions. For example, the black region $H(s_2, s_3, s_1)$ is contained in the high order Voronoi region $\text{VOR}(s_2, s_3, s_1)$ (i.e., the set of all points whose first nearest neighbor is s_2, then s_3 and finally s_1.) Finally, SIC-SINR diagrams are also shown to be related to *hyperplane arrangements* [10], a plane-tessellation formed by the intersection of the set of $\binom{n}{2}$ hyperplane of pairs in S.

Applications. The connections to both high-order Voronoi diagram and hyperplane arrangements play a key role in [5] where it used to provide to (1) the topological characterization of SIC-SINR regions and (2) algorithms for these diagrams. Specifically, these connections are used for establishing a bound of $O(n^{2d+1})$ on the number of components in d-dimensional n-station networks. In addition, they yield algorithmic applications for drawing and maintaining SIC-SINR diagram as well as for answering efficiently point-location queries.

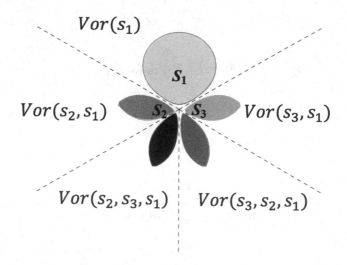

Fig. 5. The relation between the SIC-SINR diagram and the high-order Voronoi diagram.

6   Final Note: Towards Wireless Computational Geometry

A major long-term goal of the study of SINR diagrams is to develop the area of "wireless computational geometry" in which SINR diagrams play a role that is similar to that of Voronoi diagrams in computational geometry. Indeed, this review aimed at highlighting the intimate connections between these models which encourages further study of SINR diagrams in more realistic settings (e.g., adding obstacles, directional antennas, etc.) and their Voronoi counterparts.

References

1. Andrews, J.G.: Interference cancellation for cellular systems: a contemporary overview. IEEE Wirel. Commun. **12**(2), 19–29 (2005)
2. Andrews, M., Dinitz, M.: Maximizing capacity in arbitrary wireless networks in the SINR model: complexity and game theory. In: Proceedings of the INFOCOM (2009)
3. Aronov, B., Katz, M.J.: Batched point location in SINR diagrams via algebraic tools. CoRR, abs/1412.0962 (2014)
4. Aurenhammer, F., Edelsbrunner, H.: An optimal algorithm for constructing the weighted voronoi diagram in the plane. Pattern Recogn. **17**, 251–257 (1984)
5. Avin, C., Cohen, A., Haddad, Y., Kantor, E., Lotker, Z., Parter, M., Peleg, D.: SINR diagram with interference cancellation. In: Proceedings of the SODA, pp. 502–515 (2012)
6. Avin, C., Emek, Y., Kantor, E., Lotker, Z., Peleg, D., Roditty, L.: SINR diagrams: convexity and its applications in wireless networks. J. ACM **59**(4), 18:1–18:34 (2012)

7. Avin, C., Lotker, Z., Pignolet, Y.-A.: On the power of uniform power: capacity of wireless networks with bounded resources. In: Fiat, A., Sanders, P. (eds.) ESA 2009. LNCS, vol. 5757, pp. 373–384. Springer, Heidelberg (2009)

8. Baccelli, F., Blaszczyszyn, B.: Stochastic geometry and wireless networks volume 1: theory. Found. Trends Network. 3, 249–449 (2009)

9. Costa, M., El-Gamal, A.: The capacity region of the discrete memoryless interference channel with strong interference. IEEE Trans. Inf. Th. 33, 710–711 (1987)

10. Edelsbrunner, H.: Algorithms in Combinatorial Geometry. Springer, Heidelberg (1987)

11. Edelsbrunner, H., Kirkpatrick, D., Seidel, R.: On the shape of a set of points in the plane. IEEE Trans. Inf. Th. 29(4), 551–559 (1983)

12. Etkin, R.H., Tse, D.N.C., Wang, H.: Gaussian interference channel capacity to within one bit. IEEE Trans. Inf. Th. 54(12), 5534–5562 (2008)

13. Goldsmith, A.: Wireless Communications. Cambridge University Press, Cambridge (2005)

14. Goussevskaia, O., Wattenhofer, R., Halldórsson, M.M., Welzl, E.: Capacity of arbitrary wireless networks. In: Proceedings of the INFOCOM, pp. 1872–1880 (2009)

15. Gupta, P., Kumar, P.R.: The capacity of wireless networks. IEEE Trans. Inf. Th. 46(2), 388–404 (2000)

16. Halldórsson, M.M., Wattenhofer, R.: Wireless communication is in APX. In: Albers, S., Marchetti-Spaccamela, A., Matias, Y., Nikoletseas, S., Thomas, W. (eds.) ICALP 2009, Part I. LNCS, vol. 5555, pp. 525–536. Springer, Heidelberg (2009)

17 Kantor, E., Lotker, Z., Parter, M., Peleg, D.: The topology of wireless communication. In: Proceedings of the STOC (2011)

18. Moscibroda, T.: The worst-case capacity of wireless sensor networks. In: Proceedings of the IPSN, pp. 1–10 (2007)

19. Okabe, A., Boots, B., Sugihara, K., Chiu, S.N.: Spatial Tesselations. Princeton University Press, Princeton (1992)

20. Ozgur, A., Leveque, O., Tse, D.: Hierarchical cooperation achieves optimal capacity scaling in ad hoc networks. IEEE Trans. Inf. Th. 53, 3549–3572 (2007)

21. Sato, H.: The capacity of the gaussian interference channel under strong interference. IEEE Trans. Inf. Th. 27(6), 786–788 (1981)

22. Tse, D., Viswanath, P.: Fundamentals of Wireless Communication. Cambridge University Press, Cambridge (2005)

23. Viterbi, A.J.: Very low rate convolution codes for maximum theoretical performance of spread-spectrum multiple-access channels. IEEE J. Sel. Areas Commun. 8(4), 641–649 (1990)

Computations by Luminous Robots

Paola Flocchini[✉]

School of Electrical Engineering and Computer Science,
University of Ottawa, Ottawa, Canada
flocchin@site.uottawa.ca

Abstract. The study of computability issues by a system of simple, autonomous, oblivious, mobile robots, operating in the plane in LOOK-COMPUTE-MOVE cycles, has been the object of intensive investigations. These robots do not have explicit communication mechanisms, but they implicitly cooperate towards a common goal.

This paper focuses on *luminous robots*, a recently introduced model where the robots are equipped with a light that can take a constant number of different colors. The light is visible to the observing robots and stays lit from a computation cycle to the next. The availability of lights, which provides little communication and memory, has clearly a great impact on the system of robots. We review the recent results, highlighting the many open problems and research directions.

1 Introduction

Consider a set of autonomous mobile computational entities, called *robots*, which operate in the Euclidean plane initially occupying distinct points. The robots are provided with (possibly different) local coordinate systems centered in themselves, and with sensors that allow them to perceive the positions of the other robots. They operate in cycles of LOOK-COMPUTE-MOVE activities: when active, a robot observes the others, it computes a destination point, and it moves towards it. Once a cycle is completed, however, a robot *forgets* any previous information, and it starts the next cycle from scratch, making a computation solely on the basis of its current observation. Such forgetful behaviour is called *obliviousness*. Robots operate synchronously (FSYNC), if they perform simultaneously all their *Look-Compute-Move* activities in synchronized rounds; semi-synchronously (SSYNC), if only subsets of robots are activated in synchronized rounds; or asynchronously (ASYNC), if there is no synchronization and each robot operates at its own pace.

Autonomous oblivious robots have been extensively investigated, considering a variety of assumptions on their characteristics (e.g., limited vs. full visibility, level of synchrony, availability of a common coordinate system, etc.) and studying how these assumptions impact the robots' computational power on basic coordination tasks. Typical tasks in this setting are, for example, *Pattern Formation*, where the robots must place themselves so to form a given shape (e.g., see [13,16,21,24]), *Gathering*, where the robots must move to the same point

© Springer International Publishing Switzerland 2015
S. Papavassiliou and S. Ruehrup (Eds.): ADHOC-NOW 2015, LNCS 9143, pp. 238–252, 2015.
DOI:10.1007/978-3-319-19662-6_17

(e.g., see [1,2,8,14,18]), *Scattering*, where robots have to move from each other so to cover the space (e.g., see [3,7,17]). For a recent account of the current investigations, see [12].

Obliviousness is very limiting for the robots, which can rely only on the environment to decide their next step towards a common goal. On the other hand, it is a very desirable characteristic because it provides a form of fault-tolerance and self-stabilization.

Recently, to partially overcome the limits of obliviousness while maintaining some of its advantages, a stronger model has been introduced where oblivious robots carry a "light" that can take different colors (a constant number of them), can be seen by the observing robots, and stays on from cycle to cycle. Clearly, the availability of lights drastically changes the computational power of the robots, which have now the means for a little memory (by coding information to be stored into the light), and a little communication (by using the light to transmit information to the other robots). Although the idea was suggested some time ago [19], the actual model has been formalized only recently [6], and has already generated several interesting results.

The first natural question is to understand the impact that the availability of lights has on the computability power of the robots depending on their level of synchrony. This capability is quite strong; in fact, ASYNC luminous robots can perform any task solvable by SSYNC non-luminous ones. Moreover, in contrast with the situation in the classical setting, in the luminous realm there is no computational difference between the ASYNC and the SSYNC models. Several results in this regards have been derived [6], but the full computational picture describing the impact of lights on the various models is still to be completed.

Another interesting issue is to de-couple the concept of lights into two quite distinct components: the one that provides some memory to the robots (an internal state that persists from cycle to cycle), and the one that allows some form of communication (an external visible sign). Understanding the computational power of each component versus the power of their combination is a challenging and important issue. A first step in this direction has been done by studying the impact of the two individual capabilities on the rendezvous problem [15] and leaving a wealth of open problems still to be addressed.

This paper reviews the recent discoveries on luminous robots indicating the research directions emerged by the study of this new model.

2   The Robots

Let $\mathcal{R} = \{r_1, r_2, \cdots, r_n\}$ be a set of *robots*, operating in $\mathbb{R}^2$. Each robot is the centre of its own coordinate system and the robots have no agreement on the orientation of the system, on its handedness, or on their unit of distance. We denote by $r(t) \in \mathbb{R}^2$ the position occupied by robot $r \in \mathcal{R}$ at time t (for description purposes, the positions are expressed in a global coordinate system, which is unknown to the robots). Two robots r and s are said to *collide* at time t if $r(t) = s(t)$.

The robots are provided with sensors that allow them to perceive the positions of the other robots. If they can sense the whole environment, we say that they have *full visibility*, if they are able to perceive other robots only within a certain visibility radius, we say that they have *limited visibility*.

The robots are autonomous (i.e., without any external control), anonymous (i.e., without internal identifiers), indistinguishable (i.e., without external markings), without any direct means of communication.

At any time, robots can be active or inactive, and initially they are all inactive. When activated, a robot performs a LOOK-COMPUTE-MOVE sequence of operations: it first obtains a snapshot of the positions, expressed in its local coordinate system, of all visible robots (LOOK); using the last obtained snapshot as an input, the robot executes an algorithm (the same for all robots) to compute a destination point $x \in \mathbb{R}^2$ (COMPUTE); finally, it moves towards x (MOVE). It then stays inactive until the next activation.

The robots are *oblivious* in the sense that, when a robot becomes inactive, all its local memory is reset. In other words, upon becoming active again, a robot has no memory of past computations and snapshots.

A *luminous* robot r is a robot that, in addition to the above capabilities, is endowed with a persistent and externally visible state variable $Light[r]$, called *light*, whose values are from a finite set C of *colors*. The value of $Light[r]$ (i.e., its color) can be changed in each cycle by r at the end of its COMPUTE operation. A light is *externally visible* in the sense that its color at time t is visible to all robots that perform a LOOK operation at that time in its visibility radius. A light is *persistent* in the sense that, while r is oblivious and forgets all other information from previous cycles, the color is not automatically reset at the end of a cycle.

With regards to the activation and timing of the robots, there are two basic settings: *semi-synchronous* (SSYNC) and *asynchronous* (ASYNC). In SSYNC, the time is discrete; at each time instant t (a *round*) a subset of the robots is activated and performs its operations atomically, ending at time $t + 1$. At any given round, any subset of robots may be activated. In particular, if all robots are activated at every round, the system is *fully synchronous* (FSYNC). At the opposite spectrum, in ASYNC, there is no common notion of time; each robot is activated independently, the LOOK operation is instantaneous, but the COMPUTE and MOVE operations can take an unpredictable (but finite) amount of time, unknown to the robot.

The choice of the activations is done by an *adversary*, which, for fairness, activates each robot infinitely often. In ASYNC, the adversary choses also the (finite) duration of each operation. The adversary might or might not have the power to interrupt the movement of a robot before it reaches its destination in the MOVE operation. If it does, the system is said to be NON-RIGID; the only constraint on the adversary is that there exists a constant $\delta > 0$ such that, if interrupted before reaching its destination, a robot moves at least a distance δ, not known to the robot itself. Notice that, otherwise, the adversary would be able to prevent a robot from reaching any given destination in a finite number of turns. If movements are not under the control of the adversary, and every robot reaches its destination at every turn, the system is said to be RIGID.

3   Luminous Robots in FSYNC, SSYNC, and ASYNC

3.1   The Setting

A natural research investigation when considering luminous robots is to understand the impact of lights on the computability power of the robots, with respect
to the different synchrony levels of their scheduler. It is well known that, in
absence of lights, the three models form a strict hierarchy in that there exist
problems solvable in FSYNC but not in SSYNC (for oblivious robots [21]), as well
as problems solvable in SSYNC but not in ASYNC (for non-oblivious robots [20]).
The first question is whether such a strict dominance exists also in the context
of luminous robots. Interestingly, it turns out that the availability of lights does
not preserve it: the difference between asynchrony and semi-synchrony disappears making the two model equally powerful [6].

In the following, when robots are luminous we will refer to luminous ASYNC
(resp. luminous SSYNC, luminous FSYNC) models, and we denote by $ASYNC^m$ (resp.
$SSYNC^m$, $FSYNC^m$) luminous models using m colors.

3.2   Luminous ASYNC vs. SSYNC

This Section describes the relationship between luminous $ASYNC^{O(1)}$ versus non-
luminous SSYNC, as well as the one between luminous $ASYNC^{O(1)}$ and $SSYNC^{O(1)}$.

(a) **Luminous ASYNC is at least as powerful as SSYNC.** First of all, synchronous systems equipped with lights (with a constant number of colors) are at least
as powerful as semi-synchronous systems without lights. In fact, as shown below,
any problem solvable in SSYNC without lights is also solvable by asynchronous
luminous robots.

Given an algorithm $\mathcal{P}$ that solves a problem in SSYNC, there is a simulation
protocol for luminous robots in ASYNC in which every execution is equivalent to
a SSYNC execution of $\mathcal{P}$. The lights used by the simulation protocol can have five
colors: T(rying), M(oving), S(topped), F(inished), W(aiting). At the beginning, all
lights are set to T. The protocol is a sequence of Mega-Cycles, each of which starts
with all robots trying to execute protocol $\mathcal{P}$ (with color T) and ends with all robots
finishing the Mega-Cycle (with color F) having executed $\mathcal{P}$ once. All robots with
light F eventually turn their lights to T and when this process is completed, a new
Mega-Cycle starts. The protocol is designed in such a way that, during each Mega-
Cycle, every robot executes exactly one LOOK-COMPUTE-MOVE step of algorithm
$\mathcal{P}$, changing light to M when performing the move prescribed by algorithm $\mathcal{P}$. The
execution of the simulation is semi-synchronous because robots allowed to execute
the step concurrently in the same Mega-Cycle are guaranteed to have observed
the same snapshot (i.e., while nobody was moving). The transitions describing
the change of color of the robots in a Mega-Cycle are depicted in Fig. 1.

Theorem 1. *[6] Let $\mathcal{P}$ be an algorithm that solves a problem in SSYNC. There
exists an algorithm $\mathcal{S}(\mathcal{P})$ in luminous ASYNC in which every execution is equivalent to a SSYNC execution of $\mathcal{P}$.*

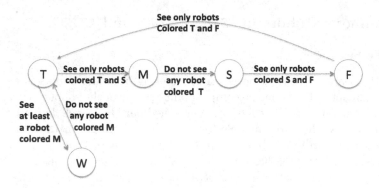

Fig. 1. ASYNC simulation of a SSYNC algorithm (executing $\mathcal{P}$ on the current snapshot only when transitioning from T to M). (1) The content of a circle represents the color of the computing robot; (2) the caption on an arrow describes a condition on the colors seen by the robot; (3) the transition between two colors indicates the local change of color corresponding to the given condition.

(b) Luminous ASYNC is more powerful than SSYNC. Luminous robots in ASYNC turn out to be actually *more powerful* than robots without lights in SSYNC. In fact, there exist problems that can be solved by asynchronous luminous robots with $O(1)$ colors, but that are unsolvable by semi-synchronous robots without colors. One such a problem is rendezvous of two oblivious robots where the robots, initially located in different points of the plane, need to gather exactly in the same point. In fact, it is well known that rendezvous is unsolvable in SSYNC [21]. The impossibility in the classical model without colors is due to the fact that the adversary could schedule the activations in such a way that the robots are forced to oscillate and never meet. However, there exists a simple solution in luminous ASYNC that uses four colors for anonymous, oblivious robots with non rigid movement and no common coordinate systems [6]. A luminous robot can exploit the lights to decide when to move towards the midpoint and when instead to move towards the other robot, without ever risking to switch position. The solution makes use of four colors: $\{a, b, c, d\}$, and it is depicted in Fig. 2.

Theorem 2. *[6]* ASYNC$^{O(1)}$ *is more powerful than* SSYNC.

(c) Luminous ASYNC$^{O(1)}$ is as powerful as luminous SSYNC$^{O(1)}$. Finally, in contrast with the dominance of SSYNC versus ASYNC observable without lights, the difference between asynchrony and semi-synchrony disappears when they are both enhanced with lights. This can be seen by showing that, given an algorithm $\mathcal{P}$ designed for SSYNC^k, there is an execution of the simulation protocol described earlier, which uses precisely $O(k)$ colors (thus works in ASYNC$^{O(k)}$). The result then follows.

Theorem 3. *[6]* ASYNC$^{O(1)}$ *is as powerful as* SSYNC$^{O(1)}$.

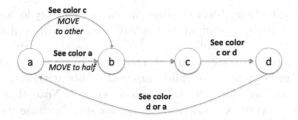

Fig. 2. Rendezvous for luminous robots in ASYNC. (1) The color in a circle represents the color of the computing robot; (2) a transition between a color and another indicates a change in the robot's color; (3) the caption on top of a transition (if any) indicates what the computing robot sees; (4) the caption below a transition describes the corresponding movement (if any) prescribed by the algorithm.

3.3 Luminours ASYNC vs. FSYNC

For the case of fully synchronous system the only known result is that there exist problems that robots cannot solve without lights, even if they are fully-synchronous, but that can be solved by asynchronous luminous robots with $O(1)$ colors. One such problem is the OSCILLATING POINTS PROBLEM requiring two robots, x and y, initially in distinct locations, to alternately come closer and move further from each other.

Theorem 4. *[6] FSYNC is not more powerful than* $\text{ASYNC}^{O(1)}$.

3.4 Open Problems

First and foremost, it is still unknown whether or not there are problems solvable by fully-synchronous robots but not by asynchronous luminous robots; a positive answer would imply that $\text{ASYNC}^{O(1)}$ and FSYNC are incomparable, while a negative one would imply that $\text{ASYNC}^{O(1)}$ is more powerful than FSYNC.

Moreover, it is known that the availability of a snapshot renders asynchronous luminous robots more powerful than regular fully-synchronous one [6]; whether a property weaker than a snapshot would suffice to achieve the same result is still open.

4 Computing by Luminous Robots: Mutual Visibility

4.1 The Setting and the Problem

In this section we consider a variant of the model, where robots cannot see through other robots; in other words, r can *see* another robot s (equivalently, s is *visible* to r) at time t if and only if no other robot lies in the segment $r(t)s(t)$ at that time. Moreover, collisions are not permitted. In such a setting, the color of robot r at time t can be seen by all robots visible by r at that time.

The Mutual Visibility problem requires the robots, initially located in different position, to terminate in a configuration where they are still in distinct locations with no three of them being collinear.

Let $\mathcal{H}(t)$ denote the convex hull of $\{r_1(t), r_2(t), \cdots, r_n(t)\}$ at time t. The robots lying on its boundary are called *external robots* at time t, while the ones lying in its interior are the *internal robots* at time t. Note that a robot may not know where the convex hull's vertices are located, because its view may be obstructed by other robots. However, it can easily determine whether it is an external or an internal robot.

4.2 Solutions

We describe two solutions, Algorithm *Shrink* and Algorithm *Contain* whose goal is to allow the robots to position themselves at the vertices of a convex polygon, thus solving Mutual Visibility. These algorithms are based on different strategies, and are tailored for different situations. Protocol *Shrink* uses two colors and requires rigid movements, while protocol *Contain* uses more colors but operates also with non-rigid movements [9–11].

(a) Algorithm *Shrink*

Consider RIGID robots in SSYNC. The main idea of Algorithm *Shrink* is to make the external robots move towards the inside of the convex hull formed by the robots, so to shrink it (see Fig. 3). Initially the robots have a pre-defined colour *Off* and they become *Vertex* to signal termination. Eventually, all the robots reach a strictly convex configuration, they all see each other with colour *Vertex*, and they terminate.

More precisely, the behaviour of the robots is as follows: a vertex robot moves inside the triangle formed by itself and its own two neighbors on the convex hull's boundary (note that they are necessarily visible). The goal of the move is to make the convex hull shrink, and possibly to increase the number of vertex robots. The destination inside the triangle is carefully calculated so to avoid collisions with other robots that may be moving at the same time, and to prevent the moving robot to become a non-vertex robot. A few other technicalities are required to treat the special case when the initial configuration is a line and to avoid deadlocks when all robots visible to an internal one are coloured *Vertex* [10].

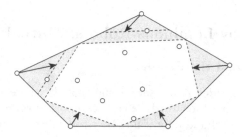

Fig. 3. From [10]: Combined motion of all vertex robots in Algorithm *Shrink*

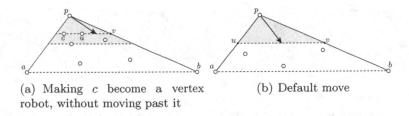

(a) Making c become a vertex   (b) Default move
robot, without moving past it

Fig. 4. From [10]: Move of an external robot p in two cases (robots' locations are indicated by small circles), where a and b are the locations of p's two neighbors on $\mathcal{H}$.

Algorithm *Shrink* correctly terminates in RIGID SSYNC using two colors. Moreover, it is possible to slightly modify it so to solve Mutual Visibility also when the two colors are not available, but robots have knowledge of n (the total number of robots in the system).

Theorem 5. *[10] Protocol* Shrink *always solves* Mutual Visibility *by* RIGID *robots in* SSYNC *with 2 colors, or with* no *colors if the robots know their number,* n.

(b) Algorithm *Contain*

Consider NON-RIGID robots in SSYNC. Algorithm *Contain* consists of two phases: an *interior depletion* phase and a *vertex adjustments* phase, to be executed in succession. In the first phase, the internal robots move towards the boundary of the convex hull signalling that they become external by changing their light, and in the second phase the robots (who are now all external) make small adjustments to finally reach a strictly convex configuration.

More precisely, let $\mathcal{H}'(t)$ be the convex hull of the positions of the internal robots at time $t \in \mathbb{N}$. When a robot r understands that it lies on a vertex of $\mathcal{H}'$, it moves towards the boundary of $\mathcal{H}$, part of which is identifiable by r. The destination point depends on the position of r within $\mathcal{H}'$ (whether r is the only internal robot, or $\mathcal{H}'$ is a line segment and r occupies one endpoint, or $\mathcal{H}'$ is a non-degenerate polygon and r it lies on one of its vertices). Some of these cases are depicted in Fig. 5. It can be shown that eventually all robots become *External* and the interior of the convex hull is depleted.

The vertex adjustments phase proceeds as follows: when a robot lies at a vertex of $\mathcal{H}$ and it sees only robots with light sets to *External*, it makes the "default move" depicted in Fig. 4b. It also sets its light to *Adjusting*, to remember it already adjusted. When the adjustment is done, the robots at a and b are guaranteed to occupy vertices of $\mathcal{H}$. Each external robot becomes a vertex robot at some point, then it adjusts its position while remaining a vertex, possibly making its adjacent robots become vertices as well, and it terminates. When all robots have terminated, the configuration is strictly convex, and therefore Mutual Visibility is solved.

Theorem 6. *[11] Protocol* Contain *always solves* Mutual Visibility *by* NON-RIGID *robots in* SSYNC *with 3 colors.*

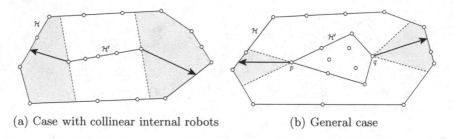

(a) Case with collinear internal robots   (b) General case

Fig. 5. From [11]: Interior depletion phase of Algorithm *Contain*.

Slight variations of protocol *Contain* can solve the problem under a variety of combination of conditions and knowledge.

Theorem 7. *[11] In* SSYNC, Mutual Visibility *can be solved by* NON-RIGID *robots with no colors, if the robots know δ (the minimum distance traversed by a robot) and their number n, and with 2 colors, if the robots know δ. In* ASYNC, Mutual Visibility *can always be solved with 3 colors in* ASYNC *by* RIGID *robots, and in* ASYNC *by* NON-RIGID *robots, if they agree on the direction of one coordinate axis.*

An interesting issue that has been investigated in regards to the Mutual Visibility problem in FSYNC is the time complexity of a solution. Indeed, a RIGID logarithmic time algorithm has been proposed [22] and such an algorithm would improve the time complexity of protocol *Contain* if it were to be executed in a synchronous system with rigid movements: note that *Contain* is designed for the weaker models of SSYNC (NON-RIGID) or ASYNC (RIGID).

4.3   Open Problems

Some questions follow immediately from the above solutions: Can Mutual Visibility be solved in ASYNC without additional assumptions (like agreement on direction)? What is the power of rigidity? Is it necessary in ASYNC? Is there a NON-RIGID solution that employs less than 3 colors? Can Mutual Visibility be solved without colors and without additional information? Note that the two colors of protocol *Shrink* are used only for termination purposes, and they could be traded by knowledge of n.

5   Partial Luminosity: Communication Versus Memory

5.1   The Setting and the Problem

All the results discussed so far assume the availability of lights that are externally visible to robots performing their LOOK operation, as well as locally persistent from one computation cycle to the next. Visibility of lights is mostly used by the robots to *communicate* pieces of information, while persistence is providing them with limited *memory* in an otherwise oblivious system. As seen in the previous Sections, the combination of these capabilities is understandably quite powerful.

To better understand the power of communication versus the one of memory in robots' computation, in [15] the two capabilities have been considered separately. More precisely, two robots' models have been proposed. For silent *finite-state* robots (FSTATE), the light of a robot is visible only to the robot itself. For *finite-communication* robots (FCOMM), a robot's light is visible only to the other robots. The two settings have been studied in relation to the classical rendezvous problem.

Rendezvous consists of having two robots meet in the same point. It is well known that in the oblivious robots' model without any light, the problem is solvable only in FSYNC [21]. The main problem is "conscious" symmetry breaking, which cannot be achieved under SSYNC and ASYNC schedulers. On the other hand, it is solvable for luminous robots (with full lights); in fact there is a 2-colors algorithm for SSYNC, which is optimal [23], and a 4 colors algorithm in ASYNC (described in Sect. 3.2) [6].

When considering FSTATE versus FCOMM robots, several components have emerged. For example, it has become more evident that the power of memory versus the one of communication is highly connected to other capabilities of the robots. In particular, the rigidity of their movements, and the level of synchrony of the scheduler. It is indeed the combined use of these features that allows rendezvous to be solved.

5.2 FSTATE **Rigid Robots in** SSYNC

When the lights provide only constant memory, an algorithm using six colors (internal states) has been devised for rigid robots in SSYNC.

The main idea of the algorithm is to make use of the (possibly different) units of distance of the robots to have them gather either as a result of a series of "symmetric" rules, or by exploiting any accidental symmetry breaking that might happen due to lack of synchrony or to disagreements on the unit distance. The role of internal states is to allow the robots to recognize such symmetry breaking, should it occur.

Intuitively, the attempted behaviour of the robots is the following: they try to reach a configuration in which they both observe the other robot at distance greater than (or equal to) their own unit. They then try to switch position and, when this happens, they attempt to gather by meeting in the mid-point between them. Note that the internal state can be used to differentiate the various stages of this attempted behaviour and, in particular, to memorize the other's position (left/right) and thus detect a switch. It can be shown that if the attempted behaviour is indeed reached, gathering is eventually achieved. On the other hand, if the attempted behaviour is not reached (because of disagreement on the unit distance or asymmetries in the activation scheduling), a detectable symmetry breaking occurs, which can be immediately exploited to gather anyway.

Theorem 8. *[15] In* SSYNC, *Rendezvous of two* FSTATE *rigid robots is solvable with six internal states.*

5.3   FCOMM **Rigid Robots in** ASYNC

Also in ASYNC for FSTATE robots, the algorithm uses the local unit distance as a computational tool, but in a rather different way, since a robot cannot remember its own color and has to infer information by observing the other robot's light.

As before, the robots attempt to coordinate a behaviour that would eventually result in gathering if their schedule happens to be synchronized and/or their unit distances are the same, but that would achieve gathering also for any break of symmetry given by a deviation from the attempted behaviour. Intuitively, the two robots try to reach a configuration in which they both see each other at distance lower than their unit distance. At this point they try to compare their unit distance, in order to break symmetry. They do that by attempting the creation of a configuration where their distance is equal to the sum of the respective unit distances. When this is accomplished, each robot can infer the other's unit and compare it with its own. If a robot has a smaller unit, it moves towards its partner, which waits. Otherwise, if their units are equal, as soon as a robot wakes up, it moves towards the mid-point and orders its partner to stay still. If both robots do so, they gather in the middle. If one robot is delayed due to asynchrony, it acknowledges the order to stay still and tells the other robot to come.

The coordination required to accomplish this overall attempted behaviour relies on the communication through the external lights, as well as on some level of synchrony that may or may not occur depending on the scheduler. If it does not occur, i.e., if the robots find themselves in different states, symmetry is broken and it can be shown that gathering is guaranteed anyway.

Theorem 9. *[15] In* ASYNC, *Rendezvous of two rigid* FCOMM *robots is solvable with twelve colors.*

5.4   FCOMM **Robots in** SSYNC

The situation of FCOMM Robots in SSYNC is quite different and much simpler. Indeed, there exist an algorithm using 3 colors, which is optimal and does not even require rigidity of movement. The robots can assume three colors: a, b, and c. If a robot sees the other colored a, it colors itself b and it moves to the mid-point between them, if a robot sees the other colored b, it colors itself c and it stay still, finally, if a robot sees the other colored c, it colors itself a and moves to the other robot's position (the algorithm is depicted in Fig. 6). It is easy to see that such a simple algorithm solves the problem.

Theorem 10. *[15] In* ASYNC, *Rendezvous of two* FSTATE *robots is solvable with three colors.*

5.5   **Open Problems**

The above results open more questions than they close. The general question on the relationship between the computability power of external lights versus

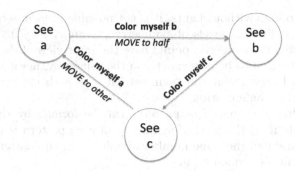

Fig. 6. SSYNC rendezvous in FCOMM. (1) the content of a circle represents what the computing robot sees; (2) the caption on top of a transition describes the change of the robot's color; (3) the caption below a transition describes the corresponding movement (if any) prescribed by the algorithm.

internal lights, and thus on whether it is better to communicate or to remember is still open. The striking difference between the simplicity of the SSYNC solution for the FCOMM model and the complexity of the one for the FSTATE model, as well as the inability to find a SSYNC solution for FSTATE, seem to indicate that FCOMM is more powerful that FSTATE.

Another parameter that plays an important role in robots' computational power is represented by rigidity of movement, which appears to be crucial to be able to find solutions in FSTATE and in ASYNC in the FCOMM model. Is rigidity necessary?

6   Computing by Luminous Robots: Sequences of Patterns

6.1   The Setting and the Problem

Pattern formation by oblivious robots is one of the basic problems and it has been studied extensively in various settings (e.g., see [13,16,21,24]). A pattern is given to each robots as a set of point in its own coordinate system, the robots must place themselves in the plane so to form the pattern, possibly scaled, rotated or translated.

Forming a sequence of patterns is the natural next step. The main challenge is to have the oblivious robots create some form of memory in the environment to be able to move from one pattern to the next. Let $S = <S_0, \ldots, S_{m-1}>$ be the sequence of patterns to be formed, and let Γ be the initial configuration of the robots. A set of robots are said to form S if they form the infinite periodic sequence $S^\infty = \langle S_0, S_2, \ldots, S_{m-1} \rangle^\infty$, starting from an arbitrary pattern $P \in S$.

6.2   Solutions

Not surprisingly, both with and without lights, the sequences of patterns formable by the robots highly depend on symmetry considerations. Intuitively, in the model of robots without lights, it is not possible to form sequences where the patterns have different levels of symmetry ("symmetricity"). The reason is that robots located in symmetric points have the same view of the environment and they can be forced to behave exactly in the same way, never breaking their symmetry class. In particular the "symmetricity" of each pattern must divide that of the starting configuration.

Without lights, a sequence of patterns can be formed by rigid robots in SSYNC if and only if (i) the level of symmetry of each pattern is the same; (ii) the patterns contain all the same number of points; (iii) no pattern can appear more than once in the sequence [4].

For luminous robots, a quite different characterization holds for the weaker ASYNC model. The "symmetricity" of each pattern must still divide the one of the starting configuration. However, in striking contrast with the previous case, a sequence of patterns can be formed by luminous robots also allowing: (i) the level of symmetry of each pattern to change; (ii) the patterns to contain different numbers of points (*contractions*); (iii) the sequence to contain repeated patterns (*repetitions*) [5].

The availability of lights is clearly very powerful, and it is exploited mainly to allow symmetry breaking in various situations: for example, to permanently identify a set of leaders, to color robots that need to move on the same point for forming a pattern and later need to move away from each other, to color differently instances of a same pattern that repeats in the sequence. Interestingly, the necessary number of colors $f(S)$ for forming sequence S, although difficult to express in a closed formula, is easily computable and bounded as follows:

$$\mu(S)^{\frac{1}{\alpha_M(S)}} \cdot (\lceil \alpha_M(S)/\alpha_m(S_i) \rceil) > f(S) \geq \max\{\mu(S)^{\frac{1}{\alpha_M(S)}}, \lceil \alpha_M(S)/\alpha_m(S) \rceil\}$$

where $\mu(S)$ is the maximum number of occurrences of some pattern in S, and $\alpha_M(S)$ (resp $\alpha_m(S)$) is the maximum (resp. minimum) number of classes of robots with the same view of the environment ("symmetricity") in some pattern in S.

Theorem 11. *[5] Any sequence of patterns S can be performed by robots with $O(f(S))$ colors.*

7   Conclusions

In this paper we have reviewed the existing results on computations by luminous robots. The investigations have just started and the current results leave many new research directions and open problems.

First of all, the computational picture of the power of luminous robots is not fully understood because some gaps still exist: for example, the relationship between ASYNC luminous robots and FSYNC non-luminous ones is still unknown. It is clear that the availability of lights highly increases the robots' capabilities; what is not clear is the impact that other assumptions have in combination with lights. For example, rigidity of movement seems to be crucial in some settings for the robots to exploit the presence of lights: this is especially true for lights that provides only internal states (FSTATE) and lights that are instead only visible to others (FCOMM), but a formal proof is still missing. An other important question left unanswered is whether there exists a computational difference between FSTATE and FCOMM: is it more useful for the robots to remember or to communicate?

Acknowledgements. I would like to thank the researchers who have collaborated with me in the investigations on luminous robots: Shantanu Das, Giuseppe Di Luna, Sruti Gan Chaudhuri, Federico Poloni, Giuseppe Prencipe, Nicola Santoro, Giovanni Viglietta and Masafumi Yamashita.

References

1. Cieliebak, M., Flocchini, P., Prencipe, G., Santoro, N.: Distributed computing by mobile robots: gathering. SIAM J. Comput. **41**(4), 829–879 (2012)
2. Cohen, R., Peleg, D.: Convergence properties of the gravitational algorithm in asynchronous robot systems. SIAM J. Comput. **34**, 1516–1528 (2005)
3. Cohen, R., Peleg, D.: Local spreading algorithms for autonomous robot systems. Theor. Comput. Sci. **399**, 71–82 (2008)
4. Das, S., Flocchini, P., Santoro, N., Yamashita, M.: Forming sequences of geometric patterns with oblivious mobile robots. Distrib. Comput. **28**(2), 131–145 (2015)
5. Das, S., Flocchini, P., Prencipe, G., Santoro, N.: Synchronized dancing of oblivious chameleons. In: Ferro, A., Luccio, F., Widmayer, P. (eds.) FUN 2014. LNCS, vol. 8496, pp. 113–124. Springer, Heidelberg (2014)
6. Das, S., Flocchini, P., Prencipe, G., Santoro, N., Yamashita, M.: The power of lights: synchronizing asynchronous robots using visible bits. In: 32nd International Conference on Distributed Computing Systems (ICDCS), pp. 506–515 (2012)
7. Dieudonné, Y., Petit, F.: Scatter of weak mobile robots. Parallel Process. Lett. **19**(1), 175–184 (2009)
8. Dieudonné, Y., Petit, F.: Self-stabilizing gathering with strong multiplicity detection. Theor. Comput. Sci. **428**(13), 47–57 (2012)
9. Di Luna, G.A., Flocchini, P., Gan Chaudhuri, S., Santoro, N., Viglietta, G.: Robots with lights: overcoming obstructed visibility without colliding. In: Felber, P., Garg, V. (eds.) SSS 2014. LNCS, vol. 8756, pp. 150–164. Springer, Heidelberg (2014)
10. Di Luna, G.A., Flocchini, P., Poloni, F., Santoro, N., Viglietta, G.: How oblivious mobile robots can achieve mutual visibility. In: Proceedings of 26th Canadian Computational Geometry Conference (CCCG) (2014)
11. Di Luna, G.A., Flocchini, P., Gan Chaudhuri, S., Poloni, F., Santoro, N., Viglietta, G.: Mutual visibility by luminous robots without collisions (2015). http://arxiv.org/abs/1503.04347
12. Flocchini, P., Prencipe, G., Santoro, N.: Distributed Computing by Oblivious Mobile Robots. Morgan & Claypool, San Rafeal (2012)
13. Flocchini, P., Prencipe, G., Santoro, N., Widmayer, P.: Arbitrary pattern formation by asynchronous oblivious robots. Theor. Comput. Sci. **407**(1–3), 412–447 (2008)
14. Flocchini, P., Prencipe, G., Santoro, N., Widmayer, P.: Gathering of asynchronous robots with limited visibility. Theor. Comput. Sci. **337**(1–3), 147–168 (2005)
15. Flocchini, P., Santoro, N., Viglietta, G., Yamashita, M.: Rendezvous of two robots with constant memory. In: Moscibroda, T., Rescigno, A.A. (eds.) SIROCCO 2013. LNCS, vol. 8179, pp. 189–200. Springer, Heidelberg (2013)

16. Fujinaga, N., Yamauchi, Y., Kijima, S., Yamashita, M.: Asynchronous pattern formation by anonymous oblivious mobile robots. In: Aguilera, M.K. (ed.) DISC 2012. LNCS, vol. 7611, pp. 312–325. Springer, Heidelberg (2012)

17. Izumi, T., Potop-Butucaru, M.G., Tixeuil, S.: Connectivity-preserving scattering of mobile robots with limited visibility. In: Dolev, S., Cobb, J., Fischer, M., Yung, M. (eds.) SSS 2010. LNCS, vol. 6366, pp. 319–331. Springer, Heidelberg (2010)

18. Lin, J., Morse, A.S., Anderson, B.D.O.: The multi-agent rendezvous problem. Parts 1 and 2. SIAM J. Control Optim. **46**(6), 2096–2147 (2007)

19. Peleg, D.: Distributed coordination algorithms for mobile robot swarms: new directions and challenges. In: Pal, A., Kshemkalyani, A.D., Kumar, R., Gupta, A. (eds.) IWDC 2005. LNCS, vol. 3741, pp. 1–12. Springer, Heidelberg (2005)

20. Prencipe, G.: The effect of synchronicity on the behavior of autonomous mobile robots. Theor. Comput. Syst. **38**(5), 539–558 (2005)

21. Suzuki, I., Yamashita, M.: Distributed anonymous mobile robots: formation of geometric patterns. SIAM J. Comput. **28**, 1347–1363 (1999)

22. Vaidyanathan, R., Busch, C., Trahan, J., Sharma, G., Rai, S.: Logarithmic-time complete visibility for robots with lights. In: 29th IEEE International Parallel and Distributed Processing Symposium (IPDPS) (2015)

23. Viglietta, G.: Rendezvous of two robots with visible bits. In: Flocchini, P., Gao, J., Kranakis, E., der Heide, F.M. (eds.) ALGOSENSORS 2013. LNCS, vol. 8243, pp. 286–301. Springer, Heidelberg (2014)

24. Yamashita, M., Suzuki, I.: Characterizing geometric patterns formable by oblivious anonymous mobile robots. Theor. Comput. Sci. **411**(26–28), 2433–2453 (2010)

Online Lower Bounds and Offline Inapproximability in Optical Networks

Shmuel Zaks[✉]

Department of Computer Science, Technion, Haifa, Israel
zaks@cs.technion.ac.il

Abstract. We present lower bounds and inapproximability results for optimization problems that originated in studies of optical networks. They include offline and online scenarios, and concern problems that optimize the use of components in the optical networks, specifically Add-Drop Multiplexers (ADMs) and regenerators.

First we discuss the online version of the problem of minimizing the number of ADMs in optical networks. In this case lightpaths need to be colored such that overlapping paths get different colors, path that share an endpoint can get the same color, and the cost is the total number endpoints (=ADMs); the key point is that an endpoint shared by two same-colored paths is counted only once. Following [19] (where we showed tight competitive ratios for several networks), we present in this paper a $\frac{3}{2}$ lower bound on the competitive ratio for a path network.

We next present problems that deal with the use of regenerators in optical networks. Given a set of lightpaths in a network G and a positive integer d, regenerators must be placed in such a way that in any lightpath there are no more than d hops without meeting a regenerator. We first discuss the online version of the problem of optimizing the number of locations where regenerators are placed, following [17]. When there is a bound on the number of regenerators in a single node, there is not necessarily a solution for a given input. We distinguish between feasible inputs and infeasible ones. For the latter case our objective is to satisfy the maximum number of lightpaths. For a path topology we consider the case where $d = 2$, and show a lower bound of $\sqrt{l}/2$ for the competitive ratio (where l is the number of internal nodes of the longest lightpath) on infeasible inputs, and a tight bound of 3 for the competitive ratio on feasible inputs.

Last we study the problem where we are given a finite set of p possible traffic patterns (each given by a set of lightpaths), and our objective is to place the minimum number of regenerators at the nodes so that each of the traffic patterns is satisfied (that is, regenerators are placed such that in any lightpath there are no more than d hops without meeting a regenerator). We prove - following [16] - that the problem does not admit a PTAS for any $d, p \geq 2$.

Some of these problems have interesting implications to problems stated within scheduling theory.

© Springer International Publishing Switzerland 2015
S. Papavassiliou and S. Ruehrup (Eds.): ADHOC-NOW 2015, LNCS 9143, pp. 253–270, 2015.
DOI:10.1007/978-3-319-19662-6_18

1   Introduction

1.1   Background and Problem Definition

Background: Optical wavelength-division multiplexing (WDM) is the most promising technology today that enables us to deal with the enormous growth of traffic in communication networks, like the Internet. Optical fibers using WDM technology can carry around 80 wavelengths (colors) in real networks and up to few hundreds in testbeds. As satisfactory solutions have been found for various coloring problems, the focus of studies shifts from the number of colors to the hardware cost. These new measures provide better understanding for designing and routing in optical networks.

A communication between a pair of nodes is done via a *lightpath*, which is assigned a certain wavelength. In graph-theoretic terms, a lightpath is a simple path in the network, with a color assigned to it. We concentrate on the hardware cost, in terms of ADMs and regenerators.

ADMs: Each lightpath uses two Add-Drop Multiplexers (ADMs), one at each endpoint. If two adjacent lightpaths, i.e. lightpaths sharing a common endpoint, are assigned the same wavelength, then they can use the same ADM, provided their concatenation is a simple path. An ADM may be shared by at most two lightpaths. The total cost considered is the total number of ADMs. For a detailed technical explanation see [13].

Stated in graph-theoretic terms, we are given a set of paths $\mathcal{P}$, and need to assign them colors, such that two edge-intersecting paths must get different colors. (The issue of vertex-intersecting paths will not be discussed here.) Such a color assignment is termed a *legal coloring*. Path that share an endpoint can get the same color. The cost is measured in terms of the total number of ADMs. As each path is using two ADMs, one at each endpoint, the total number of ADMs is $2|\mathcal{P}|$. When two paths that share an endpoint get the same color, we save one ADM, so the cost is $2|\mathcal{P}|$ minus the number of these savings. The goal is to minimize the total number of ADMs. We thus the following minimization problem:

ADM MINIMIZATION (MINADM)

Input: A graph $G = (V, E)$, a set $\mathcal{P}$ of simple paths in G.
Output: A legal coloring of $\mathcal{P}$.
Objective: Minimize the number of ADMs.

Regenerators: The energy of the signal along a lightpath decreases and thus amplifiers are used every fixed distance. Yet, as the amplifiers introduce noise into the signal there is a need to place a regenerator every at most d hops.

The *length* of a lightpath is the number of edges it contains. The *internal vertices* (resp. *edges*) of a lightpath or a path ℓ are the vertices (resp. edges) in ℓ different from the first and the last one. Given an integer d, a lightpath ℓ is *d-satisfied* if there are no d consecutive internal vertices in ℓ without a regenerator. A set of lightpaths is *d-satisfied* if each of its lightpaths is d-satisfied.

Given p sets of lightpaths $L_1, \ldots, L_p$, with $L_i = \{\ell_{i,j} \mid 1 \leq j \leq x_i\}$ (that is, x_i is the number of lightpaths in the set L_i), we consider the union of all lightpaths in the p sets $\cup L_i = \{\ell_{i,j} \mid 1 \leq i \leq p, 1 \leq j \leq x_i\}$. An *assignment* of regenerators is a function $reg : V \times \cup L_i \rightarrow \{0,1\}$, where $reg(v,\ell) = 1$ if and only if a regenerator is used at vertex v by lightpath ℓ. When one set of lightpaths is given, the number of regenerators put at node v is $reg(v)$. We present two problems.

The first problem deals with the case whre there is a limit imposed by the technology on the number of regenerators that can be placed in a network node [5,11]. We denote this limit by k and refer to the case where this limit is not likely to be reached by any regenerator placement as $k = \infty$. When $k = \infty$ we consider the *regenerator location problem* (RLP) where the objective is to minimize the number of nodes that are assigned regenerators. When k is bounded there are inputs for which there is no feasible regenerator placement that satisfies both conditions. For example, consider the case $d = 2$ and $k = 1$, and three identical lightpaths $u - v - w - x$. Each of these lightpaths must have a regenerator either at v or w, and this is clearly impossible. In this case we consider the *Path Maximization Problem* (MaxPATH) that seeks for regenerator placements that serve as many lightpaths as possible, as follows:

PATH MAXIMIZATION (MaxPATH)

Input: An undirected graph $G = (V, E)$, a set $\mathcal{P}$ of paths in G, $d, k \geq 1$.

Output: A regenerator assignment reg for which $reg(v) \leq k$ for every node $v \in V$.

Objective: Maximize the number of paths of $\mathcal{P}$ that are d-satisfied.

The second problem considers a scenario where we are given a finite set of p possible traffic patterns (each given by a set of lightpaths), and our objective is to place the minimum number of regenerators at the nodes so that each of the traffic patterns is satisfied. Thus, given $p \geq 1$ sets of lightpaths, and a distance $d \geq 1$, we need to determine the smallest number of regenerators that d-satisfy each of the p sets. Formally, for two fixed integers $d, p \geq 1$, the optimization problem we study is defined as follows.

(d,p)-TOTAL REGENERATORS $((d,p)$-TR)

Input: An undirected graph $G = (V, E)$ and p sets of lightpaths $\mathcal{L} = \{L_1, \ldots, L_p\}$.

Output: A function $reg : V \times \cup L_i \rightarrow \{0,1\}$ s.t. each lightpath in $\cup L_i$ is d-satisfied.

Objective: Minimize $\sum_{v \in V} reg(v)$ $(reg(v) = \max_{1 \leq i \leq p} \sum_{\ell \in L_i} reg(v, \ell))$.

When $p = 1$ (that is, when there is a single set of requests) the problem is trivially solvable in polynomial time, as the regenerators can be placed for each lightpath independently. The case $d = 1$ is not interesting either, as for each internal vertex $v \in V$ and each $\ell \in \cup L_i$, $reg(v, \ell) = 1$, so there is only one feasible solution, which is optimal.

Online Algorithms: An online minimization algorithm is said to be c-competitive if for any input, it produces a solution that is at most c times that used by an optimal offline algorithm (see [4]).

The motivation for the online scenario in the context of optical networks stems from the need to utilize the cost of use of the optical network. We assume that the switching equipment (ADMs or regenerators) is already installed in the network. Once a lightpath arrives, we need to assign it a color, and our target is to optimize the objective function.

Approximation Algorithms: Given an NP-hard minimization problem Π, we say that a polynomial-time algorithm $\mathcal{A}$ is an α-approximation algorithm for Π, with $\alpha \geq 1$, if for any instance of Π, algorithm $\mathcal{A}$ finds a feasible solution with cost at most α times the cost of an optimal solution. In complexity theory, the class APX (Approximable) contains all NP-hard optimization problems that can be approximated within a constant factor. The subclass PTAS (Polynomial Time Approximation Scheme) contains the problems that can be approximated in polynomial time within a ratio $1 + \varepsilon$ for *any* fixed $\varepsilon > 0$. In some sense, these problems can be considered to be *easy* NP-hard problems. Since, assuming $P \neq NP$, there is a strict inclusion of PTAS in APX (for instance, MINIMUM VERTEX COVER $\in$ APX \ PTAS), an APX-hardness result for a problem implies the non-existence of a PTAS unless $P = NP$.

1.2   Previous Work

We list below some previous works regarding optimization problems for ADMs and regenerators.

ADMs: Minimizing the number of ADMs in optical networks is a main research topic in recent studies. The problem was introduced in [13] for the ring topology. An approximation algorithm for the ring topology with approximation ratio of $\frac{3}{2}$ was presented in [7], and was improved in [9,20] to $\frac{10}{7} + \epsilon$ and $\frac{10}{7}$, respectively.

For general topology [8] described an algorithm with approximation ratio of $\frac{8}{5}$. The same problem was studied in [6] and an algorithm with an approximation ratio of $\frac{3}{2} + \epsilon$ was presented. This algorithm is further analyzed in [12].

The problem of online path coloring is studied in earlier works, such as [15]. The problem studied in these works has a different objective function, namely the number of colors.

Regenerators: Placement of regenerators in optical networks has become an active area in recent years. Most of the researches have focused on the technological aspects of the problems. Moreover, heuristics and simulations heve been performed in order to reduce the number of regenerators are performed in (e.g., [5,10,14,18,21,23,24]). The regenerator location problem (RLP) was shown to be NP-complete in [5], followed by heuristics and simulations. In [11] theoretical results for the offline version of RLP are presented. The authors study four variants of the problem, depending on whether the number k of regenerators per node is bounded, and whether the routings of the requests are given. Regarding the complexity of the problem, they present polynomial-time algorithms and NP-completeness results for a variety of special cases.

We note that while considering the path topology, RLP has implications for the following scheduling problem: Assume a company has n cars and that car i needs to be serviced within every at most d days between day a_i and b_i. Furthermore, assume that the garage can serve at most k cars per day and charges a certain cost each time the garage is used. The objective is to service the cars in the fewest number of days and hence minimizing the number of times the garage is used.

Other objective functions have also been considered in the context of regenerator placement. e.g., in [16] the problem of minimizing the total number of regenerators is studied under other settings.

1.3 Summary of Results

This paper focuses on impossibility results concerning the above-mentioned problems. The first two are lower bounds on competitive ratios of online settings, which turn out to be best possible; in one of the cases we also show an algorithm whose competitive ratio matches the lower bound. The third result is an inapproximability result for an offline one.

1. Online ADM minimization

   Regarding the problem of minimizing the number of ADMs in the online setting, in [19] we showed a competitive ratio of $\frac{7}{4}$ for any network topology, including rings of size at least four, $\frac{5}{3}$ for a triangle network, and $\frac{3}{2}$ for a path topology, and showed that these results are best possible.

   In this paper we present the lower bound of $\frac{3}{2}$ for a path topology; this result is presented in Sect. 2.

2. Online Path maximization

   In [17] we studied the MaxPATH problem. We considered the case where G is a path, $d = 2$, and at most one regenerator can be put in a single location. In this case there is not necessarily a solution, and the measurement is the maximum number of lightpaths that can be satisfied. We distinguish between feasible inputs (for which there is a solution) and infeasible ones, on a path topology. We proved a lower bound of $\sqrt{l}/2$ for the competitive ratio for general instances which may be infeasible (where l is the number of internal nodes of the longest lightpath), and a tight bound of 3 for the competitive ratio of deterministic online algorithms for feasible instances. In addition we studied in [17] the case where there is no bound on the number of regenerators that can be used in any single node, and the graph G and the value of d are arbitrary. In this case there is a solution for every input, and the measurement is the number of locations in which regenerators are placed. We showed an $O(\log |X| \cdot \log d)$-competitive randomized algorithm for any network topology, that can be made deterministic (with the same competitive ratio) for some cases including tree topology networks, where X is the set of all paths of length d in G, and a deterministic lower bound of $\Omega\left(\frac{\log(|E|/d) \cdot \log d}{\log(\log(|E|/d) \cdot \log d)}\right)$, where E is the edge set of G.

   In this paper we present the above-mentioned lower bounds of $\sqrt{l}/2$ for the competitive ratio for general instances which may be infeasible, and the tight bound of 3 for the competitive ratio of deterministic online algorithms for feasible instances; these results are presented in Sect. 3.

3. (d,p)-total regenerators

In [16] we provided hardness results and approximation algorithms for the (d,p)-TOTAL REGENERATORS problem $((d,p)$-TR for short). We proved that for any fixed integers $d, p \geq 2$, (d,p)-TR does not admit a PTAS unless P = NP, even if the underlying graph G has maximum degree at most 3, and the lightpaths have length at most $2d$. We complemented this hardness result with a constant-factor approximation algorithm with ratio $\min\{p, H_{d \cdot p} - 1/2\}$, where $H_n = \sum_{i=1}^{n} \frac{1}{i}$ is the n-th harmonic number. We proved that (d,p)-TR is polynomial-time solvable in paths when all the lightpaths share the first (or the last) edge, as well as when the maximum number of lightpaths sharing an edge is bounded.

In this paper we show that (d,p)-TR does not admit a PTAS; this result is presented in Sect. 4.

2   Online ADMs Minimization

Theorem 1. *For any $\epsilon > 0$, there is no $(\frac{3}{2} - \epsilon)$-competitive deterministic algorithm for path topology.*

Proof. Let G be a path with $2k$ nodes $u_1, v_1, u_2, v_2, ..., u_k, v_k$, $u_1 < v_1 < u_2 < v_2 < ... < u_k < v_k$. Let ALG be any deterministic algorithm. The value of k will be determined later.

The adversary works in two phases. In the first phase the input is $a_1, a_2, ..., a_k$ where $\forall i, a_i = (u_i, v_i)$. In the second phase the input depends on the decisions made by ALG during the first phase. Let $w(a_i)$ be the color assigned by the algorithm to path a_i. For every $1 \leq i < k$, if $w(a_i) = w(a_{i+1})$ then the input contains two paths $b_i = (u_1, u_{i+1})$ and $b'_i = (v_i, v_k)$, otherwise the input contains one path $c_i = (v_i, u_{i+1})$.

Let $0 \leq x \leq k-1$ be the number of times $w(a_i) = w(a_{i+1})$ is satisfied. Then $w(a_i) \neq w(a_{i+1})$ is satisfied $k - 1 - x$ times.

During the first phase the algorithm uses $2k$ ADMs, one for each node.

For the paths b_i and b'_i, let $\lambda = w(a_i)(= w(a_{i+1}))$. λ is not feasible neither for b_i nor for b'_i. Then the algorithm assigns other colors to b_i and b'_i, and it uses 4 ADMs, for a total of $4x$ ADMs.

For the path c_i, let $\lambda = w(a_i)$ and $\lambda' = w(a_{i+1})(\neq \lambda)$, coloring c_i with one of these colors ALG uses one ADM, otherwise it uses 2 ADMs. Therefore for the paths c_i, ALG uses at least $k - 1 - x$ ADMs.

Summing up, we get that ALG uses at least $2k + 4x + (k-1-x) = 3(k+x) - 1$ ADMs.

On the other hand the following solution is possible. For any consecutive paths $c_i, c_{i+1}, ..., c_{i+j}$ color such that $w(b_{i-1}) = w(a_i) = w(c_i) = w(a_{i+1}) = w(c_{i+1}) = ... = w(c_{i+j}) = w(a_{i+j+1}) = w(b'_{i+j+1})$. This solutions use $2k + 2x$ ADMs, one ADM at each u_i, v_i, x additional ADMs at u_1, and x additional ADMs at v_k.

Therefore the competitive ratio of ALG is at least $\frac{3(k+x)-1}{2(k+x)} = \frac{3}{2} - \frac{1}{2(k+x)} \geq$ $\frac{3}{2} - \frac{1}{2k}$. For any $\epsilon > 0$ we can choose $k > \frac{1}{2\epsilon}$, so that the competitive ratio of ALG is bigger then $\frac{3}{2} - \epsilon$. $\qquad\square$

3   Online Path Maximization

We consider the simple instance of the MAXPATH problem, i.e. the case where the network is a path, $d = 2$, and at most one regenerator can be place in one location.

We say that an instance is *feasible*, if there is a regenerator assignment that d-satisfies all the paths in $\mathcal{P}$, and *infeasible* otherwise. We show that if the input instance is infeasible, no online algorithm (for MAXPATH) has a small competitive ratio; precisely, we show that no online algorithm is better than $\sqrt{l}/2$-competitive, where l is the length of the longest path in the input. We then focus on feasible instances.

For infeasible instances we have the following result:

Lemma 1. *Consider the path topology. For $k = 1$ and $d = 2$, any deterministic online algorithm for MAXPATH has a competitive ratio at least $\sqrt{l}/2$, where l is the number of internal vertices of the longest path.*

Proof. The adversary first releases a path of length $l + 1$ with l internal vertices. The online algorithm has to satisfy this path, otherwise, the competitive ratio is unbounded. Then the adversary releases $\sqrt{l}$ paths along the first path each with $\sqrt{l}$ (disjoint) internal vertices. If the online algorithm does not satisfy any of these paths, the competitive ratio is at least $\sqrt{l}$ and we are done. Suppose x of these paths are satisfied. In order to make the first path and these x paths 2-satisfied, there is one regenerator placed in each node along these x paths. For each of these x paths P, the adversary releases $\sqrt{l}/2$ paths along P each with two (disjoint) internal vertices. The online algorithm is not able to satisfy any of these short paths and the total number of 2-satisfied paths is $x + 1$. On the other hand, the optimal offline algorithm satisfies all the paths except the first path of length l, i.e., $\sqrt{l} + x\sqrt{l}/2$ paths. As a result, the competitive ratio of the online algorithm is $\frac{(x+2)\sqrt{l}}{2(x+1)} > \sqrt{l}/2$. $\qquad\square$

We now consider feasible instances, that is, instances, where there exists a placement of regenerators such that all paths are satisfied. We will prove that, for feasible instances, there is a tight bound of 3 for the competitive ratio. That is, we provide an online algorithm Algorithm 1 with competitive ratio 3, and we show a lower bound of 3 for the competitive ratio of every deterministic online algorithm for feasible instances.

Note that a regenerator assignment 2-satisfies a path P if and only if it constitutes a vertex cover of the edges of P, except its first and last edges. Therefore, in this section, for simplicity we assume that the leftmost and rightmost edges of the paths have been removed and a regenerator assignment is a vertex cover of the edges of the paths.

Algorithm 1 adopts a greedy approach and satisfies a newly presented path whenever possible. When a path P_i is presented, it checks whether there exist two consecutive internal vertices of P_i that are already assigned regenerators for previous paths. If yes, this means it is impossible (under the current assignment) to satisfy P_i. Otherwise, the algorithm satisfies P_i, as follows. There are two possible locations for the leftmost regenerator of P_i, namely, either its leftmost internal node, or the internal node adjacent to it. Among these two alternatives we choose the alternative that uses the smaller number of regenerators by trying the following regenerator allocation process. Suppose we put a regenerator at a certain internal node v of P_i. We check whether the node at distance 2 from v already has a regenerator; if not, we put a regenerator there and continue; if yes, we put a regenerator at the node at distance 1 from v^1. This continues until P_i is 2-satisfied.

Algorithm 1. Online algorithm for a path-topology, $k = 1$ and $d = 2$.

1: When the path P_i with endpoints s_i, t_i is presented:
2: **if** $reg(v) = reg(v') = 1$ for two consecutive internal nodes v, v' of P_i **then**
3:      leave P_i unsatisfied;
4: **else**
5:      $last \leftarrow s_i$
6:      **while** P_i is not 2-satisfied **do**
7:          **if** $reg(last + 2) = 0$ **then**
8:              $last \leftarrow last + 2$
9:          **else**
10:             $last \leftarrow last + 1$
11:         **end if**
12:         $reg(last, P_i) \leftarrow 1$
13:     **end while**
14: **end if**

Theorem 2. *Algorithm 1 is 3-competitive for* MAXPATH *for feasible inputs in path topologies, when* $k = 1$ *and* $d = 2$.

Proof. See Appendix A.   □

Theorem 3. *Any deterministic online algorithm for* MAXPATH *has a competitive ratio at least* 3 *even when the instance is restricted to feasible ones on path topologies and* $k = 1, d = 2$.

Proof. We will prove that, for every $\varepsilon > 0$, there exist infinitely many inputs such that every algorithm has competitive ratio at least $3 - \varepsilon$. Choose an integer n, such that $\frac{2}{n+1} < \varepsilon$. The adversary provides initially a path P_0 with $13n - 2$ edges. The algorithm must satisfy the path P_0, since otherwise the adversary

---

[1] The node at distance 1 must have no regenerator, else there are two consecutive internal nodes with regenerators and the algorithm would have rejected the path.

stops and the competitive ratio is infinite. We divide P_0 into n subpaths P_i, $i = 1, 2, \ldots, n$, with 11 edges each, where between two consecutive subpaths there exist two edges.

Consider any such subpath P_i, $i = 1, 2, \ldots, n$. Suppose that there exist two edges ab and cd of P_i, where $\{a, b\} \cap \{c, d\} = \emptyset$, such that $reg(a, P_0) = reg(b, P_0) = reg(c, P_0) = reg(d, P_0) = 1$. Then the adversary provides next the paths $P_{i,1} = (a, b)$ and $P_{i,2} = (c, d)$. These two paths $P_{i,1}$ and $P_{i,2}$ cannot be satisfied, since each of the vertices a, b, c, d has a regenerator for path P_0. So the competitive ratio of the algorithm is at least 3.

We thus can assume that there do not exist such edges ab and cd for any of the P_i's. That is, there exist at $most$ three consecutive vertices u_1, u_2, u_3 of P_i, such that $reg(u_1, P_0) = reg(u_2, P_0) = reg(u_3, P_0) = 1$, while for every other edge uu' of P_i, there exists a regenerator for P_0 either on vertex u or on vertex u'. Then, there exist five consecutive vertices $v_1^i, v_2^i, v_3^i, v_4^i, v_5^i$ of P_i, such that $reg(v_1^i, P_0) = reg(v_3^i, P_0) = reg(v_5^i, P_0) = 1$ and $reg(v_2^i, P_0) = reg(v_4^i, P_0) = 0$.

The adversary now provides the path $P_i' = (v_2^i, v_3^i, v_4^i)$. Thus, since $reg(v_3^i, P_0) = 1$ and $reg(v_2^i, P_0) = reg(v_4^i, P_0) = 0$, the only way that the algorithm can satisfy P_i' is to place regenerators for P_i' at the vertices v_2^i and v_4^i (that is, $reg(v_2^i, P_i') = reg(v_4^i, P_i') = 1$).

The adversary proceeds as follows. In the case where the algorithm chooses not to satisfy the path P_i', the adversary does not provide any other path that shares edges with P_i. Otherwise, if the algorithm satisfies P_i', then the adversary provides the paths $P_i'' = (v_1^i, v_2^i)$ and $P_i''' = (v_4^i, v_5^i)$ (see Fig. 1). In this case, $reg(v_2^i, P_i') = reg(v_4^i, P_i') = 1$ and $reg(v_1^i, P_0) = reg(v_5^i, P_0) = 1$, and thus the paths P_i'' and P_i''' remain unsatisfied by the algorithm.

We now show that this instance is indeed feasible. Actually, we show that even the instance that includes P_0, all the paths $P_i' = (v_2^i, v_3^i, v_4^i)$, all the paths $P_i'' = (v_1^i, v_2^i)$ and all the paths $P_i''' = (v_4^i, v_5^i)$ is feasible. To see this, we put regenerators at the nodes v_1^i, v_3^i, v_5^i, that will satisfy $P_i'' = (v_1^i, v_2^i)$, $P_i' = (v_2^i, v_3^i, v_4^i)$ $P_i''' = (v_4^i, v_5^i)$, respectively. We then put regenerators at all other nodes (including the nodes v_2^i, v_4^i), which clearly satisfies P_0.

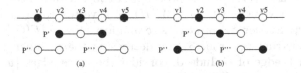

Fig. 1. Adversary for Lemma 3. (a) The online assignment where P'' and P''' cannot be satisfied. (b) The optimal assignment where all paths are satisfied.

Denote by h the number of subpaths P_i, for which the adversary adds the path P_i', P_i'' and P_i'''. Thus the number of subpaths P_i, for which the adversary adds P_i', but not P_i'' or P_i''', is $n - h$. The total number of paths that the adversary provided is thus $1 + 3h + (n - h) = 1 + n + 2h$. The number of paths satisfied by the algorithm is $1 + h$. That is, the competitive ratio of the algorithm is $\frac{1+n+2h}{1+h} = 3 + \frac{n-2-h}{1+h}$. Therefore, since $h \leq n$, it follows that the competitive ratio of the algorithm is at least $3 - \frac{2}{1+n} > 3 - \varepsilon$. Since this holds for every $\varepsilon > 0$, it follows that any deterministic online algorithm has competitive ratio at least 3. This completes the proof of the lemma.   □

4 (d, p)-total Regenerators

In this section we prove that, unless $\text{P} = \text{NP}$, (d,p)-TR does not admit a PTAS for any $d, p \geq 2$, even if the underlying graph G has maximum degree at most 3 and the lightpaths have length $\mathcal{O}(d)$. Before this, we need two technical results to be used in the reductions.

MINIMUM VERTEX COVER is known to be APX-hard in cubic graphs [2]. By a simple reduction, we prove in the following lemma that MINIMUM VERTEX COVER is also APX-hard in a class of graphs with degree at most 3 and high girth, which will be used in the sequel.

Lemma 2. MINIMUM VERTEX COVER *is* APX-*hard in the class of graphs* $\mathcal{H}$ *obtained from cubic graphs by subdividing each edge twice.*

Proof. Given a cubic graph G, let H the graph obtained from G by subdividing each each twice. That is, each edge $\{u, v\}$ gets replaced by 3 edges $\{u, u_e\}$, $\{u_e, v_e\}$, and $\{v_e, v\}$, where u_e, v_e are two new vertices. We now claim that

$$\text{OPT}_{\text{VC}}(H) = \text{OPT}_{\text{VC}}(G) + |E(G)|, \tag{1}$$

where OPT_{VC} indicates the size of a minimum vertex cover. Indeed, let $S_G \subseteq V(G)$ be a vertex cover of G. We proceed to build a vertex cover S_H of H of size $|S_G| + |E(G)|$. First, include in S_H all the vertices in S_G. Then, for each 3 edges $\{u, u_e\}$, $\{u_e, v_e\}$, and $\{v_e, v\}$ of H corresponding to edge $\{u, v\} \in E(G)$, the edge $\{u_e, v_e\}$ is *not* covered by S_G, and at least one of $\{u, u_e\}$ and $\{v_e, v\}$ is covered by S_G. Therefore, adding either u_e or v_e to S_H covers the three edges $\{u, u_e\}$, $\{u_e, v_e\}$, and $\{v_e, v\}$. This procedure defines a vertex cover of H of size $|S_G| + |E(G)|$. Conversely, let $S_H \subseteq V(H)$ be a vertex cover of H, and let us construct a vertex cover S_G of G of size at most $|S_H| - |E(G)|$. We shall see that we can construct S_G from S_H by decreasing the cardinality of S_H by at least one for each edge of G. Indeed, consider the three edges $\{u, u_e\}$, $\{u_e, v_e\}$, and $\{v_e, v\}$ of H corresponding to an edge $e = \{u, v\} \in E(G)$. Note that at least one of u_e and v_e belongs to S_H. If both $u_e, v_e \in S_H$, add either u or v to S_G if none of u, v was already in S_G. Otherwise, if exactly one of u_e and v_e (say, u_e) belongs to S_H, then at least one of u and v must also belong also to S_H, and do not add any new vertex to S_G.

Note that as G is cubic, each vertex in a solution S_G covers exactly 3 edges, so $|E(G)| \leq 3 \cdot \text{OPT}_{\text{VC}}(G)$.

In order to prove the lemma, assume for contradiction that there exists a PTAS for MINIMUM VERTEX COVER in $\mathcal{H}$. That is, for any $\varepsilon > 0$, we can find in polynomial time a solution $S_H \subseteq V(H)$ such that $|S_H| \leq (1+\varepsilon) \cdot \text{OPT}_{\text{VC}}(H)$. By the above discussion, we can find a solution $S_G \subseteq V(G)$ such that

$$
\begin{aligned}
|S_G| &\leq |S_H| - |E(G)| \\
&\leq (1+\varepsilon) \cdot \text{OPT}_{\text{VC}}(H) - |E(G)| \\
&= (1+\varepsilon) \cdot (\text{OPT}_{\text{VC}}(G) + |E(G)|) - |E(G)| \\
&= (1+\varepsilon) \cdot \text{OPT}_{\text{VC}}(G) + \varepsilon \cdot |E(G)| \\
&\leq (1+\varepsilon) \cdot \text{OPT}_{\text{VC}}(G) + 3\varepsilon \cdot \text{OPT}_{\text{VC}}(G) \\
&= (1+4\varepsilon) \cdot \text{OPT}_{\text{VC}}(G),
\end{aligned}
$$

where we have used Eq. (1) and the fact that $|E(G)| \leq 3 \cdot \text{OPT}_{\text{VC}}(G)$. That is, the existence of a PTAS for MINIMUM VERTEX COVER in the class of graphs $\mathcal{H}$ would imply the existence of a PTAS in the class of cubic graphs, which is a contradiction by [2] unless P = NP. □

It is known that the edges of any cubic graph can be two-colored such that each monochromatic connected component is a path (of any length) [1]. In fact, solving a conjecture of Bermond *et al.* [3], Thomassen proved [22] a stronger result: the edges of any cubic graph can be two-colored such that each monochromatic connected component is a path of length at most 5 (see Fig. 2(a) for an example). In addition, the aforementioned colorings can be found in polynomial time [1,22]. Note that in such a coloring of a cubic graph, each vertex appears exactly once as an endpoint of a path, and exactly once as an internal vertex of another path. We next show that these results can be easily strengthened for the family of graphs $\mathcal{H}$ defined in Lemma 2.

Lemma 3. *Let $\mathcal{H}$ be the class of graphs obtained from cubic graphs by subdividing each edge twice. The edges of any graph in $\mathcal{H}$ can be two-colored in polynomial time such that each monochromatic connected component is a path of length at most 2.*

Proof. Let $H \in \mathcal{H}$ be a graph obtained from a cubic graph G by subdividing each each twice. That is, edge $\{u, v\}$ of G gets replaced by 3 edges $\{u, u_e\}$, $\{u_e, v_e\}$, and $\{v_e, v\}$ in H. Find a two-coloring of the edges of G such that each

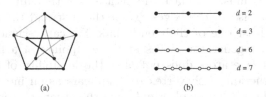

(a)   (b)

Fig. 2. (a) A two-coloring of the edges of the Petersen graph (grey and black) such that each monochromatic component is a path of length at most 5. (b) Construction of the lightpaths from a path of length 2 for several values of d, in the proof of Theorem 5. Full dots correspond to vertices of the VERTEX COVER instance (called *black* in the proof).

monochromatic connected component is a path, using [1] or [22]. To color the edges of H, do the following for each edge $\{u, v\}$ of G: color $\{u, u_e\}$ and $\{v_e, v\}$ with the same color as $\{u, v\}$, and color $\{u_e, v_e\}$ with the other color. It is then easy to check that each monochromatic connected component of the obtained two-coloring of H is a path of length at most 2.   □

We now present the main results of this section. For the sake of presentation, we first present in Theorem 4 the result for the case $d = p = 2$, and then we show in Theorem 5 how to extend the reduction to any fixed $d, p \geq 2$.

Theorem 4. $(2, 2)$-TR *does not admit a* PTAS *unless* P $=$ NP, *even if G has maximum degree at most 3 and the lightpaths have length at most 4.*

Proof. The reduction is from MINIMUM VERTEX COVER (VC for short) in the class of graphs $\mathcal{H}$ obtained from cubic graphs by subdividing each edge twice, which does not admit a PTAS by Lemma 2 unless P $=$ NP. Note that by construction any graph in $\mathcal{H}$ has girth at least 9. Given a graph $H \in \mathcal{H}$ as instance of VC, we proceed to build an instance of $(2, 2)$-TR. We set $G = H$, so G has maximum degree at most 3.

To define the two sets of lightpaths L_1 and L_2, let $\{E_1, E_2\}$ be the partition of $E(H)$ given by the two-coloring of Lemma 3. Therefore, each connected component of $H[E_1]$ and $H[E_2]$ is a path of length at most 2. Each such path in $H[E_1]$ (resp. $H[E_2]$) will correspond to a lightpath in L_1 (resp. L_2), which we proceed to define. A key observation is that, as the paths of the two-coloring have length at most 2, if any endpoint v of such a path P had one neighbor in $V(P)$, it would create a triangle, a contradiction to the fact that the girth of H is at least 9. Therefore, as the vertices of H have degree 2 or 3, any endpoint v of a path P has at least one neighbor in $V(H) \setminus V(P)$.

We are now ready to define the lightpaths. Let P be a path with endpoints u, v, and let u' (resp. v') be a neighbor of u (resp. v) in $V(H) \setminus V(P)$, such that $u' \neq v'$ (such distinct vertices u', v' exist because P has length at most 2 and H has girth at least 9; in fact we only need H to have girth at least 5). The lightpath associated with P consists of the concatenation of $\{u', u\}$, P, and $\{v, v'\}$. Therefore, the length of each lightpath is at most 4. This completes the construction of the instance of $(2, 2)$-TR. Observe that since we assume that $d = 2$, regenerators must be placed in such a way that all the internal edges of a lightpath (that is, all the edges except the first and the last one) have a regenerator in at least one of their endpoints. We can assume without loss of generality that no regenerator serves at the endpoints of a lightpath, as the removal of such regenerators does not alter the feasibility of a solution. Note that in our construction, each vertex of G appears as an internal vertex in at most two lightpaths, one (possibly) in L_1 and the other one (possibly) in L_2, so we can assume that $reg(v) \leq 1$ for any $v \in V(G)$.

We now claim that $\mathrm{OPT}_{\mathrm{VC}}(H) = \mathrm{OPT}_{(2,2)-\mathrm{TR}}(G, \{L_1, L_2\})$.

Indeed, let first $S \subseteq V(H)$ be a vertex cover of H. Placing one regenerator at each vertex belonging to S defines a feasible solution to $(2, 2)$-TR in G with cost $|S|$, as at least one endpoint of each internal edge of each lightpath contains a regenerator. Therefore, $\mathrm{OPT}_{\mathrm{VC}}(H) \geq \mathrm{OPT}_{(2,2)-\mathrm{TR}}(G, \{L_1, L_2\})$.

Conversely, suppose we are given a solution to (2,2)-TR in G using r regenerators. Since E_1 and E_2 are a partition of $E(G) = E(H)$ and the set of internal edges of the lightpaths in L_1 (resp. L_2) is exactly E_1 (resp. E_2), the regenerators placed at the endpoints of the internal edges of the lightpaths constitute a vertex cover of H of size at most r. Therefore, $\text{OPT}_{\text{VC}}(H) \leq \text{OPT}_{(2,2)-\text{TR}}(G, \{L_1, L_2\})$.

Summarizing, since $\text{OPT}_{\text{VC}}(H) = \text{OPT}_{(2,2)-\text{TR}}(G, \{L_1, L_2\})$ and any feasible solution to $\text{OPT}_{(2,2)-\text{TR}}(G, \{L_1, L_2\})$ using r regenerators defines a vertex cover of H of size at most r, the existence of a PTAS for $(2,2)$-TR would imply the existence of a PTAS for VERTEX COVER in the class of graphs $\mathcal{H}$, which is a contradiction by Lemma 2, unless P = NP. □

The extension of the result for general values of d and p follows:

Theorem 5. (d,p)-TR *does not admit a PTAS for any* $d \geq 2$ *and any* $p \geq 2$ *unless* P = NP, *even if the underlying graph* G *satisfies* $\Delta(G) \leq 3$ *and the lightpaths have length at most* $2d$.

Proof. See Appendix B. □

A   Proof of Theorem 2

Theorem 2 Algorithm 1 is 3-competitive for MAXPATH for feasible inputs in path topologies, when $k = 1$ and $d = 2$.

Proof. Let S and U denote the sets of paths that have been satisfied and unsatisfied by the algorithm, respectively. We prove the theorem by showing that $|U| \leq 2|S|$. Then, the competitive ratio of Algorithm 1 is $\frac{|P|}{|S|} = \frac{|U|+|S|}{|S|} \leq \frac{2|S|+|S|}{|S|} = 3$, i.e., Algorithm 1 is 3-competitive. In the sequel we prove that $|U| \leq 2|S|$ by associating with every path in U some paths of S, and showing that each path in S is associated with at most two paths in U.

Note also that, since the instance is assumed to be feasible, for every edge uv there exist *at most* two paths P_i, P_j, such that $uv \in P_i$ and $uv \in P_j$ (indeed, otherwise there would exist at least one path that is unsatisfied on the edge uv). Suppose that a path P_i presented at iteration i is unsatisfied, i.e., when P_i arrives, it cannot be satisfied by placing new regenerators. Then, there exists an edge $ab \in P_i$, where both a and b already have regenerators of paths that have been previously satisfied by the algorithm. We distinguish now two cases regarding the regenerators on vertices a and b.

Case 1: The regenerators on vertices a and b belong to two different paths P_j and P_h which have been satisfied previously by the algorithm, i.e., $reg(a, P_j) = reg(b, P_h) = 1$, with $j, h < i$ and $j \neq h$.

We first consider the cases where $ab \in P_j$ or $ab \in P_h$. Suppose that $ab \in P_j$. Then, since also $ab \in P_i$ by assumption, it follows that $ab \notin P_h$, since the instance is feasible. That is, b is an endpoint of P_h. In this case, associate the unsatisfied path P_i to the satisfied path P_h. Suppose now that $ab \in P_h$. Then it follows similarly that $ab \notin P_j$, and thus a is an endpoint of P_j. In this case, associate the unsatisfied path P_i to the satisfied path P_j.

Suppose now that $ab \notin P_j$ and $ab \notin P_h$, i.e., a is an endpoint of P_j and b is an endpoint of P_h. If there exists another path P_ℓ that is left unsatisfied by the algorithm, such that $ab \in P_\ell$, then associate the unsatisfied paths $\{P_i, P_\ell\}$ to the satisfied paths $\{P_j, P_h\}$. Otherwise, if no such path P_ℓ exists, then associate the path P_i to either P_j or P_h.

Case 2: The regenerators on a and b belong to the same path P_j which has been satisfied previously by the algorithm, i.e., $reg(a, P_j) = reg(b, P_j) = 1$, where $j < i$.

The edge $ab \in P_j$. Furthermore, neither a nor b is an endpoint of path P_j, since otherwise Algorithm 1 would not place a regenerator on both vertices a and b of path P_j. That is, there exist two vertices d, c of P_j, such that (d, a, b, c) is a subpath of P_j. Moreover, since a and b are consecutive vertices of P_j, according to the algorithm there must exist two other satisfied paths P_h, P_ℓ, such that $reg(d, P_h) = reg(c, P_\ell) = 1$. (Note: Here we simplify the discussion slightly by assuming that the path P_i does not contain a chain of two internal edges that both do not belong to any other paths because the algorithm can simply assign regenerators to alternate internal nodes without conflicting any other paths and this would not affect the number of paths that can be satisfied by the algorithm.)

Note also that $ab \notin P_h$ and $ab \notin P_\ell$, since the instance is feasible, and since $ab \in P_i$ and $ab \in P_j$. That is, d or a is an endpoint of P_h, while b or c is an endpoint of P_ℓ.

We claim that there exist *at most* two different unsatisfied paths P_i and $P_{i'}$ that include at least one of the edges da, ab, bc. Suppose otherwise that there exist three such unsatisfied paths P_i, $P_{i'}$, $P_{i''}$. Recall that $ab \in P_i$ and that $da, ab, bc \in P_j$. Therefore, since the instance is assumed to be feasible, it follows that, either $da \in P_{i'}$ and $bc \in P_{i''}$, or $bc \in P_{i'}$ and $da \in P_{i''}$. Since these cases are symmetric, we assume without loss of generality that $da \in P_{i'}$ and $bc \in P_{i''}$. In any optimal (i.e., offline) solution, at least one of $\{a, b\}$ has a regenerator for path P_j; assume without loss of generality that $reg(b, P_j) = 1$ (the other case $reg(a, P_j) = 1$ is symmetric). Then, it follows that $reg(a, P_i) = 1$. Then, since the edge da must be satisfied for both paths P_j and $P_{i'}$, it follows that $reg(d, P_j) = reg(d, P_{i'}) = 1$. This is a contradiction, since every vertex can have at most one regenerator. Therefore there exist at most two different unsatisfied paths P_i, $P_{i'}$ that include at least one of the edges da, ab, bc.

In the case that P_i is the only unsatisfied path that includes at least one of the edges da, ab, bc, associate the unsatisfied path P_i to either the satisfied path P_h or to the satisfied path P_ℓ. Otherwise, if there exist two different unsatisfied paths P_i, $P_{i'}$ that include at least one of the edges da, ab, bc, associate the unsatisfied paths $\{P_i, P_{i'}\}$ to the satisfied paths $\{P_h, P_\ell\}$.

We observe that by the above associations of unsatisfied paths to satisfied ones, that at most two unsatisfied paths are associated to every satisfied path P (i.e., at most one to the left side and one to the right side of P, respectively). This gives $|U| \leq 2|S|$ and the theorem follows. $\square$

B   Proof of Theorem 5

Theorem 5 (d, p)-TR does not admit a PTAS for any $d \geq 2$ and any $p \geq 2$ unless P = NP, even if the underlying graph G satisfies $\Delta(G) \leq 3$ and the lightpaths have length at most $2d$.

Proof. The case $d = p = 2$ was proved in Proposition 4. We next prove the result for $p = 2$ and arbitrary $d \geq 2$. Again, the reduction is from VERTEX COVER in the class of graphs $\mathcal{H}$ defined in Lemma 2. Given a graph $H \in \mathcal{H}$ as instance of VERTEX COVER, we partition $E(H)$ into E_1 and E_2 according to the two-coloring given by Lemma 3.

In order to build G, we associate a *parity* to the edges of H as follows. Recall that the vertices of H have degree 2 or 3. From the set of paths $\mathcal{P}$ given by Lemma 3, we build a set of paths $\mathcal{P}'$ as follows. If a vertex v appears in $\mathcal{P}$ as an endpoint of two paths P_1 and P_2 (necessarily, of different color), we merge them to build a new longer path, and add it to $\mathcal{P}'$. We orient each path $P \in \mathcal{P}'$ arbitrarily, and define the parity of the edges of P accordingly (the first edge being odd, the second even, and so on). This defines the parity of all the edges in $E(H)$.

We now subdivide the edges of H as follows. We distinguish two cases depending on the value of d. For each $P \in \mathcal{P}$:

- If $d \geq 2$ is even, we subdivide $\frac{d}{2} - 1$ times each edge of P (that is, we introduce $\frac{d}{2} - 1$ new vertices for each edge of P).
- If $d \geq 3$ is odd, we subdivide $\frac{d-1}{2}$ times each odd edge of P, and $\frac{d-3}{2}$ times each even edge of P.

This completes the construction of G. Note that $\Delta(G) \leq 3$. We call the vertices of G corresponding to vertices of H *black*, the other ones being *white*. An example of this construction is illustrated in Fig. 2(b) for several values of d in a path P with 2 edges. We now have to define the two sets of lightpaths. Again, each path of $H[E_1]$ (resp. $H[E_2]$) will correspond to a lightpath in L_1 (resp. L_2), but now we have to be more careful with the first and last edges of the lightpaths. Namely, we will construct the lightpaths in such a way that the parities of the corresponding edges of H alternate.

Let P be a path of the two-coloring of $E(H)$ with endpoints u and v. We will argue about u, and the same procedure applies to v. We distinguish two cases according to the degree of u in H. In both cases, we will associate a vertex $u' \in V(H)$ with u. First, if u has degree 2 in H, let u' be the neighbor of u in $V(H) \backslash V(P)$ (recall that $u' \in V(H) \backslash V(P)$ as H has girth at least 9). Note that by the definition of the parity of the edges of H, the edge $\{u', u\}$ has different parity from the edge of P containing u. Otherwise, u has degree 3 in H, and let u' and u'' be the two neighbors of u in $V(H) \setminus V(P)$. By the properties of the two-coloring given by Lemma 3, $\{u', u\}$ and $\{u'', u\}$ are two consecutive edges in a path of the two-coloring, hence they have different parity. Let without loss of generality $\{u', u\}$ have different parity from the edge of P containing u. Equivalently, the same discussion determines another vertex $v' \in V(H)$ associated with v (note that $u' \neq v'$ due to the high girth of H).

Then the lightpath associated with P consists of the concatenation of the edges in G corresponding to $\{u', u\}$, P, and $\{v, v'\}$. This completes the construction of the instance of $(d, 2)$-TR. Note that for both d even or odd, the length of the lighpaths is at most $2d$. Note also that the case $d = 2$ is consistent with the proof of Proposition 4. As in the case $d = 2$, we now claim that $\text{OPT}_{\text{VC}}(H) = \text{OPT}_{(d,2)-\text{TR}}(G, \{L_1, L_2\})$.

Let $S \subseteq V(H)$ be a vertex cover of H. Place one regenerator at each black vertex of G corresponding to a vertex in S; this defines a feasible solution to $(d, 2)$-TR with cost $|S|$. Indeed, at least one of every 2 consecutive black vertices of each lightpath ℓ hosts a regenerator, so the maximum distance in a lightpath without meeting a regenerator is bounded by the distance between the first and the last black vertex in a sequence of 3 consecutive black vertices, which is exactly d for both d even and odd (see Fig. 2(b)). Therefore, $\text{OPT}_{\text{VC}}(H) \geq \text{OPT}_{(d,2)-\text{TR}}(G, \{L_1, L_2\})$.

Conversely, given a solution to $(d, 2)$-TR in G using r regenerators, we perform the following transformation to each lightpath ℓ: let v_1 and v_2 be two consecutive black vertices in ℓ, and assume that v_1 is on the left of v_2 in the chosen orientation of the path corresponding to ℓ. If there are any regenerators at the white vertices between v_1 and v_2, we remove them and put a regenerator at v_1, if there was no regenerator before. We perform this operation for any two consecutive black vertices of each lightpath, inductively from right to left. This defines another feasible solution to $(d, 2)$-TR in G using at most r regenerators, since there is no lightpath with two consecutive black vertices without a regenerator. Indeed, if there were two consecutive black vertices without a regenerator after the described transformation, it would imply that the original solution was not feasible, a contradiction. The latter property implies that the regenerators at the black vertices constitute a vertex cover of H of size at most r. Therefore, $\text{OPT}_{\text{VC}}(H) \leq \text{OPT}_{(d,2)-\text{TR}}(G, \{L_1, L_2\})$.

That is, the existence of a PTAS for $(d, 2)$-TR would imply the existence of a PTAS for VERTEX COVER in the class of graphs $\mathcal{H}$, which is a contradiction by Lemma 2, unless $\text{P} = \text{NP}$.

For $p \geq 2$, it suffices to further refine in an arbitrary way the partition of $E(H)$ given by Lemma 3 into p sets of edges, which correspond to the p sets of lightpaths. For instance, if $p = 5$, we can partition $E(H_1)$ (resp. $E(H_2)$) into 2 (resp. 3) sets of paths. Then, the same proof presented above carries over to any $p \geq 2$. □

References

1. Akiyama, J., Chvátal, V.: A short proof of the linear arboricity for cubic graphs. Bull. Lib. Arts Sci. Nippon Med. Sch. **2**, 1–3 (1981)
2. Alimonti, P., Kann, V.: Some APX-completeness results for cubic graphs. Theor. Comput. Sci. **237**(1–2), 123–134 (2000)
3. Bermond, J.-C., Fouquet, J.L., Habib, M., Péroche, B.: On linear k-arboricity. Discret. Math. **52**, 123–132 (1984)

4. Borodin, A., El-Yaniv, R.: Online Computation and Competitive Analysis. Cambridge University Press, Cambridge (1998)
5. Chen, S., Ljubic, I., Raghavan, S.: The regenerator location problem. Networks **55**(3), 205–220 (2010)
6. Călinescu, G., Frieder, O., Wan, P.-J.: Minimizing electronic line terminals for automatic ring protection in general wdm optical networks. IEEE J. Sel. Area Commun. **20**(1), 183–189 (2002)
7. Călinescu, G., Wan, P.-J.: Traffic partition in wdm/sonet rings to minimize sonet adms. J. Comb. Optim. **6**(4), 425–453 (2002)
8. Eilam, T., Moran, S., Zaks, S.: Lightpath arrangement in survivable rings to minimize the switching cost. IEEE J. Sel. Area Commun. **20**(1), 172–182 (2002)
9. Epstein, L., Levin, A.: Better bounds for minimizing SONET ADMs. In: Persiano, G., Solis-Oba, R. (eds.) WAOA 2004. LNCS, vol. 3351. Springer, Heidelberg (2005)
10. Fedrizzi, R., Galimberti, G.M., Gerstel, O., Martinelli, G., Salvadori, E., Saradhi, C.V., Tanzi, A., Zanardi, A.: Traffic independent heuristics for regenerator site selection for providing any-to-any optical connectivity. In: Proceedings of IEEE/OSA Conference on Optical Fiber Communications (OFC) (2010)
11. Flammini, M., Marchetti-Spaccamela, A., Monaco, G., Moscardelli, L., Zaks, S.: On the complexity of the regenerator placement problem in optical networks. IEEE/ACM Trans. Netw. **19**(2), 498–511 (2011)
12. Flammini, M., Shalom, M., Zaks, S.: On minimizing the number of ADMs in a general topology optical network. In: Dolev, S. (ed.) DISC 2006. LNCS, vol. 4167, pp. 459–473. Springer, Heidelberg (2006)
13. Gerstel, O., Lin, P., Sasaki, G.: Wavelength assignment in a wdm ring to minimize cost of embedded sonet rings. In: INFOCOM 1998, Seventeenth Annual Joint Conference of the IEEE Computer and Communications Societies (1998)
14. Kim, S.W., Seo, S.W.: Regenerator placement algorithms for connection establishment in all-optical networks. IEEE Proc. Commun. **148**(1), 25–30 (2001)
15. Leonardi, S., Vitaletti, A.: Randomized lower bounds for online path coloring. In: Rolim, J.D.P., Serna, M., Luby, M. (eds.) RANDOM 1998. LNCS, vol. 1518, pp. 232–247. Springer, Heidelberg (1998)
16. Mertzios, G.B., Sau, I., Shalom, M., Zaks, S.: Placing regenerators in optical networks to satisfy multiple sets of requests. IEEE Trans. Netw. **20**(6), 1870–1879 (2012)
17. Mertzios, G.B., Shalom, M., Wong, P.W.H., Zaks, S.: Online regenerator placement. In: Fernàndez Anta, A., Lipari, G., Roy, M. (eds.) OPODIS 2011. LNCS, vol. 7109, pp. 4–17. Springer, Heidelberg (2011)
18. Pachnicke, S., Paschenda, T., Krummrich, P.M.: Physical impairment based regenerator placement and routing in translucent optical networks. In: Optical Fiber Communication Conference and Exposition and The National Fiber Optic Engineers Conference, p. OWA2. Optical Society of America (2008)
19. Shalom, M., Wong, P.W., Zaks, S.: Optimal on-line colorings for minimizing the number of adms in optical networks. J. Discret. Algorithms **8**(2), 174–188 (2010)
20. Shalom, M., Zaks, S.: A 10/7 + ε approximation scheme for minimizing the number of adms in sonet rings. In: First Annual International Conference on Broadband Networks, pp. 254–262, San-José, California, USA, October 2004
21. Sriram, K., Griffith, D., Su, R., Golmie, N.: Static vs. dynamic regenerator assignment in optical switches: models and cost trade-offs. In: Workshop on High Performance Switching and Routing (HPSR), pp. 151–155 (2004)

22. Thomassen, C.: Two-coloring the edges of a cubic graph such that each monochromatic component is a path of length at most 5. J. Comb. Theory, Ser. B, **75**(1), 100–109 (1999)
23. Yang, X., Ramamurthy, B.: Dynamic routing in translucent WDM optical networks. In: Proceedings of the IEEE International Conference on Communications (ICC), pp. 955–971 (2002)
24. Yang, X., Ramamurthy, B.: Sparse regeneration in translucent wavelength-routed optical networks: architecture, network design and wavelength routing. Photonic Netw. Commun. **10**(1), 39–53 (2005)

Efficient, Reliable, and Secure Smart Energy Networks

A Modular and Flexible Network Architecture for Smart Grids

Stefano Tennina[1]([✉]), Dionysis Xenakis[2], Mattia Boschi[1], Marco Di Renzo[3],
Fabio Graziosi[1], and Christos Verikoukis[4]

[1] WEST Aquila S.r.l., L'Aquila, Italy
{tennina,mattia.boschi,fabio.graziosi}@westaquila.com
[2] Department of Informatics and Telecommunications,
University of Athens, Athens, Greece
nio@di.uoa.gr
[3] Laboratory of Signals and Systems (L2S), CNRS - SUPELEC - University
Paris-Sud, Paris, France
marco.direnzo@lss.supelec.fr
[4] Department of Electronic, University of Barcelona,
Barcelona, Spain
cverikoukis01@ub.edu

Abstract. The nowadays power grid deployed and used in every Country worldwide has served relatively well in providing a seamless unidirectional power supply of electricity. Nevertheless, today a new set of challenges is arising, such as the depletion of primary energy resources, the diversification of energy generation and the climate change. This paper proposes recent advancements in this field by introducing SMART-NRG, a Marie Curie project which involves academic and industrial partners from three EU Countries. The project aims to propose new technologies to meet the specific requirements of smart grids applications. In particular, in this paper, a modular and flexible system architecture is presented to face with the challenges imposed by the different application scenarios.

Keywords: Smart grids · Wireless networks · System requirements

1 Introduction

Conventional power grids lack the capability of (real-time) monitoring and controlling the various appliances to provide optimal full system-wide energy usage. Therefore, the existing power grid is evolving into a more intelligent, more responsive, more efficient, and more environmentally friendly system, known as the Smart Grid (SG) [1]. To fully exploit the power of such new SG paradigm, efficient, reliable and secure two-way information flow among the customers and the utility companies is mandatory. In this perspective, the smart meters are new devices with the necessary intelligence to enable such two-way communication, since they are capable of collecting and delivering punctual power consumption information to remote utilities much more efficiently than conventional

© Springer International Publishing Switzerland 2015
S. Papavassiliou and S. Ruehrup (Eds.): ADHOC-NOW 2015, LNCS 9143, pp. 273–287, 2015.
DOI:10.1007/978-3-319-19662-6_19

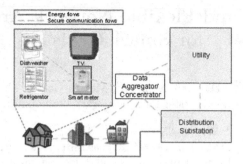

Fig. 1. An example of the Smart Energy Network [3].

meters [2]. By using this information, utility companies are able to setup Smart Energy Networks (SENs) to better monitor the energy consumption, to advice on changes of users' behavior and to adapt the generation of energy accordingly, as sketched in Fig. 1 [3]. For instance, in a home environment, the smart meter collects the power consumption information of the dishwasher, TV and the refrigerator, and also sends the control commands to them, if necessary. The data generated by the smart meters in different buildings is transmitted to a data aggregator. This aggregator could be an access point or gateway. This data can be further routed to the electric utility or the distribution substation.

At present, much effort is being made by the utility companies to install these smart meters into millions of homes across Europe. European regulations require member nations to ensure that 80 % of residential households are fitted with a smart meter by 2020 [4]. With the help of these densely deployed smart meters, SENs offer numerous benefits for consumers, operators, and the community as a whole. As for the consumers, SENs will help moderating their energy usage to reduce waste, lower their monthly bill, and use power in a more sustainable way. As for the operators/industry, SENs will help preventing outages, shorten the response time to problems, reduce cost and increase efficiency, and allow operators to resolve issues remotely. It will also integrate renewable energy and reduce carbon emissions on a macro and micro level. As for the community as a whole, SENs will help creating a safer, more secure, more reliable grid, and will reduce dependency on foreign energy supplies, thus reducing carbon emissions.

Three functionalities are instrumental for the long-term success of SENs: (*i*) Efficient and reliable networking protocols, to enable two-way communications between the smart meters and the utility companies in order to ensure that the power control centers are timely informed of the energy consumption throughout the electricity grid. (*ii*) Robust and real-time smart energy/management algorithms, to guarantee the adequate generation of energy and to minimize power outages by timely responding to the demands of the customers and by taking into account the stochastic availability of renewable energy sources and future electric vehicles. (*iii*) Secure and trusted mechanisms and algorithms, to guarantee that sensing, communications, computing, and control operations performed over the entire SEN are robust against eavesdroppers, intruders/jammers, and cyber-threats, which can compromise energy generation, distribution and control, as well as the privacy of customers' data. Accordingly, SENs can be regarded as an electric system, which uses two-way networking technologies, cyber-secure communications technologies, computational intelligence and control in an integrated fashion across electricity generation, transmission, substations, distribution and consumption, with the aim of providing a new electricity grid that is clean, safe, secure, reliable, resilient, efficient and sustainable.

In terms of the system architecture, SENs will include many devices that are not present in the current power grid (e.g., smart meters, data collectors, data centers, user terminals, etc.). Thus, its architecture must consider the integration among heterogeneous (sub-)systems and reflect the evolution of the Internet, where the customers can manage their energy bills in the same way as they manage on-line services. Heterogeneity leads to a necessary level of flexibility of the architecture to seamlessly handle different requirements for security, timeliness and bandwidth for different scenarios, e.g., homes and industrial facilities. Thus, SENs can be well regarded as a system of many systems, whose design challenges, requirements and expectations can only be achieved through a holistic analysis, design and optimization of every component. To leverage the potential benefits of SENs and to meet its design challenges, in this paper we aim to overview the SMART-NRG project [5] and the main requirements characterizing different operating scenarios, which will guide us to the definition of an adequate and flexible system architecture, where innovative communications and networking, smart energy management, and security and protection techniques will be allocated. Consequently, the remaining of this paper is organized as follows. Section 2 overviews the key performance indicators (KPIs) and requirements considered in the frame of the SMART-NRG project. Section 3 presents the four application scenarios considered. Section 4 sketches the system architecture scheme and Sect. 5 concludes the paper.

2 Key Performance Indicators

Three sets of KPIs are presented for each one of the main functional domains of interest in SMART-NRG: *communication*, *security* and *energy management*.

2.1 Communication KPIs

In terms of communications and networking, a list of requirements and performance targets can be identified by focusing on the two functional domains more relevant to this project: the advanced metering infrastructure (AMI) [6] and the Demand Response (DR) [7] domains.

The AMI lies in between the customer's premises and the nearest substation of the utility company and is required to support functions such as: (*i*) remote meter reading for billing, (*ii*) remote connect/disconnect capabilities, (*iii*) outage detection and management, (*iv*) tamper/theft detection, (*v*) short interval energy readings, which serve as the basis for market-based energy rates, and

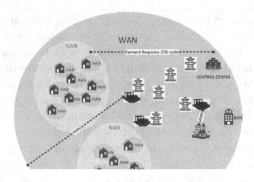

Fig. 2. Network domains in SMART-NRG.

(*vi*) distributed generation, monitoring and management. On the other hand, the DR cycle aims at reducing the energy consumption of the customers in response to an increase in the price of electricity or heavy burdens on the system, spanning all communication networks in between the smart meters at the customer's premises and the data center at the utility company. Accordingly, the communications requirements for the DR cycle vary with respect to the complexity of the system. For instance, it can be used (*i*) to shut-off an appliance, in case of low bandwidth requirements, (*ii*) perform load balancing, upon which the responsiveness of the system may not be critical and latency can be higher, and (*iii*) address imminent emergencies, such as overload conditions, requiring low latency in the message traveling across the network.

To better understand the communication networks that reside in between the area of interest in SMART-NRG, i.e., the end-to-end communication network from the customer's premises to the data center that is responsible for metering data processing, security processing and energy management, at the utility company, we consider four main types of network domains (Fig. 2): home area networks (HANs) at the user end, building area networks (BANs) at the building feeder, neighborhood area networks (NANs) to reach the local energy manager of the utility company, and wide area networks (WANs) to reach the central premises of the utility company. A WAN covers multiple NANs, a NAN is formed by a number of BANs, while a BAN is formed by a number of HANs. Depending on the scenario under scope, the BAN may not be present, e.g., in rural areas. In any case, we emphasize on the complete architecture that additionally considers data aggregation at BANs, e.g., the basement of a building, as it is relevant to the SMART-NRG project.

Aiming to provide a clear yet comprehensive specification of the communication requirements for the different network domains, in the following we overview some of the main metrics that will form the basis for the design, analysis and evaluation of the communication solutions developed in the SMART-NRG project.

– Throughput: the number of bits per second that can be successfully transmitted/received in a typical connection between two tagged nodes.

- Capacity: the maximum number of bits per second that can be successfully transmitted/received in a typical connection between any nodes.
- Spatial and time availability: the percentage of the area/time within which two communication nodes can achieve a prescribed mean Signal to Interference plus Noise Ratio (SINR) threshold or throughput target.
- Latency: the average time required to successfully transmit/receive a packet of a specific traffic type for the communication technology under scope.
- Jitter: the deviation from a predefined periodicity of a packet transmission
- Energy-efficiency: the energy consumption per bit sent from the communication infrastructure measured in a prescribed time period.
- Communication reliability: the number of communication outages identified by the communication infrastructure over those reported by the customers.
- Interoperability: the capability of two or more networks, systems, devices, applications, or components to exchange and readily use information securely, effectively, and with little or no inconvenience to the user.

2.2   Security KPIs

In terms of security in the context of smart metering, the objectives are different from other smart grid operations, where priority is often given to guarantee data availability. Current literature identifies a plethora of security requirements/objectives for the Smart Grid. Nevertheless, confidentiality, data integrity, authentication, availability of resources and time-responsiveness of the power equipment are considered critical performance indicators for the future smart grid. Given the specific domain of interest, e.g., utility servers, HAN, BAN, NAN, and WAN, in the following we overview some of the main metrics that will form the basis for the design, analysis and evaluation of the security solutions developed in the SMART-NRG project.

- Number of security attacks/breaches: this metric measures the number of security attacks attempted or breaches identified on a predefined time period.
- Number and percentage of affected customers: this metric measures the number of customers that have been affected by a specific attack that has been successfully identified on a predefined time period.
- Damages suffered from a security breach: this metric is a list of the damages following from a successfully identified security breach.
- Type of the attack: the type of the attack, i.e. physical or cyber attack.
- Service interruptions due to a security breach: this metric measures the duration or the number of service interruptions due to the presence of a security breach that has been successfully identified.
- Number of customers or authorized third parties accessing energy usage information remotely: the number of customers accessing energy usage information is a good indication for detecting inadvertent access to the smart meters.

2.3   Energy Management KPIs

In terms of energy management, current literature includes a plethora of solutions that can be classified based on the control mechanism of the DR procedure [8], on the motivations offered to customers to reduce or shift their demands [9], or on the DR decision variable [10]. In the following we overview some of the main metrics that will form the basis for the design, analysis and evaluation of the energy management solutions developed in the SMART-NRG project.

- System average interruption duration and frequency: how often and the total number of minutes per year of sustained outage per customer measured in a system-wide basis.
- Momentary average interruption frequency: the number of momentary outages per customer per year in a system-wide basis.
- Capacity factor of transmission/distribution system: total annual energy transmitted by the transmission or distribution system divided by the total annual energy capacity of the transmission system.
- Energy efficiency of the transmission/distribution system: energy delivered to the distribution grid/end use customers divided by energy entering the transmission system.
- Interoperability: the capability of two or more networks, systems, devices, applications, or components to exchange and readily use information, securely, effectively, and with little or no inconvenience to the user.
- Energy consumption ratio in peak/off-peaks periods: energy delivered to the distribution grid divided by energy entering the transmission system during averaged over the peak/off-peaks periods.
- Peak demand: largest power demand at any time instant.
- Electricity cost: cost of electricity paid by consumers.
- Total load covered by microgrids: the total load that is served by local energy resources (microgrids) in a specific geographical region of interest.
- Peak load shaving from the DR cycle: the peak load shaving following from the employment of DR programs in a specific geographical region of interest.
- Total decrease in demand by customer in peak/off-peak periods: the decrease in load demands from the customers due to the employment of DR programs, measured either at peak or off-peak periods.
- Access and frequency to real-time pricing/usage information: indicates the capability of a customer to access real-time pricing information on his/her energy usage and is an indicator of the active participation in the DR cycle.
- Number of outages: the number of outages due to malfunctions measured within a prescribed time period and geographical area.
- Prediction accuracy: a measure of the accuracy of the load/energy demand predictions achieved by the developed information processing algorithms.

3   SMART-NRG Scenarios

3.1   Dense Urban Scenario

In the first SMART-NRG scenario, we consider a scenario of an urban area that is dominated by household users (Fig. 3). All appliances in each household are connected to smart meters, which are all concentrated in Energy Consumption Controllers (ECC) in the basement, and report the power demands of each appliance to a Central Load Controller (CLC). In this context, we are interested in studying the efficiency of the proposed protocols in order to define the maximum number of appliances installed in the urban area under study that ensures that the outage probability is below a predefined threshold. We are also interested in studying the appropriate communication technology that is used for the communication of the smart meters to the ECC and the communication of the ECCs to the CLC that ensures secure transmission of the users' data, without inducing a considerable amount of management and computational overhead.

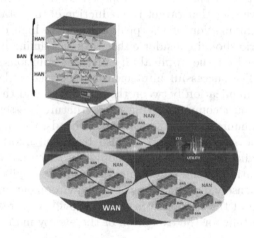

Fig. 3. Scenario 1: Dense urban.

In terms of *communication requirements*, the proposed solutions should account for the type of spectrum bands utilized by the ubiquitous networking infrastructure in order to sustain smooth coexistence of the HAN, BAN and NAN networks. The spatial and time availability of the communication infrastructure should also be taken into account, given the increased number of hops characterizing the network layout and the low power levels required to mitigate the radio interference. Another key performance requirement should be to maximize the energy-efficiency of the communication systems, given that a vastly overlapping coverage is expected in Scenario 1, due to the high network density of sensors and smart meters within a small geographical area. Smart energy saving protocols and low-power communications should be thus employed to minimize the energy overhead following from the concurrent operation of the different smart grid networking domains. The strategic deployment of smart meters is also expected to play a key role in mitigating intra-tier interference in the NAN network and allow seamless exchange of energy measurements/commands towards the ECC at the basement of the buildings. In terms of *security requirements*, the proposed solutions should exhibit increased reliability and robustness to cope with the openness of the communication infrastructure between the neighboring BANs. Data integrity, user privacy and non repudiation, should be the key pillars for safeguarding the smooth operation of the smart grid communication infrastructure while keeping the required communication and processing overheads at low levels. Emphasis should be given in effectively identifying the type of ongoing attacks and the number of customers that can be affected by a security breach in the ECC at the basement of each building. To this end, the proposed security protocols should perform anomaly detection and key management techniques in a timely yet effective manner, so as to preserve the privacy of the customers that access their energy consumption information through web interfaces. In terms of *energy management requirements*, the proposed solution should apply demand side management schemes that target the reduction of the total power consumption and/or the minimization of the peak demand in high demand hours. To this end, the scenario should consider either task scheduling or energy management methods, based on the applicability of these solutions on the nature of consumers' loads. The successful implementation of such schemes is based on the continuous communication between the consumers and the ECC; therefore, the presence of a robust communication infrastructure is essential. However, in order to avoid communication and security overheads, the application of these energy management schemes should be also based on information processing and load/price prediction models that should be executed at the ECC. This centralized coordination of the scheduling of the users-appliances' operation is essential in such a densely deployed networking environment and requires a powerful processing unit at the ECC that is able to provide near real-time processing data that are important for the execution of the energy management scheme.

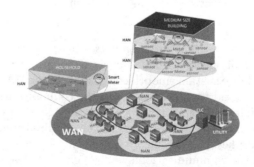

Fig. 4. Scenario 2: Sparse urban/dense rural.

3.2 Sparse Urban/Dense Rural Scenario

In the second SMART-NRG scenario, we aim to address small towns, villages or outskirts of bigger towns, where we expect a mix of medium-size buildings and households (Fig. 4). Compared to Scenario 1, the smart meters are further separated from each other due to the lower density of households per square area. In terms of communication domains, the networking layout is mainly characterized by a medium number of HANs grouped in BANs (medium-size buildings) and a number of HANs that are not grouped in BANs (individual households). In this scenario, we are interested in studying the efficiency of the proposed protocols and determine the optimum topology of the smart meter network that ensures minimum transmission delay and maximum power capacity utilization.

In terms of *communication requirements*, the proposed solutions should be able to sustain service connectivity between the (distant) smart meters without severely degrading its energy-efficiency due to high power transmissions. To achieve this, the proposed solutions should trade-off between employing multihop communications within the NANs or using the WAN to reach the central controller at the utility company. On the other hand, different from Scenario 1, a malfunctioning communication device would have a severe impact on the smart grid performance, due to the comparably lower networking density characterizing the Scenario 2. To this end, the proposed communication solutions should be robust against such malfunctioning operation of the (comparably) sparser networking infrastructure and capable of mitigating the impact of momentary outages of intermediate relays. In terms of *security requirements*, resource availability and authentication should be two of the key design principles for the developed security solutions, since denial of service type attacks and inadvertent access to the communication equipment may have a severe impact on the robustness of the smart grid communication network. To this end, KPIs such as the service interruption due to a security breach and the number of authorized third parties accessing customer energy usage information should be integral part of the optimization process in the proposed solutions. In terms of *energy management requirements*, as for Scenario 1, the ECC is responsible for the coordination of the consumers' appliances, in order to effectively execute the applied energy management methods. However, the employment of information processing and load prediction is expected to have minor impact on the overall performance of the energy management process; therefore these functions can be executed at the ECC, without the need for a powerful processing unit.

3.3 Large-Scale Grid Scenario

In the third SMART-NRG scenario, we consider an area comprising a potentially very large number of nodes. Usually, two approaches are applicable to design large scale systems: top-down and bottom-up. In the former, a high level framework of the system is modeled by a powerful grid operator and each subsystem is then refined in greater details. This approach needs a powerful operator to design at the beginning the whole architecture, which is not an easy task [3].

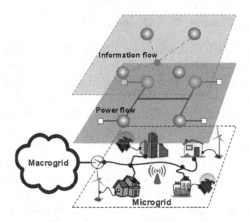

Fig. 5. Scenario 3: Large-scale grid [3].

The latter approach revolves around the definition of small systems to be composed together to form a larger one. Although it does not require a powerful operator to design the whole architecture from scratch, the final system grows up from many individually formed subsystems, which are not the output of optimized planning or centralized (and potentially optimal) management.

In SMART-NRG, we focus on the design of microgrids that could be formed in a dense urban area or in a sparse urban/dense rural area, as exemplified in Fig. 5. The lower layer shows a physical structure of the microgrid, whose main features include buildings, renewable energy generators and one access point (AP). These buildings and generators exchange power using power-lines, while they exchange information via an AP-based wireless network. The blue (top) layer shows the information flow within the microgrid and the red (middle) layer shows the power flow. With the help of such microgrids, the large scale deployment task is divided into several small basic tasks, provided that appropriate interfaces among the sub-elements are well defined. In this way, the complexity of building large scale smart grids is significantly reduced. Accordingly, in SMART-NRG we study the applicability of the proposed optimization algorithms in order to derive globally optimal decisions for the large-scale grid, under different conditions. To do so, given the intrinsic quasi-random nature of the nodes's distribution, the Stochastic Geometry [11] approach can efficiently solve the network modeling problem, since it has been widely used to compute spatial averages that capture the key dependencies of the network performance, e.g., the interference, spectral efficiency and transmission capacity, as a function of a relatively small number of system parameters.

Under the above viewpoints, the proposed *communication solutions* in Scenario 3 should account for the different coverage radius, spatial availability, and communication reliability of the individual microgrids. Nevertheless, a holistic and over-the-top joint optimization of the individual micro-grids should be developed to allow smooth coexistence and seamless interconnection in a system-wide fashion. Communication KPIs, such as latency, throughput and spatial availability, should also be part of the optimization process in a system-wide level. In terms of *security requirements*, the integrity of data, the level of authorization, as well as the authentication between the different networking entities in the individual microgrid systems should be carefully planned to avoid inadvertent equipment behaviors among the different sub-systems. To achieve this, the usage of semi-distributed security solutions should be considered, by taking into account the impact of the attack in individual microgrids against the robustness of the macrogrid as a whole and the effort required to patch a security breach in a specific microgrid. The percentage of affected customers should also be considered, since a security breach in a smaller geographical region may cause instability on the exchange of energy usage and control data information among neighboring micro-grids. In terms of *energy management requirements*, for both approaches (top-down and bottom-top) distributed energy management schemes are needed. In the top-down approach, the energy management decisions are made by the central coordinator, which are based on the information provided by the subsystems. Under this approach, a central coordinator with a powerful processing unit is necessary, in order to execute the functions of the applied energy management program in near real time. On the other hand, in the bottom-up approach, each subsystem is responsible for executing the functions of the energy management program through the direct coordination of consumers with each other, in order to achieve an aggregated power reduction.

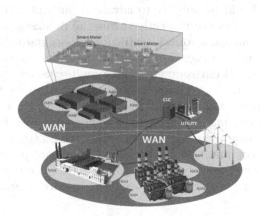

Fig. 6. Scenario 4: Industrial.

3.4 Industrial Scenario

In the fourth SMART-NRG scenario, we consider the presence of big buildings with very specific metering needs in terms of power and type of measurements (Fig. 6). Each building is also expected to have many metering points that can be either strategically or randomly deployed. Depending on the size of the industrial premises or its location, different sub-scenarios can be depicted. In SMART-NRG, we consider two main types of industrial premises. One with very big industries covering large areas and with many metering points (this case would also cover big shopping malls), which is similar to the case of large scale grid scenario. On the other hand, the case of small or medium industries, e.g., in industrial parks, is very similar to Scenario 2. Both these cases are typically characterized by harsh radio wave propagation conditions [12].

Consequently, in terms of *communication requirements*, the developed communication protocols should exhibit a high degree of communication reliability and network autonomy in order to reduce the effort required from the industrial parties to actively participate in the identification of communication outages. Energy-efficiency is also of critical importance, since the employed communication protocols should sustain a prescribed spatial and time availability of the underlying infrastructure, while reducing the operational costs required for communications in the smart grid. In terms of *security requirements*, the development of an advanced Intrusion Detection System (IDS) is of paramount importance to safeguard the power grid infrastructure and the smooth operation of the industrial equipment [13]. The number of security breaches should be kept at the minimum possible level, even if this is achieved at the cost of increased processing and signaling complexity for the effective detection of inadvertent behavior of the industrial equipment. To this end, the number of authorized third parties accessing customer energy usage information should be kept at the minimum possible level, whereas the access of the customers or other third parties in such information should be subject to advanced authentication and restricted mainly within the area of the respective industrial premises. In terms of *energy management requirements*, a robust central load coordinator is essential for the execution of the energy management programs, in order to provide reliable data, since in many industrial environments millisecond-scale monitor and control is vital. Furthermore, due to the fact that many industries cannot tolerate load adjustments, the energy management model should consider incentive-based DR programs. These programs are based on agreements between the industries and the power utilities; it is therefore necessary that the central coordinator should be equipped with computational mechanisms that can guarantee an efficient calculation of the agreement terms (based on the total power consumption or load/price predictions) that guarantees optimal resource management.

4   System Architecture

The current trend for modeling SG-specific architectures is the employment of multi-layer architectures for the SG communication infrastructure [14]. Such a layering approach enables the network engineers to reduce computing, memory and system complexity requirements for the intermediate nodes, to aggregate control/metering traffic, prioritize emergency messages, and also utilize data fusion and network coding techniques. Nevertheless, the adoption of those architectures also requires region-based DR optimization and the strategic deployment of gateways to reduce costs while simultaneously allowing the efficient and timely exchange of metering data. Such an optimization on the deployment of gateways can be efficiently achieved through the Stochastic Geometry approach [11], which takes into account the layout of the equipment as a function of the scenario under scope. Furthermore, in the classical store-and-forward paradigm used in most multi-hop relay networks, packet sent from a receiver traverses nodes in a network unchanged until it reaches the intended destination. Thus, the nodes in the network do not modify the data in the transmitted packet, hence the name store-and-forward. Ahlswede et. al. [15] demonstrated that in some topologies the max-flow min-cut bound [16] cannot be obtained by existing routing protocols that are based on the premise of store-and-forward. To solve this problem they proposed Network Coding (NC), which breaks with this paradigm by enabling nodes in the network in between the source and the destination to modify packets in order to make them more suitable for transmission. Hence, NC follows the compute-and-forward paradigm providing new ways to increase throughput, reliability, performance, security, error correction and operation.

Accordingly, the main technologies for the communication system that constitutes the reference architecture for SMART-NRG are depicted in Fig. 7. In particular, short range communication technologies (e.g., ZigBee/IEEE 802.15.4) are used in home and office indoor environments, with the goal of reaching the smart meter from every electrical equipment with the least number of intermediate hops to minimize the communication delays. Then, to reinforce the reli-

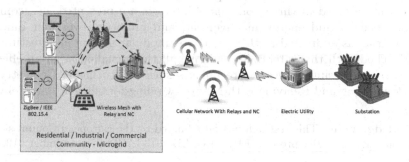

Fig. 7. An example of the SMART-NRG system (inspired by [3]).

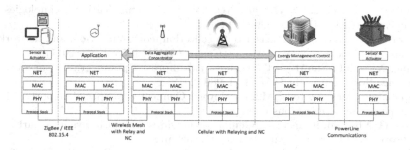

Fig. 8. Communication protocol architecture.

ability of the system, a wireless mesh based network is built among the smart meters to reach the access point for medium-long range communications (e.g., WLAN/IEEE 802.11). The mesh network intrinsically ensures that in the presence of nodes or links faults, the presence of multi-path guarantees that information is delivered from the source(s) to the access point, which will play the role of data aggregator. Moreover, the possibility to implement Relay-aided and NC schemes further helps improving reliability and scalability of the system. Then, long range communications (e.g., cellular and LTE based networks), once again powered by Relay-aided NC schemes, implement the two-ways connection of the microgrid with the electric utility (and then the substation). This three tiers network scheme is modular and flexible in the sense that moving from, e.g., Scenario 1 to Scenario 2, the middle tier of short range communications can be skipped wherever the nodes are far apart from each other, without compromising the overall system. Finally, Fig. 8 sketches the composition of these communication protocol stacks into the full picture of the system architecture, as currently under development in our System Level Simulator, built upon the OMNeT++ framework [17].

5 Conclusions

In this paper we have briefly outlined the design challenges of a Smart Grid system as part of the SMART-NRG project. In particular, a set of application requirements divided in three domains of interests are presented, i.e., communication, security and energy management, and projected over four different application areas addressed in this project. A modular system architecture has been sketched with the potential to address all those challenges in an efficient way, by encompassing innovative techniques such as Stochastic Geometry for network planning and Relaying with Network Coding for efficient data delivery.

Acknowledgement. This research has been funded by the European Commission as part of the SMART-NRG project (FP7-PEOPLE-2013-IAPP Grant number 612294).

References

1. Gungor, V.C., et al.: Smart grid technologies: communication technologies and standards. IEEE TII **7**(4), 529–539 (2011)
2. Kayastha, N.: Smart grid sensor data collection, communication, and networking: a tutorial. Wireless Comms. Mobile Comp. **14**(11), 1055–1087 (2014)
3. Fang, X., et al.: Smart grid" the new and improved power grid: a survey. IEEE Comms. Surv. Tutorials **14**(4), 944–980 (2012)
4. Communications, E. (ed.) M2M - Beyond Connectivity Special Report, vol. Q3. European Communications (2012)
5. SMART-NRG: an industry-academia partnerships and pathways for smart energy networks (2014). http://gain.di.uoa.gr/smart-nrg
6. Wenpeng, L.: Advanced metering infrastructure. South. Power Sys. Tech. **3**(2), 6 (2009)

7. Palensky, P., et al.: Demand side management: demand response, intelligent energy systems, and smart loads. IEEE TII **7**(3), 381–388 (2011)
8. Alagoz, B., et al.: A user-mode distributed energy management architecture for smart grid applications. Energy **44**(1), 167–177 (2012)
9. Wissner, M.: The smart grid-a saucerful of secrets? Appl. Energy **88**(7), 2509–2518 (2011)
10. Dong, Q., et al.: Distributed demand and response algorithm for optimizing social-welfare in smart grid. In: IEEE IPDPS, pp. 1228–1239 (2012)
11. Xenakis, D., et al.: Mobility management for femtocells in lte-advanced: key aspects and survey of handover decision algorithms. IEEE Comms. Surv. Tutorials **16**(1), 64–91 (2014)
12. Gungor, V.C., et al.: Opportunities and challenges of wireless sensor networks in smart grid. IEEE TII **57**(10), 3557–3564 (2010)
13. Kolias, C., Kambourakis, G., Stavrou, A., Gritzalis, S.: Intrusion detection in 802.11 networks: empirical evaluation of threats and a public dataset. IEEE Comms. Surv. Tutorials **PP**, 1 (2015). doi:10.1109/COMST.2015.2402161
14. Gharavi, H., Hu, B.: Multigate communication network for smart grid. Proc. of the IEEE **99**(6), 1028–1045 (2011)
15. Ahlswede, R., et al.: Network information flow. IEEE Trans. Inf. Theor. **46**(4), 1204–1216 (2000)
16. Elias, P., et al.: A note on the maximum flow through a network. IRE Trans. Inf. Theor. **2**(4), 117–119 (1956)
17. Omnet++ 4.x discrete event simulator (2015). www.omnetpp.org

A Linear Programming Approach for K-Resilient and Reliability-Aware Design of Large-Scale Industrial Networks

Béla Genge[(✉)], Piroska Haller, and István Kiss

"Petru Maior" University of Tîrgu Mureş,
N. Iorga, No. 1, 540088 Tîrgu Mureş, Mureş, Romania
bela.genge@ing.upm.ro, phaller@upm.ro, istvan.kiss@stud.upm.ro

Abstract. The profound transformation of large-scale Industrial Control Systems (ICS), e.g., smart energy networks (Smart Grids), from a proprietary and isolated environment to a modern architecture brings several new challenges. Nowadays, ICS network designers need to accommodate a variety of devices and communication media/protocols with industry-specific requirements pertaining to real-time delivery of data packets, reliability, and resilience of communication networks. Therefore, this work proposes a novel network design methodology formulated as a Mixed Integer Linear Programming (MILP) problem. The developed problem accounts for different data flows routed across an overlay network of concentrators and embodies traditional ICS design requirements defined as linear constraints. Furthermore, the MILP problem defines a K-resilience factor to ensure the installation of K back-up paths, and a linear reliability constraint adapted from the field of fuzzy logic optimization. Experimental results demonstrate the efficiency and scalability of the proposed MILP problem.

Keywords: Industrial Control Systems · Energy networks · Smart Grid · Linear Programming · Network design · Resilience · Reliability

1 Introduction

Industrial Control Systems (ICS) have undergone a profound transformation from a traditional, isolated environment to a modern industrial architecture. Nowadays, ICS encompass a wide variety of devices, software, and communication protocols which paved the way towards the implementation of new features and operational paradigms such as smart electricity networks, more generally known as Smart Grid.

This ongoing transformation is a consequence of recognizing the importance of these infrastructures which are essential to the smooth functioning of our society. Due to their significant role, large-scale infrastructures such as smart electricity grids, transportation systems, oil and gas pipelines, are often characterized as Critical Infrastructures (CI). Without any doubt, one of the most debated

© Springer International Publishing Switzerland 2015
S. Papavassiliou and S. Ruehrup (Eds.): ADHOC-NOW 2015, LNCS 9143, pp. 288–302, 2015.
DOI:10.1007/978-3-319-19662-6_20

and globally recognized CI is the electricity grid. In this direction the importance of upgrading the current infrastructure and its transformation towards the next electricity grid has been publicly acknowledged by academia, policy makers, and private companies as well. In this respect the US Department of Energy (DoE) [1] expressed the need to accelerate the deployment of advanced communication, control, and generally speaking the integration of modern Information and Communication Technologies (ICT), aiming at the modernization of the electricity grid. Subsequently, the DoE has set several performance targets for 2030, among which the reduction of peak energy demands, the efficient utilization of assets, and the integration of renewable energy resources. In Europe, on the other hand, we find several programs aiming at upgrading the current infrastructure and integrating novel Smart Grid applications [2,3].

In light of these advancements, however, ICS network designers are faced with new challenges which need to accommodate a variety of devices and communication protocols on one hand, and industry-specific requirements on the other hand. In this respect, in the list of common ICS-specific design requirement parameters we find real-time delivery of data packets, reliability of communications, and communications resilience. These parameters are well-defined by international regulators/standardization bodies in the field of ICS. The National Institute of Standards and Technology (NIST) working group on Smart Grid [4], for instance, has identified the requirements for each of the aforementioned ICS parameters in the context of specific Smart Grid applications. Examples in this sense include the reliability of communications between 98 % and 99.5 %, and a latency of less than 10 s for electric vehicle charging applications.

Based on these issues this work proposes a comprehensive methodology to design communication networks for large-scale ICS. The network design is formulated as a Mixed Integer Linear Programming (MILP) problem that accounts for data flows between different cyber assets, as well as traditional ICS design requirements pertaining to real-time traffic, reliability, and resilience. The developed methodology aims at harmonizing reliability and resilience aspects of ICS network design with well-known quality of service prerequisites, e.g., network latency, in order to deliver a modern communication infrastructure. By doing so, ICS network designers may integrate various design requirements into a methodology that minimizes the installation costs while ensuring that critical ICS-specific requirements are satisfied. It is also noteworthy that the proposed methodology may be used together with various other techniques as well. In this respect, for instance, the outcome of risk assessment given as a set of critical assets, may provide the necessary input to the methodology proposed in this paper. This way, a special emphasis may be placed on critical cyber assets in order to ensure an increased resilience and reliability factor to communications targeting such assets.

In particular, the proposed technique encompasses a *K-resilience* configuration factor to enable the deployment of K back-up communication paths for each main communication path. In this respect, for each data flow we define a *resilient communication group* (RCG), which encompasses the main data flow's communication path as well as all of its associated back-up communication paths. Reliability constraints are adopted from the traditional field of system safety, and are adapted to the present problem through the linear transformations proposed by Chiang *et al.* in the field of *fuzzy logic* optimization [5]. The proposed MILP problem is experimentally evaluated from several perspectives. More specifically, we assess the effect of various costs, resilience, and reliability parameters, on the solutions generated by the MILP model.

The remainder of this paper is organized as follows. An overview of related work is presented in Sect. 2. The problem statement together with a detailed description of the proposed methodology are given in Sect. 3. Then, the technique is experimentally evaluated from several perspectives in Sect. 4. The paper concludes in Sect. 5.

2   Related Work

This section provides a brief overview of related studies in the field of resilience and reliability-aware network design for ICS.

In [6], Carro Calvo *et al.* developed a genetic algorithm for optimal ICS network partitioning. The approach followed the traditional principles of ICS network design aimed at maximizing intra-network communications, minimizing inter-network communications, and balancing the communication over the resultant sub-networks. In [7], Zhou *et al.* formulated an optimization problem for ICS network design, which assumed a hierarchical switch-based ICS topology and incorporated costs in terms of the number of switches and of port utilization rates, traffic load balancing, as well as real-time traffic requirements expressed in terms of delay. The work of Zhang *et al.* in [8] focused on optimal ICS network design from the perspective of minimizing network delays. A relative delay metric was adopted to minimize communication delays with respect to the maximum tolerable delay. The same authors formulated in [9] an optimization problem which additionally incorporated ICS network reliability requirements in the form of probability of link failure. In [10], controllability analysis was used to reduce the size of the network design problem in water distribution systems by identifying the sub-set of nodes that fully cover the monitored physical process. Finally, the work of Zahidi *et al.* [11] showed that modern Integer Linear Programming (ILP) solvers are able to handle large network topologies. The effectiveness of ILP was demonstrated in optimizing the formation of clusters in Mobile Ad-Hoc Networks (MANETs).

Despite the variety of existing approaches the design problem of large-scale ICS networks has not been properly addressed yet. In the domain of the aforementioned studies, the approach presented in this paper is mostly related to the work of Zhang *et al.* [8,9]. However, our methodology is more suited to large-scale ICS design due to the special emphasis on traffic flows and concentrator site installations. Additionally, we propose a K-resilience factor, which is not available in related work, and we adapt the non-linear reliability constraints to the present linear problem by applying a transformation originating from the field of *fuzzy logic* optimization [5].

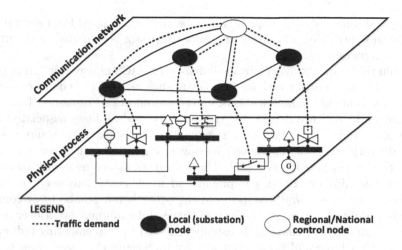

Fig. 1. Cyber-physical architecture of large-scale Industrial Control Systems.

3   Proposed Network Design Model

This section presents the proposed network design problem. It starts with the problem statement and it continues with the definition of basic sets and symbols. Then, it provides a description of the objective function and of MILP constraints.

3.1   Problem Statement

We assume the cyber-physical architecture of a large-scale ICS as illustrated in Fig. 1. The central focus of this architecture are data exchanges between different end-points, depicted with dotted lines. Data flows, hereinafter called *traffic demands*, have a well-defined purpose. Their role is to enable the implementation of various control loops ranging from local actuation strategies, to large-scale/global decision-making algorithms. In the specific context of Smart Grid, for example, traffic demands ensure the timely delivery of price information announcements in smart customer energy management applications, the transfer of smart meter data to the utility, as well as the implementation of complex demand response programs aiming at a more effective energy planning and ultimately to the stability of the energy grid.

In this architecture traffic demands are routed across a hybrid hierarchical-meshed topology consisting of data concentrator nodes. These are well-known elements in the architecture of smart energy networks, and are mainly destined to aggregate and forward data from/into various communication mediums. Data concentrators may be implemented in various locations to aggregate data from various data networks. In this respect, at the home level, data concentrators aggregate traffic from smart meters, and different other devices specific to Smart Grid applications. Then, for each building data is aggregated from several different home networks, while for each neighborhood, data is aggregated from several building area networks. Subsequently, data is then aggregated from several networks at substation level in order to ensure that data is available to the utility and to customers.

Within this complex cyber-physical architecture ICS network design requirements need to account for various parameters such as: real-time delivery of data packets, reliability of communications, and communications resilience. Real-time requirements in ICS are defined for the majority of data flows associated with critical control loops. For instance, a common requirement is to ensure a strict packet delivery time for high-speed messages, e.g., alerts, in a certain range, e.g., in the 2 ms–10 ms interval [12]. These limitations are necessary to ensure normal functioning of critical equipment and implementations need to ensure that these time limitations are met. On the other hand, reliability of communications is defined as the probability of successful communication over a specific time period [13]. Reliability is usually achieved by implementing redundant communication links and nodes, but also by increasing the *mean time to failure* of hardware. Finally, resilience is the ability of the system to function and to alleviate the disruptive impact of disturbances in various stress scenarios. For example, disruptive cyber attacks may cause significant packet delays and might block the delivery of critical commands to control hardware. Therefore, the resilience property might be implemented with alternative communication paths. Note that the aforementioned properties are not necessarily independent since for each resilience-assurance communication path network designers need to meet reliability and real-time requirements in order to comply with standard industry-specific prerequisites [12]. On the other hand, the difference between reliability and resilience is subtle, yet of paramount importance. In this respect, while reliability applies to normal up-time operation, resilience is formulated for conditions of stress, i.e., disruptive cyber attacks. Therefore, we believe that ICS network design methodologies need to incorporate not only reliability, but also resilience in order to build survivable ICS installations.

To accommodate these requirements, this work proposes a mathematical model of the ICS network design as a MILP problem. In the proposed MILP problem traffic demand paths, end-points, and concentrator sites are installed in such a way to minimize the costs of the infrastructure, while ensuring that constraints regarding real-time packet delivery, reliability, and K-resilience are satisfied.

3.2 Preliminary Notations

We define $I = \{1, 2, ..., i, ...\}$ to be the set of traffic demands (TD), and $J = \{1, 2, ..., j, ...\}$ the set of potential concentrator sites (CS). Then, let c_j^S to be the cost associated with installing CS j, c_{jl}^B be the cost of buying one unit of bandwidth between CSs j and l, c_{ij}^A the cost of buying one unit of bandwidth between the access end-point of TD i and CS j, and c_{ji}^E the cost of buying one unit of bandwidth between the egress end-point of TD i and CS j.

Then, let d_i denote the TD i, u_{jl} the link capacity between CSs j and l, and v_j the total capacity of combined access and egress demands for CS j. Considering the geographic location of access/egress end-points for each individual TD as well as the geographic location of CSs, the following binary connectivity parameters are defined. Let a_{ij} be a binary parameter with value 1 if the access end-point of TD i can connect to CS j, b_{jl} a binary parameter with value 1 if CS j can connect to CS l, and e_{ji} a binary parameter with value 1 if egress end-point of TD i can connect to CS j.

With respect to the ICS real-time performance requirements, the following communication latency parameters are defined. Let q_{ij}^A be the latency between access end-point of TD i and CS j, q_{ji}^E the latency between CS j and egress end-point of TD i, q_{jl}^L the latency between CSs j and l, and q_j^C the latency introduced by each CS j. Finally, we define the maximum tolerated latency for each individual TD i as q_i^M.

Next, we define the parameters concerning reliability. Let p_{jl} be the probability of failure of link j and l. p_{jl} is defined as a real number bounded between 0 and 1. The minimum desired path reliability for TD i is denoted by r_i.

The binary parameter h_{ik}, where $i, k \in I$, is defined such that $h_{ik} = 1$ if TDs i and k are part of the same RCG, and $h_{ik} = 0$, otherwise. Of particular importance for the proposed MILP problem is the configuration $h_{ii} = 1$, which reduces the complexity of constraints needed to identify TDs that are part of the same RCG. Before moving forward it is noteworthy that the proposed MILP problem assumes that primary and back-up TDs are part of the same set I. This is a significant design choice in the proposed MILP problem, which enables the applicability of the same constraints on both primary and back-up TDs. Moreover, it also facilitates the individual configuration of primary and back-up TDs which may represent a key aspect in the implementation of cost-efficient ICS communication networks.

Finally, we define g_j as a binary variable with value 1 if CS j is installed, x_{ij} as a binary variable with value 1 if the access end-point of TD i is connected to CS j, w_{ji} as a binary variable with value 1 if CS j is connected to the egress end-point of TD i, and t_{jl}^i as a binary variable with value 1 if TD i is routed on link (j, l), where $j, l \in J$.

3.3   Objective Function

The objective function aims at minimizing the cost of the installation:

$$\min \sum_{j \in J} c_j^S g_j + \sum_{j,l \in J, i \in I} c_{jl}^B t_{jl}^i d_i + \sum_{j \in J, i \in I} \left(c_{ij}^A x_{ij} + c_{ji}^E w_{ji} \right) d_i \qquad (1)$$

The objective function (1) accounts for the total installation cost of concentrators, and communication links between CSs. In particular, the term $\sum c_j^S g_j$ is the total cost of installing all selected CSs, the term $\sum c_{jl}^B t_{jl}^i d_i$ is the total cost of bandwidth for routing traffic demands between CSs, and $\sum \left(c_{ij}^A x_{ij} + c_{ji}^E w_{ji} \right) d_i$ is the total cost of bandwidth for access and egress end-points connected to CSs.

3.4 Constraints

Next, we define the constraints pertaining to traffic demand paths, to real-time packet forwarding, to resilience, and finally, to reliability.

General Demand Flow and Capacity Constraints. The following constraints are defined to ensure the identification of traffic demand communication paths and to satisfy capacity restrictions.

$$\sum_{j \in J} x_{ij} = 1, \sum_{j \in J} w_{ji} = 1, \qquad \forall i \in I \qquad (2)$$

$$x_{ij} \leq a_{ij} g_j, w_{ji} \leq e_{ji} g_j, \qquad \forall i \in I, j \in J \qquad (3)$$

$$\sum_{i \in I} d_i (x_{ij} + w_{ji}) \leq v_j, \qquad \forall j \in J \qquad (4)$$

$$\sum_{i \in I} d_i t_{jl}^i \leq u_{jl} b_{jl} g_j, \sum_{i \in I} d_i t_{jl}^i \leq u_{jl} b_{jl} g_l, \qquad \forall j, l \in J \qquad (5)$$

$$x_{ij} - w_{ji} - \sum_{l \in J} \left(t_{jl}^i - t_{lj}^i \right) = 0, \qquad \forall j \in J, i \in I \qquad (6)$$

The constraints defined in Eqs. (2) and (3) limit the number of connections between access/egress demand end-points and CSs to exactly one. Constraints defined in Eqs. (4) and (5) impose restrictions with respect to concentrators and link capacities. Finally, constraints defined in Eq. (6) are classical multicommodity flow conservation equations [14].

Real-Time Communication Constraints. The constraints formulated as Eq. (7) force the selection of routing paths that fulfill the latency requirements defined for each demand. In particular, for each demand i, the term $\sum \left(q_{ij}^A x_{ij} + q_{ji}^E w_{ji} \right)$ is the sum of latencies for access and egress links, the term $\sum t_{jl}^i \left(q_{jl}^L + q_j^C \right)$ is the sum of latencies owed to CSs and links between CSs, and the term $\sum q_j^C w_{ji}$ is the latency of the egress CS.

$$\sum_{j \in J} (q_{ij}^A x_{ij} + q_{ji}^E w_{ji}) + \sum_{j,l \in J} t_{jl}^i \left(q_{jl}^L + q_j^C \right) + \sum_{j \in J} q_j^C w_{ji} \leq q_i^M, \qquad \forall i \in I \qquad (7)$$

K-Resilience Constraints. As previously mentioned, in this work the K-*resilience* property of ICS communications is achieved by expanding set I with K back-up traffic demands for each primary TD. The association of TDs in a specific RCG is achieved by means of parameter h_{ik}. In this respect we define two distinct resilience cases. In the first case we assume the provisioning of independent communication links in a specific RCG. This ensures resilience to communication link failure, however, it does not provide resilience in case of compromised/failed concentrator nodes. Therefore, a second case is defined in which both links and concentrator nodes are distinctly chosen in each RCG.

The MILP problem at hand supports link and node independency by means of additional linear constraints. In this respect, Eq. (8) defines constraints pertaining to access and egress TD link independency, while Eq. (9) defines link independency between CSs selected in the same RCG.

$$\sum_{k \in I} x_{kj} h_{ik} \leq 1, \sum_{k \in I} w_{kj} h_{ik} \leq 1, \qquad \forall i \in I, j \in J \qquad (8)$$

$$\sum_{k \in I} (t_{jl}^k + t_{lj}^k) h_{ik} \leq 1, \qquad \forall i \in I, j, l \in J \qquad (9)$$

Conversely, the proposed MILP problem imposes concentrator node and link independency simultaneously through the linear constraints defined in Eq. (10). These include stricter conditions than the ones defined in Eq. (9). Consequently, if node independency is needed, the proposed MILP problem shall include the constraints defined in Eq. (10), instead of the ones deinfed in Eq. (9).

$$\sum_{k \in I, l \in J} (t_{jl}^k + t_{lj}^k) h_{ik} \leq 1, \qquad \forall i \in I, j \in J \qquad (10)$$

Reliability Constraints. Reliability in constrain-oriented programming is a well-known problem [15,16]. For the MILP at hand the reliability of communication path for a specific TD i is defined as:

$$\prod_{j,l \in J} (1 - p_{jl} t_{jl}^i) \geq r_i, \qquad \forall i \in I \qquad (11)$$

Unfortunately, this constraint is non-linear due to the multiplication of t_{jl}^i variables. However, based on the observation that t_{jl}^i is a binary variable, the transformation defined in [5] can be applied to derive a linear constraint. More specifically, owed to the binary range of values for t_{jl}^i, it follows that $(1 - p_{jl} t_{jl}^i) = (1 - p_{jl})^{t_{jl}^i}$. Consequently, constraint (11) can be rewritten as:

$$\prod_{j,l \in J} (1 - p_{jl})^{t_{jl}^i} \geq r_i, \qquad \forall i \in I \qquad (12)$$

Then, by applying the natural logarithm function on both sides of inequality (12) we obtain the following linear reliability constraint, which is adopted in the proposed MILP problem:

$$\sum_{j,l \in J} t_{jl}^i \ln(1 - p_{jl}) \geq \ln(r_i), \qquad \forall i \in I \qquad (13)$$

4   Experimental Results

In this section we provide experimental results on the applicability of the proposed MILP problem to the design of large-scale ICS communication networks.

4.1 Experimental Scenario and Parameters

The experimental assessment is conducted in the context of electricity grid communication networks. As a reference model we assume the IEEE 30-bus model [17] consisting of 6 generators and 30 substations.

The communication infrastructure of large-scale electricity grid networks is composed of several independently governed communication networks. In general, utilities rely on a main, internal communication network in order to supervise and control large-scale physical processes. However, regulations impose the installation of at least one back-up communication line leased from a different Internet Service Provider (ISP) with specific Quality of Service requirements [18]. In this respect, the problem at hand assumes that the communication medium between concentrator nodes (including the concentrator nodes) may be provided by two different ISPs: ISP_1 (primary communication network), and ISP_2 (secondary communication network). We further assume the following parameter values. We assume a first, primary communication network (governed by ISP_1) where CSs are installed in the premises of substations, which yield a number of 30 CSs. Next, we assume the same number of CSs in the second communication network (governed by ISP_2). We further assume 30 main traffic demands originating from each substation, which need to be routed to regional/national monitoring nodes connected exclusively to the second communication network.

In order to ensure a realistic estimation of parameters we arranged meetings with representatives of a local electricity grid operator and a local ISP provider. Discussions confirmed the validity of the aforementioned assumptions on communication infrastructure, and guided the choice of parameter values pertaining to costs and traffic magnitude. For the provisioning of CSs we initially assume that $c_j^S = 30$ monetary units (MU) for all CSs, while the cost of one bandwidth unit is assumed to be $c_{jl}^B = 3$ MU per Mb/s, and $c_{ij}^A = c_{ji}^E = 1$ MU/Mb/s. We further assume that the traffic that needs to be routed for each TD is $d_i = 100$ Mb/s, and the capacity of each CS is $v_j = 8000$ Mb/s and $u_{jl} = 1000$ Mb/s.

The initial connectivity between CSs is randomly configured through a uniform distribution function such that $b_{jl} = 1$ with a probability of 20 % if both CSs j and l are in the primary network and with a probability of 5 % if at least one CS is in the secondary network. The initial connectivity of access TD endpoint to CSs is similarly configured such that $a_{ij} = 1$ with a probability of 20 % if CS j is in the primary network, and with a probability of 5 %, otherwise. Since the egress end-point of TDs are exclusively assumed to be connected to CSs in the secondary network, we assume that $e_{ji} = 1$ with a probability of 20 %.

With respect to latency, for the sake of simplicity we assume that $q_{ij}^A = q_{ji}^E = q_{jl}^L = q_j^C = 1$ ms. Subsequently, we assume that the maximum tolerated latency is of $q_i^M = 10$ ms [12]. The probability of failure is assumed to be of 0.01 %, i.e., $p_{jl} = 0.0001$, while the minimum required reliability is assumed to be of 99.8 %, i.e., $r_i = 0.998$ [18]. Finally, the resilience communication group parameters h_{ik} are initialized according to the values of K, as discussed later in the following sections. Initially, we assume that $K = 1$.

The network design problem was implemented in AIMMS [19], where we adopted the popular CPLEX engine as a MILP solver. All experiments were performed on a Windows 7 OS host, with Pentium Dual Core CPU at 3 GHz, and 4 GB of memory.

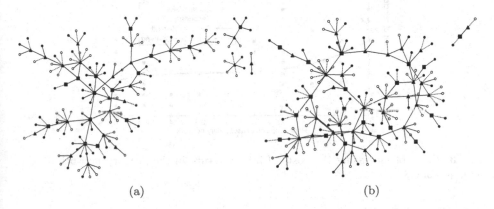

(a)   (b)

Fig. 2. Effect of costs on the generated solution: (a) equal costs in the primary and secondary networks; and (b) five times higher costs in the primary network. CSs in the primary and in the secondary network are represented with boxes and triangles, respectively. Access and egress end-points are represented by filled and empty circles, respectively.

4.2 Effect of Costs on the Generated Solution

Obviously, the focus of the objective function on costs will yield an increased sensitivity of the generated solutions on the components' costs. An example in this sense is provided in Fig. 2a and b. Here it can be seen that the increase of components' costs in the primary network, in comparison to the costs of components in the secondary network, triggers profound changes in the architecture of the generated solutions. As a result, more CSs are selected from the secondary network, which ensures minimum provisioning costs for the generated solution.

This particular trend is more visible from numerical results. In this respect, we performed several tests in which we increased the costs of components in the primary network starting from a ratio of 0.5 with respect to the cost of components in the secondary network, and up to a ratio of 5. Given the probabilistic configuration of connection options, for each cost value we ran the AIMMS solver 10 times. As illustrated in Fig. 3, the increase in the price ratio between the primary and the secondary network decreases the number of CSs allocated from the primary network, and it increases at the same time the number of CSs allocated from the secondary network.

A similar behavior was measured for the allocation of links between different network components, i.e., CSs and TDs (see Fig. 4a and b). Here (Fig. 4a) it can be seen that the increase in the price ratio will increase the preference of

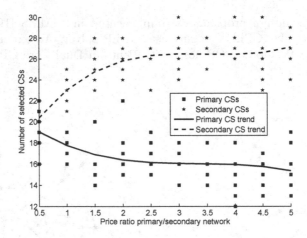

Fig. 3. Effect of increasing the costs of infrastructure in the primary network on the selection of CSs.

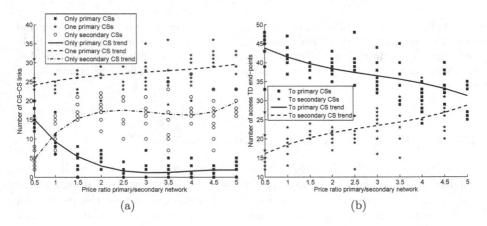

Fig. 4. Effect of increasing the costs of infrastructure in the primary network on the selection of links: (a) between CSs located in the primary and/or in the secondary network; and (b) between access TD end-points and CSs located in the primary or in the secondary network.

linking together more CSs from the secondary network and less from the primary network. The same trend is also visible for the connection of access TD end-points to CSs (see Fig. 4b).

4.3   Effect of K-Resilience Factor on the Generated Solution

Until this point we assumed the implementation of only one back-up path for each TD, i.e., $K = 1$. However, the requirement of link and node independence between different TDs of the same RCG, may yield an unfeasible linear problem

Table 1. Percentage of feasible configurations as a function of K and α

K	α [%]									
	5	10	15	20	25	30	35	40	45	50
1	0 %	0 %	0 %	81 %	91 %	100 %	100 %	100 %	100 %	100 %
2	0 %	0 %	0 %	23 %	70 %	98 %	100 %	100 %	100 %	100 %
3	0 %	0 %	0 %	0 %	33 %	81 %	90 %	100 %	100 %	100 %
4	0 %	0 %	0 %	0 %	0 %	41 %	80 %	100 %	100 %	100 %
5	0 %	0 %	0 %	0 %	0 %	15 %	52 %	91 %	97 %	100 %

for which the solver cannot provide a solution. This scenario holds especially when the configuration does not provide sufficient connectivity options in parameters a_{ij}, b_{jl}, or e_{ji}. Since the tests presented in this work assume a probabilistic configuration of connectivity parameters, this sub-section evaluates the impact of the number of back-up paths (K) and of different probability distribution values (denoted by α) on the problem's feasibility. For this purpose we assume that $\alpha \in \{5, 10, 15, 20, 25, 30, 35, 40, 45, 50\}$ (in percentage), and $K \in \{1, 2, 3, 4, 5\}$. For each (α, K) configuration we run the AIMMS solver 40 times.

The measured percentage of feasible solutions for each (K, α) configuration is given in Table 1. Here it is shown that for $\alpha \leq 15\%$ the solver is unable to find a feasible solution for any values of K. This is explained by the link independence constraint which requires that access, egress, and CS-CS links to be independent within a specific RCG. By increasing the value of α above 15 %, however, for $K = 1$, the solver finds that more than 81 % of configurations lead to a feasible solution. Nevertheless, by increasing the value of K, that is, by expanding the problem with K additional TDs in each RCG, the percentage of feasible solutions decreases dramatically. As an example in this sense, for $\alpha = 30\%$ and $K = 1$ we measure a 100 % rate of feasible solutions, while for the same α and $K = 5$ the rate of feasible solutions drops to 15 %. These results are a clear indication on the need to carefully formulate the MILP problem, since minor changes to input data may lead to significantly different outputs.

4.4 Effect of Reliability Parameters on the Generated Solution

The reliability constraints impose the identification of solutions that satisfy the minimum path reliability condition. In this section we assume a minimum reliability value of 99.8 %, that is $r_i = 0.998$ for all $i \in I$. As a result, the generated solutions guarantee that for all TDs the communication paths have a reliability of at lest 99.8 %. In Fig. 5 we illustrate the average reliability of communication paths for several solver runs. Here the value of K was fixed to 1, the value of α was fixed to 20 %, while individual link reliability was of 99.99 %. As a result, the measured path reliability for each solution was well above the minimum of 99.8 %. In fact, we measured a minimum path reliability of 99.993 %, and a maximum path reliability of 99.996 %. Obviously, the path reliability is directly

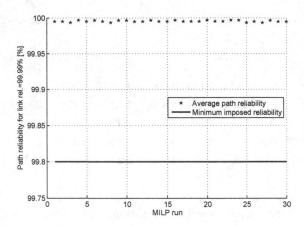

Fig. 5. The average reliability of communication paths considering that the reliability of each individual j, l link is 99.99 %.

influenced by the individual link reliability values. However, it is noteworthy that if the configuration is feasible, then the proposed MILP problem will always yield a cost-optimal solution that guarantees that reliability constraints are satisfied.

4.5   Execution Time

We performed several experiments to test the execution time of the CPLEX solver on the proposed MILP problem. As tabulated in Table 2, the execution time is influenced by the problems' dimension in terms of the number of CSs,

Table 2. Execution time (s) for $\alpha = 20\%$ and $\alpha = 40\%$ (we assume $K = 1$). Unfeasible configurations are denoted by '–'.

CS count	Total TD count including back-up					
	10	20	30	40	50	60
10	−/0.2	−/0.29	−/0.32	−/0.41	−/0.53	−/0.62
20	0.36/0.37	0.64/0.73	0.99/1.19	−/1.5	−/2.05	−/2.3
30	0.68/0.88	1.45/1.67	2.11/2.61	3.01/3.58	3.61/5.72	4.37/5.93
40	1.28/1.53	2.54/3.28	3.88/5.38	5.29/8.20	6.81/11.4	8.14/12.2
50	2.02/2.49	4.04/5.66	6.17/8.80	8.77/15.8	10.5/21.1	13.3/27.7
60	3.09/3.68	6.48/8.52	9.72/14.1	13.0/24.7	17.2/35.7	21.1/43.2
70	4.04/5.56	8.28/12.2	13.4/23.0	17.7/33.6	23.3/47.6	28.2/55.3
80	5.73/7.50	11.0/15.9	17.1/32.7	25.4/45.5	30.7/58.9	38.5/79.3
90	7.12/10.1	14.4/22.2	22.3/42.2	30.2/59.3	38.8/79.3	48.0/100.7
100	9.03/12.8	18.1/30.7	28.5/54.5	39.3/77.1	50.5/101.2	63.8/158.8

the number of TDs, and the number of configured connection options, i.e., the magnitude of α. In this respect for $\alpha = 20\,\%$ and for 10 CSs the solver determined that there are no feasible configurations available. Nevertheless, by increasing α to 40 % solutions are generated in 0.2 s for 10 CSs and 10 TDs, and in 0.62 s for 10 CSs and 60 CSs. By further increasing the number of CSs we also notice a linear increase in the solvers' execution time. Nevertheless, as depicted in Table 2 the execution time for 50 CSs and 60 TDs does not exceed 30 s, while for 100 CSs and 60 TDs the execution time is below 160 s, i.e., below 3 min. These results are a clear indication on the scalability and applicability of the proposed MILP problem to large-scale ICS networks. The linear increase of the execution time provides further confirmation on the applicability of constraint-based programming to the design of large-scale ICS network.

5   Conclusion

We proposed a novel MILP problem that accounts for the provisioning of traditional concentrator equipment, of traffic flows routed across an overlay network, and most importantly of typical ICS design requirements pertaining to real-time packet delivery, communications reliability, and resilience. The approach saved costs on investments in the system's resources, enhancing at the same time the resilience of ICS networks by a factor of K through the deployment of back-up communication paths. Experimental results proved that the technique is scalable and applicable to large-scale ICS such as modern energy network. As future work we intend to further expand the developed MILP with security requirements in order to deliver a methodology that ensures security and resilience-aware design of ICS communication networks.

Acknowledgments. This research was supported by a Marie Curie FP7 Integration Grant within the 7th European Union Framework Programme.

References

1. U.S. Department of Energy: Smart grid research and development: multi-year program plan (MYPP) 2010–2014, September 2012
2. European Commission: European smart grids technology platform, eur 22040, September 2006
3. European Commission: Horizon 2020 energy calls, January 2015
4. NIST: NIST Smart Grid Collaboration (2011). http://collaborate.nist.gov/twiki-sggrid/bin/view/SmartGrid/WebHome
5. Chiang, C., Hwang, M., Liu, Y.: An alternative formulation for certain fuzzy set-covering problems. Math. Comput. Model. **42**(34), 363–365 (2005)
6. Carro-Calvo, L., Salcedo-Sanz, S., Portilla-Figueras, J.A., Ortiz-Garca, E.: A genetic algorithm with switch-device encoding for optimal partition of switched industrial ethernet networks. J. Netw. Comput. Appl. **33**(4), 375–382 (2010)
7. Zhou, Z., Chen, B., Wang, H., Fan, Z.: Study on the evolutionary optimisation of the topology of network control systems. Enterp. Inf. Syst. **4**(3), 247–264 (2010)

8. Zhang, L., Lampe, M., Wang, Z.: A hybrid genetic algorithm to optimize device allocation in industrial ethernet networks with real-time constraints. J. Zhejiang Univ. Sci. C **12**(12), 965–975 (2011)
9. Zhang, L., Lampe, M., Wang, Z.: Multi-objective topology design of industrial ethernet networks. Frequenz **66**(5–6), 159–165 (2012)
10. Diao, K., Rauch, W.: Controllability analysis as a pre-selection method for sensor placement in water distribution systems. Water Res. **47**(16), 6097–6108 (2013)
11. Zahidi, S., Aloul, F., Sagahyroon, A., El-Hajj, W.: Optimizing complex cluster formation in manets using sat/ilp techniques. IEEE Sens. J. **13**(6), 2400–2412 (2013)
12. Institute of Electrical and Electronics Engineers: IEEE 1646–2004 standard: communication delivery time performance requirements for electric power substation automation (2004)
13. IEC62439: Industrial communication networks - high availability automation networks (2012)
14. Meixner, C., Dikbiyik, F., Tornatore, M., Chuah, C., Mukherjee, B.: Disaster-resilient virtual-network mapping and adaptation in optical networks. In: 2013, 17th International Conference on Optical Network Design and Modeling (ONDM), pp. 107–112, April 2013
15. Hsieh, Y.C.: A linear approximation for redundant reliability problems with multiple component choices. Comput. Ind. Eng. **44**(1), 91–103 (2003)
16. Billionnet, A.: Redundancy allocation for series-parallel systems using integer linear programming. IEEE Trans. Reliab. **57**(3), 507–516 (2008)
17. University of Washington - Electrical Engineering: Power Systems Test Case Archive (2015). http://www.ee.washington.edu/research/pstca/
18. Nazionale, G.R.T.: Criteri di connessione al sistema di regolazione della tensione del GRTN. Terna, Italy (2005)
19. AIMMS: Advanced Interactive Multidimensional Modeling System (2014). http://www.aimms.com/aimms/

Self-organised Key Management
for the Smart Grid

Foivos F. Demertzis[1], Georgios Karopoulos[2]($\boxtimes$), Christos Xenakis[1],
and Andrea Colarieti[3]

[1] Department of Digital Systems, University of Piraeus, Piraeus, Greece
{fdemertz,xenakis}@unipi.gr
[2] Department of Informatics and Telecommunications, University of Athens,
Athens, Greece
gkarop@di.uoa.gr
[3] WEST Aquila S.r.l., L'Aquila, Italy
andrea.colarieti@westaquila.com

Abstract. As Smart Grid deployments emerge around the world, their
protection against cyberattacks becomes more crucial. Before protective
measures are put into place, one of the main factors to be considered
is key management. Smart Grid poses special requirements compared
to traditional networks; however, the review of previous work reveals
that existing schemes are not complete. Here we propose a scalable and
distributed key management scheme for the Smart Grid based on the
Web-of-Trust concept. Our proposal is build on top of a Distributed
Hash Table for efficient lookups of trust relationships. The target of this
scheme is to create a key management system for the Smart Grid without
the need of an always available Trusted Third Party. The underlying Distributed Hash Table can be further utilised as an infrastructure to build
other Smart Grid services on top of it, like secure and/or anonymous
aggregation, billing, etc.

Keywords: Smart grid · Security · Key management · DHT · Chord

1 Introduction

In order to handle the power demand that is increasing in the last decades,
Nation states turn to renewable sources to diversify their energy mix. Since the
traditional power grid was not designed with the current situation in mind, it can
neither cope with the efficient management of diverse energy sources nor respond
effectively to events leading to blackouts; for these reasons, the Smart Grid is
considered the next step in the power grid. Information and communication technology introduction to the traditional power networks will provide advantages
like efficiency, increased reliability, resilience, distributed intelligence, and better
control of demand response. The EU has plans to replace at least 80 % of its electricity meters with smart ones by the year 2020.[1] According to a US report [1],

---

[1] http://ec.europa.eu/energy/en/topics/markets-and-consumers/smart-grids-and-
meters

© Springer International Publishing Switzerland 2015
S. Papavassiliou and S. Ruehrup (Eds.): ADHOC-NOW 2015, LNCS 9143, pp. 303–316, 2015.
DOI:10.1007/978-3-319-19662-6_21

the smart meter installations in the USA have reached 50 million of devices as of July 2014. Therefore, it is not long until large Smart Grid deployments will become true.

On the other hand, upgrading such a large and complex system, like the Smart Grid, could introduce new vulnerabilities and expose it to common cyberthreats, as have been described in relevant survey articles [2,23,28]. There are numerous possible attacks against the Smart Grid, for example attacks against the availability of the utility's control systems with varied consequences ranging from data theft to loss of human lives, against the smart meters leading to incorrect billing and personal data theft, and against the communication protocols that support the operation of the Smart Grid. Countermeasures to these attacks assume some sort of key management for supporting the cryptographic operations required for securing the Smart Grid and establishing trust relationships.

Before specifying a key management scheme that is appropriate for the Smart Grid, there are a few issues and trade-offs to be considered. These issues are mainly related to the type of cryptography, and the key provisioning system used. Typically, symmetric cryptography tends to be less computationally demanding, but more difficult in terms of key management than asymmetric; therefore, it is expected that it will be supported from Smart Grid devices. Regarding asymmetric cryptography, while appliances might have too limited computational resources for it, we argue that smart meters will be able to meet its demands. There are two alternatives here: either in software, using a crypto library, or in hardware, utilising a dedicated co-processor. Hence, we argue that smart meters will have the required capacity to implement both types of cryptography.

Key provisioning in the case of symmetric cryptography is difficult in comparison to digital certificates used in asymmetric cryptography. In the first case, each node needs to have a different secret key with every other node; in some implementations a single key is used throughout the networks, increasing the security risk substantially. Moreover, some kind of Key Distribution Center (KDC) should be employed, which creates a single point of failure, and should be online at all times; the latter is not always possible in the Smart Grid. On the other hand, digital certificates is a much more effective solution, and two alternatives exist: Public Key Infrastructures (PKIs) based on a Trusted Third Party (TTP), and Web-of-Trust which is a distributed solution. In the first case, however, usually a Certification Authority (CA) is needed, which has to be secured, creating a single point of failure; its operation introduces a significant amount of overhead as well. Moreover, PKIs are complex and might prove difficult to operate, especially for PKIs belonging to different organisations that need to interoperate. Another issue is checking the validity of a certificate which requires connectivity with the relevant server; intermittent communications as anticipated in Smart Grids will hinder key management operation. Also, trust management can be an issue when a single root CA is used; however, there are solutions to alleviate this problem by cross-signing other CAs or using bridge CAs. The Web-of-Trust concept offers a more distributed, self-organised, and scalable alternative for the

Smart Grid, which can manage trust relationships in a local level and operate even when a TTP is not always available.

Taking into account the issues found in the Smart Grid [19] and the limitations imposed by the employed cryptography type and key provisioning method, the following key management requirements can be identified:

Resilient against well known attacks. Key management solutions are expected to take previously identified attacks and weaknesses into account, whether it is in the same or in similar systems, e.g., the Internet.

Holistic key management approach. The Smart Grid should be covered as a whole by the proposed key management schemes, rather than targeting subsystems of it.

Robust against key compromise. A key management scheme intended for the Smart Grid must provide an adequate level of protection for the keys. Equally important is to afford sufficient key diversity, so that compromise of a single device does not put in risk other devices or the network.

Distributed operation. It is expected that connectivity to central servers (like CAs) will not always be available in the Smart Grid, for reasons like natural disasters and outages. Moreover, the various systems and devices found in the Smart Grid are distributed in large geographical areas and need to face intermittent communications. In such cases, key management should be flexible and less centralised in order to increase availability of authentication and authorisation services.

Upgradeability. The cryptographic modules that support cryptographic operations need to be designed carefully so that they are upgradeable. This is required because Smart Grid devices will have an average lifetime of 20 years, which is much longer than usual IT systems.

Certificate revocation. When a certificate cannot be considered trustworthy or the private key has been compromised, then there must be a proper mechanism to revoke the certificate.

Scalability. A high degree of scalability is needed since Smart Grid deployments will comprise utilities with millions of customers; this will involve the management of tens of millions of credentials and keys.

Efficiency. The devices found in the Smart Grid tend to be constrained in terms of memory, processing power and storage; moreover, there are also limitations in bandwidth and connectivity availability. Therefore, proposed key management solutions should be efficient in memory, computation, storage, and communications.

As the interest on Smart Grid security is growing, one of the main challenges is the proposal of an appropriate key management scheme that can meet the security and network requirements of Smart Grids. However, already proposed key management schemes come with several limitations. According to [23], some of the existing key management schemes are for Supervisory Control And Data Acquisition (SCADA) systems. SCADA are decision making systems used to control and monitor physical processes remotely; they provide the connection

between the cyber and the physical world and typically are closed and proprietary. While SCADA is considered a significant part of the Smart Grid, solutions
targeting SCADA do not provide an adequate solution for the Smart Grid, since
they only protect subsystems of the grid. We also show, in the related work
section, that other solutions aiming at the Smart Grid as a whole are not appropriate as well, because they lack security, robustness, efficiency, and scalability.
Hence, there is a considerable need for a key management solution that can meet
Smart Grid's special requirements.

Here, we present a distributed and scalable authentication and key management scheme, namely Self-Organised key MAnagement for the Smart grid
(SOMA-S), that can meet the special requirements that a Smart Grid has, compared to a typical communication network. Smart Grid will assume a mesh networking structure because of the advantages it can offer towards meeting Smart
Grid's requirements, and mainly the high degree of reliability, self-configuring,
and self-healing [11,27]. Hence, schemes intended for mesh networks can be
adapted to Smart Grids, taking into account their special requirements. We
utilise our previous work, a scheme called Self-Organised Mesh Authentication
(SOMA) [9,10], which is a certificate-based authentication infrastructure that
aims to create a large-scale secure authentication system for mesh networks without the need of a TTP. We adapt this framework to the Smart Grid context,
so that it can fulfil the diverse requirements set by this type of networks. Compared to related work, our proposal follows a different approach by employing
the Web-of-Trust concept found in Pretty Good Privacy (PGP) [5]; this allows
our scheme to present a few advantages over existing schemes, like scalability
and decentralisation.

In the next section we will examine related work and background on key management for the Smart Grid. Section 3 describes our key management system
called SOMA-S. Next, Sect. 4 discusses security related issues, gives a critical
overview of our proposal, and presents additional services that can be implemented for the Smart Grid, on top of the proposed infrastructure. Finally, Sect. 5
summarises important points and outlines the conclusions drawn.

2   Related Work

The first category of existing solutions focus on SCADA systems, which are
decision making systems used to control and monitor physical processes remotely,
and are considered a significant part of the Smart Grid. With the integration of
the Smart Grid with existing SCADA systems, which have been employed since
the '60s, the resulting system should have a unified key management scheme
for the secure communication of all components among them. There are quite
a few key management schemes designed especially for SCADA: [4,6–8,12,17];
however, they are not adequate because they do not take into account the rest of
the Smart Grid's components. An overview of each of these proposals is presented
in [23].

Various solutions have proved to be insecure and susceptible to different
attacks. In [24], a novel key management scheme for Smart Grids is proposed,

combining public and symmetric key cryptography; the used techniques are elliptic curve and the Needham-Schroeder authentication protocol. The authors of [25], however, prove that it is susceptible to man-in-the-middle attacks and propose their own scheme. The latter is a symmetric key distribution scheme utilising an online TTP with precomputed responses, which can be operated as an LDAP server and replicated with low cost. This scheme was also found to be vulnerable to an impersonation attack [20].

One of the easiest, yet insecure, key management methods is sharing a single symmetric key among many or even all parties. In fact, according to [19], there exist deployed systems that use the same symmetric key among all their meters. The main issue with this method is that if one node is compromised, then the whole network is at risk.

In the pursuit of efficiency, several researchers have proposed key management schemes based on shared secret keys. The scheme proposed in [13] introduces an efficient and scalable key management protocol for secure unicast, multicast, and broadcast communications in Smart Grids; its operation is based on a binary tree to manage secret keys shared among entities. This proposal does not scale well, since it requires substantial manual work in order to create the binary tree and transmit it together with the secret values to every node; moreover, as nodes join or leave the network, the whole network should update the broadcast keys. Another scheme based on shared secrets is [16], which can also support unicast, multicast and broadcast communications. The Smart Grid is divided into two levels, based on the computational resources of its devices, and each level has its own key management system. The nodes follow a binary tree arrangement and the secret key of a parent node is the hash of its children keys. The Dynamic Key Management Scheme (DKMS) [26] uses symmetric keys with frequent key updates among the nodes comprising the Smart Grid. When a node A joins the grid, a key is installed manually between node A and a bootstrapping node B. To communicate with a third node C, node A has to negotiate a new key with C using the association it has with node B. All the methods presented in this paragraph have the same issue: every node has to maintain one key for each secure connection to another node. This hinders scalability, since it involves high efforts for key management, renewal, and distribution.

Another category of key management schemes utilises ID-based cryptography. In [15], a key management scheme for Advanced Metering Infrastructure (AMI) is proposed, which is based on a key tree structure. However, the authors of [22] found that it is susceptible to de-synchronisation attacks, while it lacks scalability due to inefficient key management. For this reason, they proposed Scalable Key Management (SKM), which utilises ID-based encryption and a key graph technique for efficient multicast key management. The authors in [18] propose a key management solution occupying a CA that resides in the utility network. They use secret values shared between the CA and the smart meters together with an ID-based cryptography model. The main drawback of the above solutions, as with all ID-based systems, is that the Private Key Generator (PKG) should always be online and available; moreover, since the PKG holds the private keys of all nodes, it can be a single point of failure.

Wide-area measurement system key management (WAKE) [14] is a solution based on traditional hierarchical PKI architecture. Each participating node has a X.509 certificate and the root CA is the grid operator. Secure channels between devices are established using the Diffie-Hellman key establishment protocol. The hierarchical PKI architecture is not well suited for the Smart Grid, as stated in [3], since it does not meet the high availability requirement, because the root CA tends to be a single point of failure.

Summing up, we showed that each proposed solution presents one or more of the following limitations: (a) partial coverage of the Smart Grid, (b) vulnerability against well known attacks, (c) secret key re-use, (d) poor scalability, (e) contain a single point of failure, and (f) require high availability. On the other hand, as we will show in the following sections, SOMA-S addresses the aforementioned limitations by following a different approach, i.e. the Web-of-Trust concept. At the same time, it meets the key management requirements stressed in Sect. 1.

3   SOMA-S

3.1   Functional Components

The components that comprise a basic Smart Grid infrastructure are presented in Fig. 1; particular focus has been given to the communication links between elements rather than power delivery. Here we deliberately omit communication technologies among nodes which range from ZigBee, to WiFI, cellular, satellite, powerline communications, etc.

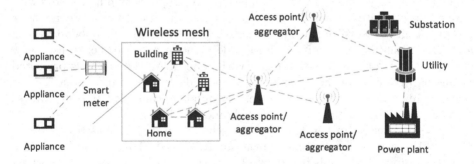

Fig. 1. Functional components of a Smart Grid

One of the basic nodes of the Smart Grid is the utility company that provides power to its customers. Here we depict a single company, but normally a Smart Grid will comprise more than one. Power generation and distribution installations do not directly communicate with the end customer, but with the utility instead. Aggregator nodes reside in between the utility and the end customers and their main purpose is to aggregate smart meter readings and forward the results towards the utility. They are considered more powerful nodes than smart meters, thus they can support more computationally intensive operations than smart meters. In each home or building there are one or more smart meters, connected in a mesh network.

3.2 Architecture

Figure 2 presents our proposal together with the Smart Grid architecture. The lowest level depicts the power delivery network over which the utility company delivers power to its customers. Above this, the ICT level resides, allowing bi-directional communications among Smart Grid elements. To support the operations of SOMA-S, we have two logical layers comprising the overlay on top of the ICT level, which we will present in the following paragraphs in more detail.

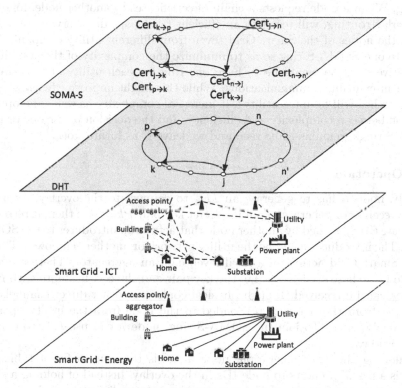

Fig. 2. DHT and SOMA-S over Smart Grid

The overlay layer, over which SOMA-S is built, provides a generic Distributed Hash Table (DHT) functionality. On a DHT we can normally store (key,value) pairs, as in every other locally stored hash table, and retrieve the value back based on the key. Our scheme is based on Stoica's Chord [21], where nodes are placed as IDs occupying a circular identifier space; in our case, every node of the Smart Grid is represented as a node of the Chord ring. Chord is scalable and efficient, taking O(log N) communication hops and keeping O(log N) state per node, where N is the total number of nodes in the system; moreover, it is robust on node joins and failures. The overlay can support on top of it diverse applications that require efficient lookup services, like authentication with SOMA-S; other services that could benefit from the overlay include billing, and secure aggregation.

In the case of SOMA-S, the overlay key of a node n is ID_n and will be derived by hashing the concatenation of its public key and a device identifier of the node $ID_n = h(Pk_n + ID_{device})$. This concatenation ensures that ID_n will be unique even when the device identifiers are not. The digital certificates, where used, follow the OpenPGP Message Format [5] or any compliant format. The overlay structure forms the framework for the transitive trust relationships and each node exchanges certificates with the nodes that is directly responsible for routing. When a node requests a chain of certificates to another node, following the overlay routing, will result to an efficient trust path discovery.

All the nodes of the Smart Grid (even from different utility companies) are put onto one SOMA-S ring so as to minimize the complexity of the system. An alternative solution would be to have one ring for each utility and an overlay ring for inter-utility communications. While this might improve efficiency (since most searches will be intra-utility), it increases complexity as well. The optimal trade-off between complexity and efficiency, and the decision to use one or more SOMA-S rings, remains to be seen and we leave it as future work.

3.3   Operation

Initially, node n has to generate an ID_n to connect to the overlay. As a first step, n receives or generates a public/secret key pair Pk_n/Sk_n; then, it receives a certificate $Cert_n$ signed by another node that acts as an introducer to the SOMA-S ring. The introducer could be the utility network or another "empowered" node of the Smart Grid acting on a local level, e.g., an aggregator. There are three ways to load these credentials, i.e. the key pair and the certificate, to each node: (a) they can be received through physical contact or other direct channels, (b) generated from the node, or (c) loaded to the node from the utility operator before deployment. To join the SOMA-S ring, node n can use ID_n to connect to the overlay.

Following the ID_n generation, node n needs to set-up a finger table [21], which is a list of pointers to node IDs in the overlay. Instead of holding a single pointer to the next node, $n.finger = ID_{next\_node}$, a list of m nodes is maintained, $n.finger(m)$, with their logical inter-node distance increasing exponentially, providing the efficient look up mechanism. Every entry in the table will be associated with an IP address; moreover, as n establishes trust relationships with other nodes from its finger table, it adds the corresponding certificate to this table entry as following. Node n can use its certificate to introduce itself to an "empowered" node n' residing in its finger table, which also holds a certificate signed by the utility. Next, n' will check the authenticity of key Pk_n, and sign it; similarly, n will sign the authenticity of $Pk_{n'}$. Following the PGP

Web-of-Trust, n' will put in its keyring the certificate $Cert_{n \to n'}$; additionally, n will hold $Cert_{n' \to n}$ in its keyring. After the bootstrapping phase, Pk_n is signed by other nodes as well, using n's Web-of-Trust. This will lead node n to set-up an authentication aware finger table. This way, n can be authenticated later, even if some of the signing nodes have been withdrawn or their certificates revoked, using signatures that are still valid. Regarding performance, SOMA-S is based on the Chord protocol, hence it follows its mathematical properties; therefore, a node would be able to find a chain of certificates in O(log N) time and in O(log N) number of certificates for any trust path given.

An example operation of SOMA-S is shown in Fig. 2. When node n wants to authenticate with p, it must have a trust chain to p. If p is already in the trust path of n, then they can communicate directly. Otherwise, node n has to follow the look-up methods provided by SOMA-S as following. Node n communicates with the closest node to p with which it has a mutual trust relationship, i.e. node j; that is the closest node to p in n's finger table, or p is an introducer for n. Node j does the same and ends up communicating with node k. Finally, k communicates with p completing the authentication between n and p. Details on joining the ring, stabilisation, and lookup procedures can be found in [10].

To further protect the Smart Grid from misbehaving nodes, we can use a reputation framework to include ratings from all the experiences between principals, in addition to the above certificate path-building method. On every transaction between a node and a finger, an outcome will be recorded and its reputation score calculated. We do not wish to claim specific parameter values as accurate ratings other than the positive and negative outcomes between events. For instance, supposing a node j was malicious and was misbehaving in routing, node n could undershoot in its finger table and therefore avoiding the problematic node as if it was faulty. After a while, intentional routing misbehaviour by specific nodes would be represented in the rest ring effectively, skipping it in their finger tables. This approach results in a reputation based path ranking that is similar to the discrete ranking of PGP Web-of-Trust [5], but also allowing for further flexibility and extensibility, and more complex representation of social interactions and structures.

Regarding certificate revocation, in a large system like the Smart Grid, a typical Certificate Revocation List (CLR) can become very lengthy creating efficiency issues. To solve this, administrators set short validity periods to certificates so that, when a previously revoked certificate expires, it is removed from the CRL. However, this creates higher operational overhead, especially when a large number of certificates needs to be frequently re-issued.

In SOMA-S a node can explicitly revoke its certificate by using a revocation certificate as described in OpenPGP [5]. The node that revoked its certificate does not need to send the revocation request certificate to all the nodes in SOMA-S, but only to its predecessor, and exchange it through Chord's stabilisation protocol with all the nodes that update their finger table to the node itself. When a node requires to check if a certificate is currently revoked, it only needs to proceed with the normal lookup operation.

4 Discussion

In this section we discuss several issues related to our scheme. First, we provide a security analysis by employing a few representative attack scenarios; for each case we study the actions taken by our proposal. Next, we review the key management requirements of Smart Grids and check on which degree SOMA-S fulfils them. Finally, we present alternative services that can be offered on top of the DHT infrastructure supporting SOMA-S.

4.1 Security Analysis

In this section we analyse the possible attack scenarios derived from the characteristics of our architecture and the desirable system properties. Attacks range from the certificate exchange mechanism, the control and use of key material and, finally, the overlay routing itself.

Node Join. A malicious party could try to implant a fake node that it controls to the SOMA-S ring. However, bootstrapping of new nodes is controlled by the utility (or delegates like aggregators) so that not even a large number of colluding malicious nodes could successfully introduce this new node to the Smart Grid.

Utility Certificate Revocation. An issue that could probably arise is what happens when the utility certificate is revoked. If the node has not already bootstrapped to the network, then it needs to acquire a new certificate signed using the new utility certificate. If the node has already bootstrapped to the network, then no action is needed; the utility certificate is used for bootstrapping only, and the node's certificate will have been signed by other nodes and considered valid until expired or revoked.

Certificate Chain. A node wanting to find a certificate chain to another node, needs to authenticate first a chain of intermediate nodes. These intermediate nodes have valid IDs and the digital certificates provide authentication and non-repudiation for each node in the certificate path. The consistent hashing is a pre-image resistant mechanism between the IP and logical address which, provides a simple defence against impersonation and Sibyl attacks. If one of the nodes misbehaves or simply creates multiple identities, it will be trivially detected, since the certificates are bound to its device identifier. Therefore, this node can simply be ignored and move on the previous node preceding the target node in the finger table. As long as a single node in the finger table follows the protocol, the authentication can proceed. Moreover, with our reputation extension we can have a threshold for misbehaviour tolerance. After this threshold a malicious node can be blacklisted or be dealt accordingly to predefined rules.

Denial of Service. If a node joins a SOMA-S ring, where the majority of the nodes are malicious, then its identity even though could not be forged, the authentication service for that node could be potentially disrupted through Denial of

Service (DoS). Against DoS attacks, SOMA-S is resilient by using ratings and hashing. With ratings as a defence mechanism, peers can blacklist and exclude malicious nodes that sent fraudulent messages after a threshold, since their identities are detectable. In combination to reputation, consistent hashing is used to distribute the logical identities of the nodes. Therefore, malicious peers would require a large majority to eclipse a node (cut off his ingoing and outgoing links). This is due to the fact that the IDs are mapped using consistent hashing, where the standard hardness assumptions for the chosen hash function apply.

Credential Exchange. An attack directly on the certificate exchange is thwarted by the use of nonces and timestamps, which ensure freshness, and prevent man-in-the-middle attacks. Additionally, the inclusion of origin and target data safeguards against certificate hijacking and all forms of impersonation.

Overlay attack. Attacks on the routing protocol, itself, can be hard to avoid if the majority of nodes are malicious. Even though impersonation is averted through the logical-IP address relationship of the public key certificates, a DoS attack could potentially disrupt the overlay as a whole. In such a case, our reputation model provides the necessary insight to marginalise or expel malicious nodes. Attacks on the network infrastructure itself could include churn attacks and potential network failures from the malicious nodes. Against such attacks, SOMA-S is resilient by requiring at least one correct node in the finger table for correct routing. In addition, instead of using a single successor, employing successor lists will provide additional routes and mitigate the effects of overlay attacks.

4.2   Critical Appraisal

In this section, we reconsider the key management requirements presented in Sect. 1 and discuss on which degree SOMA-S fulfils them.

First of all, our scheme is based on the well known PGP Web-of-Trust and asymmetric cryptography operations, so that it can be considered *resilient against well known attacks*, at least to the extent these two building blocks can be considered resilient.

SOMA-S also provides *full coverage to the Smart Grid* since all its nodes are added to the ring.

Regarding *robustness against key compromise*, while the adequate protection of keys is highly dependent on the implementation, key diversity can be supported by using key derivation techniques based on the public/private key pair of each node. Thus, we argue that this requirement is partially met, but SOMA-S provides all the necessary elements to fully achieve this target.

SOMA-S can support *distributed operation* of its nodes given that, after the bootstrapping phase, its operation is based on a Web-of-Trust and no central TTP is needed. Hence, it shows high availability even over intermittent connections or no connectivity at all with the central utility servers.

Smart Grid devices are expected to have a long lifetime, in the order of 20 years. SOMA-S fulfils *upgradeability* since its distributed nature allows digital certificates to be easily and inexpensively updated with longer key sizes.

The requirement of *certificate revocation* is covered as well, since it is a procedure provided by SOMA-S. Moreover, it is implemented in a distributed manner without the administrative burden imposed by CRLs or having availability requirements like Online Certificate Status Protocol (OCSP) based methods.

Regarding the *scalability* requirement, we argue that SOMA-S can support large numbers of devices since after bootstrapping, where the utility is involved, it is completely decentralised and there is low administrative cost.

Efficiency is highly related to the hardware that will be used. Even though SOMA-S utilizes digital certificates and asymmetric cryptography, there are ways to mitigate the performance penalty by using session keys based on these certificates.

4.3   Smart Grid Services

The overlay described previously is only used as a meta-structure providing a key management service among the Smart Grid nodes. A possible extension to our proposal would be to leverage the DHT to allow diverse and powerful services in the Smart Grid. The DHT will provide an efficient indirection layer, with the services being implemented on top of the overlay. These services will be implemented by the utility companies themselves to overcome the usual point-to-point limitations of the traditional approaches, providing robust services for data aggregation, demand-side-management, privacy protection, advanced policies for billing, two-way consumer-producer services, etc.

5   Conclusions

Due to Smart Grid's special characteristics and requirements, existing key management solutions cannot be applied as is; moreover, as we showed, existing proposals leave space for further improvements. This paper has proposed a distributed and scalable key management system for the Smart Grid without the need of a TTP with high availability. Its operation is based on a DHT for efficient discovery of trust relationships among the Smart Grid nodes. Having the utility take part during the bootstrapping phase, we ensure that it will be difficult for malicious nodes to join the Smart Grid. After this phase, trust policy is more decentralised and flexible in order to promote scalability and resilience.

In our case, the overlay is used as a meta-structure to infer trust relationships and not as the means to provide distributed directory storage. The same overlay, however, could be utilised as an infrastructure to provide other Smart Grid related services, like secure aggregation, and billing. Our future work includes a more detailed definition of the underlying overlay infrastructure, together with the description of additional services for the Smart Grid on top of it.

Acknowledgement. This research has been funded by the European Commission as part of the SMART-NRG project (FP7-PEOPLE-2013-IAPP Grant number 612294).

References

1. Utility-scale smart meter deployments: Building block of the evolving power grid. Technical Report, The Edison foundation, September 2014
2. Aloul, F., Al-Ali, A., Al-Dalky, R., Al-Mardini, M., El-Hajj, W.: Smart grid security: threats, vulnerabilities and solutions. Int. J. Smart Grid Clean Energy $1(1)$, 1–6 (2012). https://dx.doi.org/10.12720/sgce.1.1.1-6
3. Baumeister, T.: Adapting PKI for the smart grid. In: 2011 IEEE International Conference on Smart Grid Communications (SmartGridComm), pp. 249–254, October 2011
4. Beaver, C., Gallup, D., Neumann, W., Torgerson, M.: Key management for SCADA. Cryptog. Information Sys. Security Dept., Sandia Nat. Labs, Technical Report, SAND2001-3252 (2002)
5. Callas, J., Donnerhacke, L., Finney, H., Shaw, D., Thayer, R.: OpenPGP Message Format. RFC 4880 (Proposed Standard) (Nov 2007). http://www.ietf.org/rfc/rfc4880.txt, updated by RFC 5581
6. Choi, D., Kim, H., Won, D., Kim, S.: Advanced key-management architecture for secure SCADA communications. IEEE Trans. Power Delivery $24(3)$, 1154–1163 (2009)
7. Choi, D., Lee, S., Won, D., Kim, S.: Efficient secure group communications for SCADA. IEEE Trans. Power Delivery $25(2)$, 714–722 (2010)
8. Dawson, R., Boyd, C., Dawson, E., Nieto, J.M.G.: SKMA: a key management architecture for SCADA systems. In: Proceedings of the 2006 Australasian workshops on Grid computing and e-research, vol. 54, pp. 183–192. Australian Computer Society, Inc. (2006)
9. Demertzis, F., Xenakis, C.: SOMA: Self-Organised Mesh Authentication. In: Camenisch, J., Lambrinoudakis, C. (eds.) EuroPKI 2010. LNCS, vol. 6711, pp. 31–44. Springer, Heidelberg (2011). http://dx.doi.org/10.1007/978-3-642-22633-5_3
10. Demertzis, F.F., Xenakis, C.: SOMA-E: Self-organized mesh authentication-extended. Math. Comput. Model. $57(7–8)$, 1606–1616 (2013)
11. Gharavi, H., Hu, B.: Multigate communication network for smart grid. Proc. IEEE $99(6)$, 1028–1045 (2011)
12. He, W., Huang, Y., Sathyam, R., Nahrstedt, K., Lee, W.C.: SMOCK: a scalable method of cryptographic key management for mission-critical wireless ad-hoc networks. IEEE Trans. Inf. Forensics Secur. $4(1)$, 140–150 (2009)
13. Kim, J.Y., Choi, H.K.: An efficient and versatile key management protocol for secure smart grid communications. In: 2012 IEEE Wireless Communications and Networking Conference (WCNC), pp. 1823–1828. IEEE (2012)
14. Law, Y.W., Palaniswami, M., Kounga, G., Lo, A.: WAKE: Key management scheme for wide-area measurement systems in smart grid. IEEE Commun. Mag. $51(1)$, 34–41 (2013)
15. Liu, N., Chen, J., Zhu, L., Zhang, J., He, Y.: A key management scheme for secure communications of advanced metering infrastructure in smart grid. IEEE Trans. Industr. Electron. $60(10)$, 4746–4756 (2013)
16. Long, X., Tipper, D., Qian, Y.: An advanced key management scheme for secure smart grid communications. In: 2013 IEEE International Conference on Smart Grid Communications (SmartGridComm), pp. 504–509, October 2013

17. Mittra, S.: Iolus: a framework for scalable secure multicasting. In: ACM SIGCOMM Computer Communication Review, vol. 27, pp. 277–288. ACM (1997)
18. Nicanfar, H., Jokar, P., Leung, V.: Smart grid authentication and key management for unicast and multicast communications. In: 2011 IEEE PES Innovative Smart Grid Technologies Asia (ISGT), pp. 1–8, November 2011
19. NIST: Guidelines for smart grid cybersecurity: vol. 1 - smart grid cybersecurity strategy, architecture, and high-level requirements, vol. 2 - privacy and the smart grid vol. 3 - supportive analyses and references. Technical Report, NIST (2014). doi:10.6028/NIST.IR.7628r1
20. Park, J.H., Kim, M., Kwon, D.: Security weakness in the smart grid key distribution scheme proposed by Xia and Wang. IEEE Trans. Smart Grid 4(3), 1613–1614 (2013)
21. Stoica, I., Morris, R., Karger, D., Kaashoek, M.F., Balakrishnan, H.: Chord: a scalable peer-to-peer lookup service for internet applications. ACM SIGCOMM Comput. Commun. Rev. 31(4), 149–160 (2001)
22. Wan, Z., Wang, G., Yang, Y., Shi, S.: SKM: Scalable key management for advanced metering infrastructure in smart grids. IEEE Trans. Industr. Electron. 61(12), 7055–7066 (2014)
23. Wang, W., Lu, Z.: Cyber security in the smart grid: Survey and challenges. Computer Networks 57(5), 1344–1371 (2013). http://www.sciencedirect.com/science/article/pii/S1389128613000042
24. Wu, D., Zhou, C.: Fault-tolerant and scalable key management for smart grid. IEEE Trans. Smart Grid 2(2), 375–381 (2011). http://ieeexplore.ieee.org/stamp/stamp.jsp?arnumber=5743049
25. Xia, J., Wang, Y.: Secure key distribution for the smart grid. IEEE Trans. Smart Grid 3(3), 1437–1443 (2012)
26. Xiao, S., Gong, W., Towsley, D.: Dynamic key management in a smart grid. In: Dynamic Secrets in Communication Security, pp. 55–68. Springer, New York (2014). http://dx.doi.org/10.1007/978-1-4614-7831-7_5
27. Xu, Y., Wang, W.: Wireless mesh network in smart grid: modeling and analysis for time critical communications. IEEE Trans. Wireless Commun. 12(7), 3360–3371 (2013)
28. Yan, Y., Qian, Y., Sharif, H., Tipper, D.: A survey on cyber security for smart grid communications. IEEE Commun. Surv. Tutorials 14(4), 998–1010 (2012)

Information-Quality Based LV-Grid-Monitoring Framework and Its Application to Power-Quality Control

Mislav Findrik[1]([✉]), Thomas le Fevre Kristensen[2], Thomas Hinterhofer[1], Rasmus L. Olsen[2], and Hans-Peter Schwefel[1,2]

[1] Telecommunications Research Center Vienna (FTW), Vienna, Austria
{findrik,hinterhofer,schwefel}@ftw.at
[2] Aalborg University, Aalborg, Denmark
{tfk,rlo}@es.aau.dk

Abstract. The integration of unpredictable renewable energy sources into the low voltage (LV) power grid results in new challenges when it comes to ensuring power quality in the electrical grid. Addressing this problem requires control of not only the secondary substation but also control of flexible assets inside the LV grid. In this paper we investigate how the flexibility information of such assets can be accessed by the controller using heterogeneous off-the-shelf communication networks. To achieve this we develop an adaptive monitoring framework, through which the controller can subscribe to the assets' flexibility information through an API. We define an information quality metric making the monitoring framework able to adapt information access strategies to ensure the information is made available to the controller with the highest possible information quality. To evaluate the monitoring framework, an event-driven voltage controller is simulated in an LV grid. This controller utilizes the flexibility of photovoltaic (PV) panels to get the voltages into acceptable ranges when the limit is exceeded. This is done by controlling the grid periodically during the time interval that starts when a voltage limit is exceeded and ends when an acceptable voltage level is reestablished. We show how the volatile behaviour of the PV panels causes overvoltages in a baseline scenario. We then show the controller's ability to keep the voltages within their limits. Lastly, we show how control performance can be increased by optimizing information access strategies.

1 Introduction

The current electrical grid is facing increased penetration of renewable energy resources. In particular, in the low voltage (LV) grid photovoltaic (PV) panels are being widely installed on rooftops of the end customers, and electric vehicles are expected to be strongly present. Thus, the end customers are transforming from passive consumers to active "prosumers" that can locally generate and feed the power into the grid. The volatile nature of PVs may lead to over-voltage problems that can occur very rapidly in time [1]. Overcoming this challenge requires

© Springer International Publishing Switzerland 2015
S. Papavassiliou and S. Ruehrup (Eds.): ADHOC-NOW 2015, LNCS 9143, pp. 317–329, 2015.
DOI:10.1007/978-3-319-19662-6_22

control of LV grid assets in order to maintain the voltage profiles [2,3]. Reference [4] outlines the evolution steps of the low voltage grid controlling approaches, starting from "local control" and gradually moving towards "advanced control" that utilizes active management and control via communication infrastructures with limited bandwidth and availability.

We present a monitoring architecture that facilitates operations of control approaches on top of heterogeneous off-the-shelf communication infrastructure with varying network properties. The existing off-the-shelf communication infrastructures are an economically feasible solution for last-mile coverage of LV grids. However, such network infrastructures may be shared (e.g. cellular networks) and are not highly dependable in providing sufficient quality-of-service (QoS). In order to tackle these problems we present a monitoring framework acting as an adaptive communication middleware. The framework features the ability to adapt to network QoS conditions and configure information access to demands of the control approach.

The paper is organized as follows: Sect. 2 covers the related work; in Sect. 3 the LV grid controller scenario is presented together with a description of the baseline data access scenario; in addition, assumptions on the communication network are presented. In Sect. 4 the adaptive monitoring framework is defined and optimization approaches for information access are presented. Section 5 describes the simulation framework and shows the evaluation results of a monitoring framework. Finally, in Sect. 6 conclusions are made and future research directions are outlined.

2   Related Work

Recently, in the domain of Network Control Systems a new control strategy called periodic event-triggered control (PETC) was proposed [5]. PETC is suited for applications where communication resources are scarce, hence we applied the basic principle of PETC (i.e. control triggered by an event) to a voltage control scenario. For the developed controller we analysed how different information access strategies are influencing performance of the controller and how information access can be adapted based on quality metric for sensor information called mismatch probability (mmPr) [6]. MmPr covers the aspect of real-time access, impact of access delays and access strategies on information accuracy in a distributed system and it was used to optimize access to location information for communication network optimization in References [7,8]. In this paper, we applied information access scheduling techniques for PETC controllers in details described in References [9,10], and analysed benefits of such scheduling on the voltage controller.

3   Power Quality Control Scenario

The scenario of this paper assumes a centralized low voltage grid controller (LVGC) located in a secondary (medium voltage to LV) substation. When more distributed power sources penetrate into the LV grid, voltage control becomes more challenging. In particular, the voltage rise is the major issue in LV grids with high share of PVs due to active power injection and small X/R ratios [11]. The purpose of our reference controller is to keep the voltage levels within defined range by regulating grid-interfaced PV inverters upon over-voltage occurrence. As mentioned in Reference [1], the advantage of having a centralized intelligent voltage control mechanism opposed to a local one at each PV lies in its ability to do fair and optimal decisions by orchestrating all resources in LV grid.

The controller relies on real-time measurements of voltages at the grid connection points (i.e. electrical buses) and information about active power injected from PVs. This section provides a detailed overview of the voltage control algorithm, describes a baseline communication pattern in which measurements are sent by the assets, and it presents assumptions on the communication network used in this paper. Note that the development of a voltage controller is not the purpose of the paper; instead the paper will use the described controller as the example of a controller that utilizes several assets distributed across the LV grid for solving the over-voltage problem, and for which information access is provided by the adaptive monitoring framework.

3.1 Voltage Control Approach

The basic principle of the voltage controller is to be active only in the case the measured bus voltages are above a pre-defined upper voltage threshold. The limit in our studies shall be set to 5% of the nominal voltage (1.05 pu). Upon sensing the voltage violation the controller will be triggered to start running in the periodic time steps T_s. Due to radial topologies of LV grids PVs located upstream the feeder are contributing to voltage rise at Bus i. Therefore, to handle the over-voltage event at Bus i, the controller is designed not only to control injected power of PVs at Bus i, but also PVs located upstream the feeder of Bus i. The subset of upstream buses to which the control is spanning is denoted with X and cardinality of X with N. After each time step, the LVGC distributes set-points containing the maximal active power to PV inverters located at the bus where the voltage event has occurred and the upstream buses (see Fig. 1).

LVGC calculates PV set-points based on the difference between the nominal and the measured voltage at Bus i (when there is an over-voltage event at Bus i), as well as the measured active power injection of PV at Bus i and PVs of the buses in Set X. Subsequently upon the arrival, the PV inverters use these set-points for limiting the maximal power injection. The detailed voltage control algorithm is given by Algorithm 1; where K_p, K_i denote proportional and integral gains, respectively, and t_{V_m} the time when voltage was above or below the threshold.

The PV model used for evaluation purposes is adopted from the simulation framework in reference [12] (details are described in Sect. 4), where active power output in normal operation is given by:

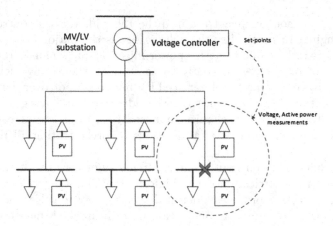

Fig. 1. An example of a LV grid with PVs and overvoltage violation at the last bus

read $V_{measured,i}$, $i \in BUS$;
if $V_{measured} > 1.05$ then
  $err = V_{measured} - 1.05$;
  $P_{max\ bus\ i} = P_{injected\_i} + K_p * err + K_i * \int_{tV_m>1.05}^{tV_m<1.05} err(\tau)d\tau$
  for $b \in X$ do
    $P_{max\ bus\ b} = P_{injected\_b} + K_p * err + K_i \int_{tV_m>1.05}^{tV_m<1.05} err(\tau)d\tau$
  end
end

Algorithm 1. PI control algorithm for stabilizing the voltage at the bus i utilizing PVs at the bus i and the upstream buses $b \in X$

$$p = \begin{cases} p_{rated}, & \textbf{if } \mu A I_{solar} \geq p_{rated} \\ \mu A I_{solar}, & \textbf{otherwise} \end{cases}$$

where μ is the efficiency of the solar cells, A the area they cover, and I_{solar} the solar irradiance.

3.2 Baseline Data Access

The described controller is located in the secondary substation and needs to receive updates from the assets in the grid. A baseline data access is defined for each grid asset: the assets send their update messages to the LVGC according to a Poisson process where the parameter MTBU (Mean Time Between Updates) describes the mean time between updates. The Poisson updates are chosen since their arrival before the controller execution are random, thus yielding the average result as if fixed updates would be randomly scheduled in time. An update contains values of the local voltage and active power injection for each asset. In the evaluation section later, all assets will use the same MTBU parameters.

The baseline data access is illustrated in Fig. 2. The figure also shows two phases. In the inactive phase, the controller receives updates from the assets but does not communicate new set-points. Only when triggered by an out-of band value of a voltage sensor (received at time t_1 at the controller in Fig. 2), the controller transits to the active phase in which it communicates set-points to the assets.

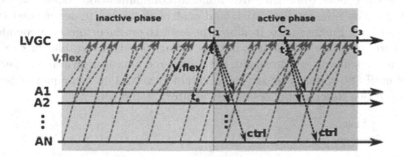

Fig. 2. Baseline data access with Poisson interval updates

3.3   Network Modelling

We assume a reliable communication network, e.g. utilizing retransmissions to compensate for losses. Therefore, the network is abstracted via a stochastic process characterizing the message delays. For the evaluations in this paper, we use independent Poisson processes for the communication between the assets and the controller. A Poisson process is chosen, as a light tailed distribution (fast decaying tail) of transmission delays resulting from assumption that number of retransmissions follows a geometric distribution. The latter is true when the errors are independent of the size of the transmitting packet (see reference [13] for the details). The upstream delay U effecting the messages from the assets to LVGC is characterized by a rate λ_U, while downstream delay from LVGC to the assets is characterized by a rate λ_D. For the purpose of reducing number of parameters for the evaluation analysis, later on we assume that upstream and downstream delays have the same rate, e.g. they are symmetric. Furthermore, in the evaluation, message re-ordering effects are neglected, since delay values are chosen such that $MTBU > 1/\lambda_U$, thus yielding a low probability of a message $i-1$ arriving before the message i. Also, it is assumed that network traffic generated by the controller and the sensors is negligible compared to other cross-traffic in the network, hence it will not have an impact on the delay distributions.

4   Adaptive Monitoring Framework

The main contribution of this paper is to show the benefits of using dynamic data access compared to the static baseline scenario. Consequently, this section contains a detailed specification of the adaptive monitoring framework.

4.1 Adaptivity Concept

This section describes architecture of the monitoring framework, together with the developed approach for adaptive data access. As already mentioned in the introduction section, we are using a grid control approach that is only active in case voltage levels exceed given boundaries. This approach of only being active if voltages are out-of-band assures minimal communication requirements (i.e. bandwidth), as no communication and control is needed under normal stable grid conditions. Furthermore, it allows the asset to produce/inject independently of the current power grid status in case voltage levels are stable. In case of a violation of a soft voltage threshold, the considered approach switches to periodic control until the voltages return to acceptable levels. As a consequence, we will investigate two different phases of the monitoring and control:

- inactive phase
- active phase

Figure 3 depicts our dynamic data access approach, where no monitoring data transfers are happening prior to the active phase (i.e. prior to the reception of the voltage violation event at time t_e). As a result, the input data for the first control step of the active phase needs to be acquired as a reaction to the voltage violation event. In this phase, we call it the *starting phase*, the controller can request its input data dynamically (e.g. subscription to the neighbouring controllable assets of the voltage event location) from the adaptive monitoring framework. After a configurable timeout (waiting time t_{wait}), the voltage controller will do its first computation step C_1 regardless of the complete reception of all asset flexibilities. In case some asset information is not available after t_{wait}, the controller cannot use it for control and the asset is considered to be a normal non-controllable load or producer. As a result, no set-point will be sent to this asset. For simplicity, the control messages are not depicted in Fig. 3. Arrows from the LVGC to the assets represent the request of data.

During the active phase, we investigate an approach that uses scheduling of updates as introduced in [9] in order to be synchronised with the voltage

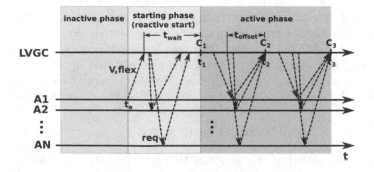

Fig. 3. Reactive start approach

controller steps C_i. As a consequence, the resulting data quality is high as the delay between the measurements and the control step is minimised.

4.2 Task Management Approach

The *task manager* is the main component of the monitoring framework and is responsible for configuring, controlling and coordinating the adaptivity concept mentioned in the last section. The basic approach for the management of the full set of controller monitoring tasks is depicted in Fig. 4. It is based on a tight interlinking with the *Quality Estimator* and the *Network QoS Manager*, which are also sub-components of the monitoring framework: (1) The task manager can request a data quality estimation for a specific variable var_{id} and a set of access configurations $cfg_{id1}, cfg_{id2}, \ldots$. A configuration cfg defines a specific access configuration (push, pull, event-based, scheduled, etc.). For this paper, we are focusing on the (scheduled) pull (i.e. request-reply) access technique. (2) After receiving a new estimation request, the quality estimator requests the communication network QoS options from the according monitoring source defined by var_{id}. (3) This request is answered with a set of QoS options $QoS_1, QoS_2, \ldots$ from the given source address. The set can be empty if there is no connectivity at all or it can contain multiple QoS objects if the connection can handle different priorities or because of the existence of multiple routes/technologies. A QoS object consists out of a delay distribution and a loss probability. For simplicity reasons, we consider only one QoS option per asset for the studies in this paper. (4) In case there is connectivity, the data quality estimation is acknowledged by returning a unique estimation id est_{id}, otherwise a NACK is returned.

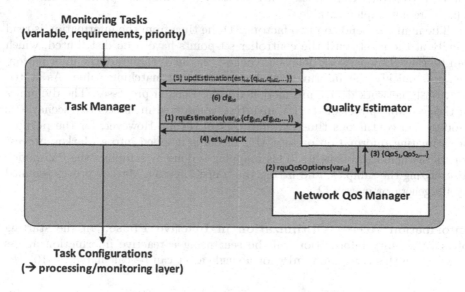

Fig. 4. Monitoring framework components

(5) Based on the available QoS options, the quality estimator can now compute the data qualities to be expected for the set of requested access configurations. The results of the data quality estimation is finally reported to the task manager by using the updEstimation() function. (6) The task manager will then decide for one access configuration, and the decision is sent to the quality estimator.

4.3   Information Quality Estimation

In this section we present the algorithms for quality estimation running inside of **Quality Estimation** component of the monitoring framework. The algorithms are used for optimizing the information access to the grid measurements (e.g. voltages and PV power outputs) for the controller described in Sect. 2. The information access is optimized separately for the two phases of the grid state; namely, for the first one in which there is no overvoltage and controller is inactive and the second one in which overvoltage has occurred and controller is running in equidistant periodic time steps T_s.

In this paper we measure information quality using the mismatch probability (mmPr) metric. Let S(t) is the state of an asset and S'(t) be the state of the asset known by the controller. S'(t) is updated through the communication network making it a step function that only changes when updates arrive, whereas S(t) can be continuous. We then define the mmPr for the time instance at which the controller runs(t_c) as:

$$Pr(S(t_c) \leq S'(t_c)) \tag{1}$$

This quality metric is used because if the asset has less active power available than the controller thinks, the controller may determine set-points for the asset which are not implementable.

The mmPr depends on two factors: (1) the time from the information is read locally at the asset until the controller set-points have been distributed, which can be caused by network delays or caches. (2) the dynamics of the information, i.e. how quickly the information changes to a mismatching value. As stated previously network delays are modelled using Poisson processes. The dynamics of the available power of the PV inverters, considered in the control scenario, is modelled in continuous time through control theory. However, for the purpose of estimating information quality, this model is mapped into a Markov process for its mathematical benefits. This mapping is done by sampling the PV model, discretizing the samples, and fitting the transitions to a Markov process defined by its generator matrix Q.

Information Access Optimization in Inactive Phase. In the starting phase, flexibility information will be sent using a reactive information access strategy. In this case, the mmPr for a single asset can be calculated as [10]:

$$Pr(mm) = \int_0^\infty Pr(mm|t) f_{mmPr}(t) dt \tag{2}$$

where $Pr(mm|t)$ is the probability of the information changing to a mismatching value during the time t, and $f_{mmPr}(t)$ is the density function of the total experienced delay. The total experienced network delay is here defined as the time from the information was read at the sensor until the controller has distributed setpoints based on the information, and is, thereby, dependent on the information access strategy.

$Pr(mm|t)$ is computed from the Markov process describing the flexibility information. Since we consider the flexibility of assets, too much flexibility will have no negative consequences for the system, and the information is only considered to be mismatching if the flexibility of the asset is in a lower state when it is utilized than it was when the information was read. For a general Markov process defined by the generator matrix $\mathbf{Q}$ this means [10]:

$$Pr(mm|t) = \sum_{i=1}^{M} \left(\pi_i \sum_{j=1}^{i-1} p_{ij}(t) \right) \tag{3}$$

where π is the stationary probabilities of the Markov Chain, $p_{ij}(t)$ is the probability of being in state j at time t given the state i at time 0, and can be calculated from standard transient Markov chain calculations [10].

$f_{mmPr}(t)$ is for the reactive information access strategy calculated in [10] as:

$$f_{\text{mmPr,rea}} = (f_u * f_d)(t) \tag{4}$$

where f_u and f_d is the density function of the network delay in the upstream and downstream respectively. In this paper we consider several assets providing flexibility information. Therefore, the controller must either wait for all answers to arrive, or determine a proper time to wait before it runs using the information available. We optimize this time by minimizing the mmPr averaged over all relevant assets. However, since we now consider the controller running at the latest after a predetermined waiting time, the stochastic upstream network delay is replaced by the deterministic waiting time. Using this and the reciprocal of the network downstream delay λ_d we get:

$$f_{\text{mmPr,rea}} = f_d(t - t_{wait}) \qquad\qquad , t > t_{wait} \tag{5}$$

Using this, the mmPr, $mmPr_{rec}$, can be calculated for a single asset assuming the response was received before the waiting period was over. If the response was not received before the end of the waiting period, there is no old information to be used resulting in a certain information mismatch. Using this and the distribution of the upstream delay $F_u(t)$ we can calculate the mmPr for a single asset as:

$$mmPr_{1Asset} = mmPr_{rec}F_u(t_{wait}) + 1 \cdot (1 - F_u(t_{wait})) \tag{6}$$

This mmPr notion can be extended to N assets by conditioning on the number of responses that is received and averaging over all assets.

$$mmPr_{tot} = \sum_{n=0}^{N} \frac{n \cdot mmPr_{1Asset} + (N - n)1}{N} \cdot Pr(\text{n responses arrive}) \tag{7}$$

Information Access Optimization in Active Phase. Instead of getting pushed updates from the assets in Poisson intervals, the task manager utilizes pull access (e.g. request-response communication patter) to retrieve the information from each controlled asset in the active phase. The goal of the task manager is to find T_{offset} value such that $mmPr$ metric is minimized and asset information arrives before the controller executions C_i (see Fig. 3). T_{offset} is a function of upstream and downstream delays (λ_U and λ_D) as well as information dynamics (**Q** matrix). In recent paper [9] we derived analytical formulas for deriving $mmPr$ curve for different data access strategies. Here, we recall the formula for case of pull data access:

$$
mmPr_{pull} = \int_0^{T_o} \int_0^{T_o-t} \sum_{i=1}^{S} \pi_i Q_{i,i}[exp(Qs)]_{i,i} ds \int_0^{T_o-t} \lambda_U exp(-\lambda_U x) dx
$$

$$
\lambda_D exp(-\lambda_D t) dt + \left(1 - \int_0^{T_o} \left(\frac{\lambda_D \lambda_U}{\lambda_U - \lambda_D} exp(-\lambda_D x) + \frac{\lambda_D \lambda_U}{\lambda_D - \lambda_U} exp(-\lambda_U x) dx\right)\right) *
$$

$$
\int_0^{T_o} \int_0^{T_o+T_s-t} \sum_{i=1}^{S} \pi_i Q_{i,i}[exp(Qs)]_{i,i} ds \lambda_D exp(\lambda_D t) dt
$$

From the equation above, T_{offset} is derived at the point where $mmPr_{pull}$ is minimal. Due to its length, the closed-form solution is omitted from the paper.

5   Simulation Framework and Evaluation Results

In order to evaluate the monitoring framework we implemented a prototype of the publisher layer inside of DiSC simulation framework [12]. DiSC is MATLAB based open-source simulation framework that contains implementation of PV panels, as well as the data of real solar irradiation and household consumption collected in Denmark. Moreover, it allows verification of different control approaches. Utilizing DiSC framework we constructed a LV grid with 7 buses, each having 15 household loads connected together with PVs with total rated power of 7 kW. DiSC tool was configured to generate data for all assets and to calculate voltages each second, which is the highest precision offered by the DiSC models. The voltage profile during a summer day was generated for all buses and shown in Fig. 5.

In normal operating conditions with no PVs in LV grid, the feeder voltage decreases as the distance from the substation increases. However, with PV installations at all buses the most far bus becomes the most sensitive one where overvoltage occurs first. Since simulating the whole monitoring stack takes significant amount of time our analysis is further focused on observing the controller performance in a time interval from $2.6 * 10^4\,s$ until $4.5 * 10^4\,s$, since in that time interval, overvoltage events are occurring at Bus 7.

Furthermore, to compare the monitoring framework solution against the baseline data access a suitable performance index has to be defined. For this purpose we define Overvoltage Surface (OS) index. The index is calculated by multiplying time in which the bus suffered overvoltage with the voltage value subtracted by critical threshold 1.05 pu (i.e. integral of voltage curve going above

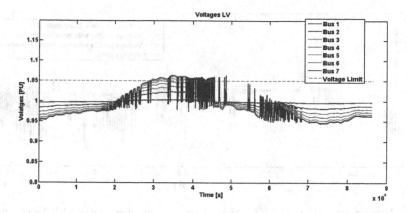

Fig. 5. Voltage profile for all buses during 86400 s (24 h)

1.05 pu). Note, that during zero delay conditions and infinite rate updates the controller would have OS index value 0 since overvoltage would never happen; worse QoS network conditions or different data access strategies would yield higher index value. The simulation configuration parameters for the controller, the baseline data access and the monitoring framework are given in Table 1.

Table 1. Simulation configuration parameters

Fixed configuration parameters	Value
The controller period T_S (active phase)	10 s
K_p (Proportional gain of the controller)	3000
K_p (Integral gain of the controller)	1000
N (# of buses upstream used for over-voltage control)	1
# of Markov states in the information model	20
MTBU for baseline access (inactive and active phase)	10 s
Variable configuration parameters	Values
T_w	[4 6 7] s
T_{offset}	[3.6 4.4 6.1] s
Symmetrical mean delays $\frac{1}{\lambda_U} = \frac{1}{\lambda_D}$	[1 2 3] s

The baseline data access is compared against the monitoring framework adaptive over three different mean delay values, namely 1, 2 and 3 s in both directions. Adaptive parameters T_w and T_{offset} are calculated for corresponding delays using equations given in previous section, which are configured before the simulation runs. The comparison of voltage profiles when using the monitoring framework opposed to the baseline data access is shown in Fig. 6. For further analysis OS indexes are calculated form voltage profiles obtained for different delays and presented in Table 2.

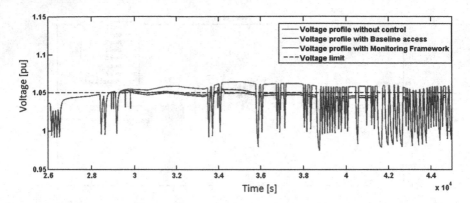

Fig. 6. Voltage profile at Bus 7 for the reference control scenario with 1 s symmetric delays

The results show that control, even with the poor data quality resulting from the baseline access scenario, is effective in reducing the OS index (hence reducing the overvoltage effect) by more than half for all network delays compared to the case without control. Moreover, the baseline scenario is significantly affected by the network delays, the increase from 1 s to 3 s causes the OS index to almost double. Using the optimized information access via the monitoring framework, the OS index is reduced for all network delays in comparison to the baseline data access. Hence, its sensitivity to changing network delays is significantly reduced.

Table 2. Comparison of data access schemes based on OS index

Simulation Scenario		OS index
Without control (worst case)		11 494
Symmetric delays 1s	Baseline Data Access	2700,8
	The monitoring framework	1308,5
Symmetric delays 2s	Baseline Data Access	3644,0
	The monitoring framework	1352,1
Symmetric delays 3s	Baseline Data Access	5219,1
	The monitoring framework	1382,4

6   Conclusion and Future Work

In this paper a novel monitoring framework for Smart Grid controllers with adaptive information access is presented. We provided a detailed description of the monitoring framework components and described the algorithms within components as well as their interaction. Furthermore, a sample controller for regulating voltage profiles in LV grid was specified and the benefits from the adaptivity provided by the monitoring framework are evaluated via the simulation with respect to the baseline data access. It is shown that control performance with the monitoring solution is significantly improved. The future work shall focus on extensive test-bed evaluation of the framework with different type of Smart Grid controllers running over heterogeneous access networks (power-line communication, cellular networks, xDSL and WiFi).

Acknowledgments. This work is part of SmartC2Net research project, supported by the FP7 framework programme under grant No318023. FTW is also supported by the Austrian Government and City of Vienna within the competence center COMET.

References

1. Ran, B., Negeri, E., Baken, N., Campfens, F.: Last-mile communication time requirements of the smart grid. In: Sustainable Internet and ICT for Sustainability (SustainIT), pp. 1–6, October 2013
2. Groenbaek, J., Bessler, S., Schneider, C.: Controlling ev charging and pv generation in a low voltage grid. In: 22nd International Conference and Exhibition on Electricity Distribution (CIRED 2013), pp. 1–4, June 2013
3. Kupzog, F., Schwalbe, R., Prggler, W., Bletterie, B., Kadam, S., Abart, A., Radauer, M.: Maximising low voltage grid hosting capacity for pv and electric mobility by distributed voltage control. Elektrotechnik und Informationstechnik **131**(6), 188–192 (2014)
4. Einfalt, A., Kupzog, F., Brunner, H., Lugmaier, A.: Control strategies for smart low voltage grids. In: Integration of Renewables into the Distribution Grid, CIRED 2012 Workshop, pp. 1–4, May 2012
5. Heemels, W., Donkers, M., Teel, A.: Periodic event-triggered control for linear systems. IEEE Trans. Autom. Control **58**(4), 847–861 (2013)
6. Bogsted, M., Olsen, R.L., Schwefel, H.P.: Probabilistic models for access strategies to dynamic information elements. Perform. Eval. **67**(1), 43–60 (2010)
7. Nielsen, J., Olsen, R., Madsen, T., Schwefel, H.: On the impact of information delay on location-based relaying: a markov modeling approach. In: Wireless Communications and Networking Conference (WCNC), pp. 3045–3050. IEEE, April 2012
8. Nielsen, J., Olsen, R., Madsen, T., Uguen, B., Schwefel, H.P.: Location quality aware policy optimization for location based relay selection in mobile networks. Wireless Netw. (2015)
9. Findrik, M., Groenbaek, J., Olsen, R.: Scheduling data access in smart grid networks utilizing context information. In: 2014 IEEE International Conference on Smart Grid Communications (SmartGridComm), pp. 302–307, November 2014
10. Kristensen, T., Olsen, R., Rasmussen, J.: Analysis of information quality in event triggered smart grid control. In: Proceedings of the 2015 IEEE 81st Vehicular Technology Conference (2015)
11. Juamperez, M., Yang, G., Kjar, S.: Voltage regulation in lv grids by coordinated volt-var control strategies. J. Mod. Power Syst. Clean Energy **2**(4), 319–328 (2014)
12. Pedersen, R., Sloth, C., Andresen, G.B., Wisniewski, R.: Disc: a simulation framework for distribution system voltage control. In: Submitted for European Control Conference (2015)
13. Bertsekas, D., Gallager, R.: Data Networks, 2nd edn. Prentice-Hall Inc, Upper Saddle River (1992)

Energy Efficient Small-Cell Discovery
Using Users' Mobility Prediction

Apollinaire Nadembega[1(✉)], Abdelhakim Hafid[1],
and Ronald Brisebois[2]

[1] Network Research Lab, University of Montreal, Montreal, Canada
{nadembea, ahafid}@iro.umontreal.ca
[2] BIBLIOMONDO, Montreal, Canada
ronald.brisebois@mondoin.com

Abstract. Deployment of small cells (i.e., picocells and femtocells) within macrocell coverage is seen as a cost-effective way to increase system capacity and to equip wireless WANs with the ability to keep up with the increasing demand for data capacity. Existing cell discovery mechanisms are tailored for homogeneous networks (macrocells only). User Equipment (UE) cannot efficiently save energy in the process of small cells detection in order to exploit offloading opportunities provided by such heterogeneous deployments. In this paper, we propose a Mobility Prediction aware Scanning Start Time Estimation (MPSTE) scheme to discover/detect small cells efficiently in terms of energy. Based on the current data on road segments (e.g., density of road segment, UEs' speeds and physical aspects of road segment) and current behaviour of UEs on the road segment, MPSTE allows deriving the time interval UE will spend in the small cell and making decision to perform handoff or no; if handoff is necessary, MPSTE derives the best time to begin the scanning process to discover small cells. Simulation results show the benefits of MPSTE over existing schemes in terms of energy saving by UEs.

Keywords: Energy efficiency · Mobility prediction · Handoff decision · Handoff time estimation · Wireless network

1 Introduction

Over the past few years, there has been significant growth in the volume of mobile data traffic following the proliferation of smart phones and new mobile devices that support a wide range of applications and services. These mobile data traffic demands have been increasing exponentially with the advent of radio access technologies; cell sizes are also becoming increasingly smaller and smaller [1–3]. Indeed, operators are increasingly resorting to deploy macro-cells for wide-area coverage and smaller pico or femto-cells for high capacity hotspots and relays to improve coverage [4]; these new types of deployments are commonly referred to as heterogeneous networks (HetNets) [5] and are currently receiving significant attention in industry [22]; macro-tier guarantees the coverage, while the overlay network allows offloading data traffic from the macro-cell network and satisfying local capacity demands; small cells can extend the network

© Springer International Publishing Switzerland 2015
S. Papavassiliou and S. Ruehrup (Eds.): ADHOC-NOW 2015, LNCS 9143, pp. 330–343, 2015.
DOI: 10.1007/978-3-319-19662-6_23

coverage and the reduced cell size leads to higher spatial frequency reuse and increased network capacity. Indeed, LTE-Advanced systems provide the option of extending the system bandwidth using carrier aggregation of up to 100 MHz using five 20 MHz carriers [6]. HetNets deployments are one of the key enablers in providing ubiquitous coverage and capacity enhancements for satisfying high data rate and Quality of Service requirements for LTE-Advanced networks [22]. User Equipments (UEs) are allowed to roam freely within the HetNet and undergo vertical handoffs between the heterogeneous layers; however, there are other requirements regarding energy/battery savings. Indeed, the study of energy efficient schemes for discovering and selecting radio nodes deployed in different frequency layers is a relevant and important problem [2].

In this paper, we propose a scheme Mobility Prediction aware Scanning Start Time Estimation (MPSTE) that allows energy saving for UEs in the process of small cell discovery in HetNets; MPSTE takes into account the UE's residence time in the small cell and the arrival time to the small cell. More specifically, MPSTE uses the current physics of traffic flows of road segments as a basis for defining probability distributions of traffic variables. Based on these probability distributions, MPSTE builds a table of transit times of each road segment; making use of these tables, we estimate the residence time of an UE in a small cell and the required time to reach this small cell; these estimates are computed only when we are certain that UE will transit the small cell. Based on the estimated residence time in the small cell, we propose the concept of threshold residence time (TRT) in the small cell for inter-frequency scanning. If the estimated residence time is below this threshold, UE should not search for or connect to small cells; in this case, the small cell connection opportunities become so short that they offer very limited benefit. Otherwise, MPSTE estimates the best time to begin scanning to discover small cells; the goal is to reduce unnecessary scanning and thus save energy. In our proposed scheme, we assume that the path of UE is known in advance (e.g., using the schemes described in [7, 8, 16, 21]).

The remainder of this paper is organized as follows. Section 2 presents related work. Section 2 gives an overview of the system model used. Section 3 describes the proposed MPSTE scheme. Section 4 reports some simulation results and discusses the performance of MPSTE. Finally, Sect. 5 concludes the paper.

2 Related Work

Even though current literature includes various studies towards understanding the main challenges of interference management in the presence of femtocells, little light has been shed on the open issues of mobility management in the two-tier macrocell-femtocell network [22]. In [9–11], the authors analyzed the additional mobility performance benefits gained from using UEs supporting inter-site carrier aggregation (CA) between macro and small cells, instead of legacy UEs supporting only data reception from a single cell. They did focus on the time when UEs should perform handoffs; these approaches did not take into account energy saving. In 3GPP standardization, there is currently active interest in mobility enhancements in HetNets [12–14, 19, 20]; researchers are investigating efficient discovery of small cells which are deployed in different carriers for offloading purposes using network-centric as well as UE assisted enhancements. A. Prasad et al. [12] proposed a scheme, using a flexible fingerprint matching region around fingerprint reference locations, that provides energy efficient small-cell discovery. The basic idea is that UE will initiate inter-frequency scanning only if the received signal strength of the current location indicates that there is a small cell in the vicinity. This scheme suffers from two key limitations: (1) UE initiates inter-frequency scanning, in order to identify a small cell in the vicinity, independently of its expected residence in the cell; and (2) the time to start scanning is not determined; UE begins the scanning as soon as it identifies a small cell in the vicinity; this causes a waste of energy especially if UE does not handoff to the small cell. A. Prasad et al. [13] proposed an energy efficient small cell discovery mechanism for HetNets; they make use of a threshold speed for inter-frequency scanning, above which UE should not search for or connect to small cells. Indeed, UE searches for inter-frequency cells much less frequently when moving slowly; a threshold speed vt is defined above which UE should stop scanning for inter-frequency cells. The threshold speed is defined such that above this speed, the time spent by UE in the small cell is smaller than a minimum time-of-stay (i.e., residence time). The authors use the current speed of UE to estimate its average time spent in small cell without considering speed variation (e.g., because of road traffic variation and signalization) along UEs' movement and the speed of UEs that are currently located in the small cell. W.-H. Yang et al. [14] proposed a handoff scheme with geographic mobility awareness (HGMA) which considers historical handoff patterns of mobile devices. They evaluated a mobility-based scanning and cell selection mechanism in HetNets consisting of WiFi (small cell) and WiMAX (macro cell) networks. HGMA prevents UE from triggering unnecessary handoffs according to its received signal strength and moving speed; indeed, when UE is around cell boundaries, due to temporal RSS dropping, HGMA checks whether UE's current speed is higher than its average speed when it is within WiFi networks; if the response is yes, UE will also trigger a handoff event because it is likely to move out of its current WiFi networks. The main limitation of this scheme is the fact that authors do not estimate the time to start scanning to discover small cells. Also, they make use of average speed, to estimate UE's residence time in the small cell, without taking into account the current state of road traffic; this may cause inaccurate estimation of UE's residence time in the cell.

We conclude that existing schemes do not estimate the best time to start the scanning process; furthermore, their estimation of residence time (of UE in small cells) may be inaccurate because they used average speed of UE without taking into account the dynamics of road traffic.

3   Proposed Scheme

In this section, we describe the proposed scheme (MPSTE) for small cell discovery. First we describe our approach to estimate aggregate transit times that are used to build transit times tables of road segments. Then, we present the process that makes use of transit times tables to decide whether to scan (or not) in order to discover small cells.

3.1   Transit Times Estimation

The aggregate transit times estimation scheme is based on the current data about the traffic flows; indeed, we compute in one time, the transit times for UEs that will transit the same road segments; it computes tables of transit times of the road segments. We assume that the road topology consists of several roads and nodes. We refer to the location where at least two roads intersect as a road intersection. We refer to the location where the border of a cell and a road intersect as handoff point. We assume that a road intersection or a handoff point is represented by a node; each node is identified by a node ID (i.e., labeled node) that is related to its geographic coordinates (i.e., latitude and longitude); we assume that the handoff point is an available information coming from the network; the handoff point location is updated when current location (i.e., location used in the model) and real location are quite distant (e.g., 50 m). We refer to the road portion between two intersections a and b as a road segment identified by (a, b) where the navigation direction a towards $b (a \rightarrow b)$ is different from the navigation direction b towards $a (b \rightarrow a)$. Based on the geographic coordinates, we build an oriented graph that represents the road topology for the proposed system; this graph is readily available and may get downloaded from the city road maps. The road segment database contains road segment ID, traffic direction of lane on the road segment, list of ID of small cells covering the road segment (or part of it), and list of ID of macro cells deployed in the same frequency layer; these tables are updated when physical aspects of road segments change (e.g., because of lane size reduction, closed lane or road work) or after a pre-defined time (e.g., climatic different seasons).

In traffic flow theory, it is common to model a vehicular flow as a continuum and represent it with macroscopic variables of flow f(t) (veh/s), density d(t) (veh/m) and velocity v(t) (m/s). The definition of a flow gives the following relation between these three variables: $f(t) = d(t) * v(t)$. Thus, we make the assumption that the state of traffic flow is fully characterized by the density d; we compute the density of a road segment in terms of the number of UEs in that specific road segment of the density of road segment S at time t is expressed as follows:

$$D(S,t) = \frac{Num(S,t)}{A_S} \tag{1}$$

where $Num(S, t)$ denotes the number of UEs in S at time t and A_S is the size of S. For example, for two traffic flow conditions (i.e., two density classes), we define di as the boundary density between (i) *lightly dense segments* (i.e.,$D(S, t) \leq d_i$) and (ii) *highly dense segments* (i.e.,$D(S, t) > d_i$).

For each density class, we estimate transit time for each road segment. To compute the transit time of road segment S for a density class, we first compute PDF of S transit/travel times by UEs. The transit time, by UE travelling on road segment S towards the destination, is mainly impacted by traffic flow conditions (i.e., density) on S [15]. The probability population consists of (1) in the congested condition, the times to transit S by UEs who have already transited S and are currently located on adjacent (to S) road segments; these times are known; (2) in the free flow condition, the times to transit S ΔT_i by the user under consideration UEi; these times are computed making use of

Eqs. 2–6 and UEi's driving behavior (i.e., acceleration, deceleration, constant velocity) on the road segments already transited, in the same traffic flow condition, just before entering S; and (3) in the under-saturated condition: the times to transit S by UEs (i.e., UE_1,..UE_i,..) who are currently on S; these times are computed making use of Eqs. 2–6 and UEi's last known driving behavior (i.e., acceleration, deceleration, constant velocity). The transit time ΔT_i, for an UE_i is computed based on velocity function; the physical expression of velocity at time is given by:

$$v(t) = \sigma \times (t_i - t_{i-1})v(t_{i-1}) \tag{2}$$

Whereby σ and $(t_i - t_{i-1})$ denote the acceleration/deceleration (depending on the movement) and the minimum time granularity, respectively. Based on Eq. 2 and the simplified driving behavior cycle (i.e., a cycle of acceleration, maintaining of a constant velocity, deceleration and finally the stopping), we derive the expression of travel time of acceleration phase Δt_i^a and deceleration phase Δt_i^d as follows:

$$\Delta t_i^a = \frac{v_m}{a} \ and \ \Delta t_i^d = \frac{v_m}{d} \tag{3}$$

where a, d and v_m denote the acceleration, the deceleration and the velocity during the constant velocity phase, respectively. Thus, we need to compute the travel distance of constant velocity phase l_i^c; it is given by:

$$
\begin{aligned}
l_i^c &= l_s - (l_i^a + l_i^d) \\
l_i^a &= \left[\frac{a}{2} \times (\Delta t_i^a)^2 \right] \\
l_i^d &= \left[\frac{d}{2} \times (\Delta t_i^d)^2 \right] + (v_m + \Delta t_i^d)
\end{aligned}
\tag{4}
$$

where l_s, l_i^a and l_i^d denote the length of the road segment, the travel distance during the acceleration and deceleration phases, respectively. Using Eq. 4, the expression of travel time of a constant velocity phase Δt_i^c Δt_i^c is as follows:

$$\Delta t_i^c = \frac{l_i^c}{v_m} \tag{5}$$

Finally, we sum the travel time of each phase of the driving behavior cycle to obtain the travel time ΔT_i on the road segment. Its expression is given by:

$$\Delta T_i = \Delta T_i^a + \Delta T_i^c + \Delta T_i^d \tag{6}$$

Let n_S denotes this population and $n_S^{\Delta T_i}$ denotes the fraction of n_S who transits S with $\Delta t_{S_i}^j$ as transit time. Along road segment S, transit time Δt_{S_i} is a random variable with distribution ψ. We derive the probability distribution ψ of transit times of road segment S as follows:

$$\psi(\Delta T_i) = \frac{n_S^{\Delta T_i}}{n_S} \tag{7}$$

Using Eq. 7, we derive the cumulative distribution function (CDF), F_S, of transit times of road segment S as follows:

$$F_S(\Delta T') = \sum \psi_S(\Delta T \leq \Delta T') \tag{8}$$

To estimate time T^S when UE will transit S, we derive the inverse function of the CDF of travel times of S $F_S()$; we select the values of probability 0.5 that determines the median of transit/travel time of S; The expressions of T^S is given by:

$$T^S = F_S^{-1}(0.5) \tag{9}$$

In order to take into account the density of road segment S and increase the estimation accuracy, the transit time is estimated for different densities of S; in other word, when density of S changes, the transit time is computed based on Eq. 9; MPSTE keeps a table of transit time of each road segment S for different densities (see Table 1).

Let $f()$ be the function that returns the transit time of a given road segment using the transit time table of the road segment; for example, $f(Sj, D(Sj, t)) = T_i^{Sj}$ if we assume $D(Sj, t) \in [d_{i-1}, d_i]$ (see Table 1). In order to compute time T to transit a path formed by subsequent transitions of road segments (e.g., road segments covered by small cells that will be transited by UE), we need only to sum the transit times of road segments that form this path. More generally, if S_1 to S_m is UE's path to transit, then, T can be computed as follows:

$$T = \sum_{q=1}^{m} \left(p_q \times \frac{f(Sq, D(Sq, t))}{Ds_q} \right) \text{ with } p_q \leq Ds_q \tag{10}$$

where Ds_q is the length of S_q and p_q is the length of S_q portion transited by UE. In this way, we compute the transit times for UEs who will transit the same road segments (i.e., from S_1 to S_m) before the next update of transit times tables of S_1 to S_m.

Table 1. Transit time table of road segment Sj

Density	Transit times
$[0, d_1]$	T_1^{Sj}
....	...
$]d_{i-1}, d_i]$	T_i^{Sj}
...	...
$]d_{n-1}, d_n]$	T_n^{Sj}

3.2 Small Cell Discovery

To achieve both energy efficiency and small cell usage efficiency, UE should perform inter-frequency measurements only when it is close to the small cell (see Fig. 1).

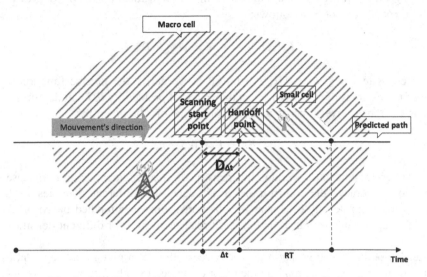

Fig. 1. Mobility prediction aware Scanning Start Point Estimation.

In Fig. 1, RT denotes the estimated residence time in the small cell, Δt denotes the time needed to complete the handoff process, and $D_{\Delta t}$ denotes distance, travelled by UE, during Δt. RT is computed using Eq. 10 while Δt is assumed to be known a priori. Let S_{hp} be the road segment where the handoff point is located, Ds_{hp}, the length of S_{hp} and $f(s_{hp}, D(s_{hp}, t))$ the time to transit Ds_{hp} when S_{hp} density is $D(s_{hp}, t)$; we compute the velocity $v(S_{hp}, t)$ at time t on S_{hp} and then the distance $D_{\Delta t}$, transited during Δt; v (S_{hp}, t) and $D_{\Delta t}$ can be computed as follows:

$$v(s_{hp}, t) = \frac{Ds_{hp}}{f(s_{hp}, D(s_{hp}, t))}$$
$$D_{\Delta t} = \Delta t \times v(s_{hp}, t) \tag{11}$$

For UEs who rapidly transit the small cell, the small cell usable time would be short; in this case, small cells offer very limited benefit. Figure 2 shows the operation of MPSTE in making the decision to scan or not to discover small cells. In MPSTE, UEs will initiate inter-frequency scanning only if the UE's predicted path will transit a small cell(s); using schemes in [7, 8, 21], we are able to predict UE's path (i.e., road segments to be traversed). For a given UE, if the predicted path includes a small cell(s), MPSTE will estimate the residence time RT of this UE in the small cell; RT represents the transit time of the road segments (or a portion of a road segment) covered by the small cell that will be transited by UE. Thus, with the knowledge of road segments that will be transited in the next small cell, we can use Eq. 10 to compute RT.

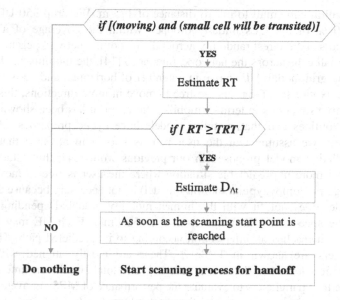

if [(moving) and (small cell will be transited)]

YES

Estimate RT

if [RT ≥ TRT]

YES

Estimate $D_{\Delta t}$

As soon as the scanning start point is reached

NO

Do nothing

Start scanning process for handoff

Fig. 2. MPSTE process

If estimated residence time RT is greater than or equal to threshold TRT, we estimate the required distance to complete a handoff process, called $D_{\Delta t}$; $D_{\Delta t}$ is computed using Eq. 11. Thus, before starting the small cell scanning for handoff process, MPSTE waits that UE reaches the scanning start point. To avoid increasing energy consumption by UEs using MPSTE, we assume that the wireless network operator administrates an entity, called Controller (similar to the controller defined in [19], that performs MPSTE process; indeed, the Controller computes the transit times tables; then, for each UE, Controller estimates RT and sends the scanning start point location when the handoff is required in the small cell.

4 Performance Evaluation

In this section, we evaluate, via simulations, the performance of MPSTE in terms of energy consumption during small cell scanning process. We compare MPSTE against the schemes described in [13] and [14], referred to as AP1 and AP2, respectively. We selected AP1 and AP2 because to the best of our knowledge (1) they represent the best work related to energy saving during small cell scanning process; and (2) they are based on UEs' mobility behavior to trigger small cell scanning; indeed, whether the UE's speed is smaller than a threshold speed (resp. average speed of UEs in the small cell), AP1 (resp. AP2) triggers small cell scanning. According to simulation scenario, the best threshold speed for AP1 is 15 km/h.

Similar to [13], we consider a heterogeneous network consisting of seven macro base station sites, with three sectors each, altogether 21 macro cells. Three small cells are randomly dropped in each macro cell, with minimum small to small distance of 60 m and minimum small to macro distance of 175 m. We drop 350 UEs in these cells, with 20 % of the UEs initially dropped within the coverage of a randomly selected hotspots and the rest randomly across the rest of the network; each UE transits at least two small cells across the network. Similar to [14], the mobility model for UEs is a Manhattan grid model; UEs move in a number of horizontal and vertical streets in an urban area. Notice that the users are free to move in more directions; this is similar to less predictive situations in term of mobility; however, it has been shown that users follow daily routines and that mobility models have cyclic properties; also, in the proposed work, we assume that the user path is known in advance thanks to the mobility prediction model proposed in our previous work [21]; thus, study the performance for a more unpredictable situation where the user is free to move in more directions (e.g., random waypoint mobility model) is not necessary because this type of mobility model does not fit with the human mobility model. Depending on traffic conditions, the speeds of UEs range from 5 to 50 km/h. Each UE may change its direction when it reaches an intersection according to its predicted path. The system-level parameters are shown in Table 2. These are mostly aligned with HetNets mobility-specific parameters defined in [17, 18]. The total simulation time is 12 h.

We define four parameters to evaluate the performance of MPSTE: average number of scanned small cells, energy consumption rate and two handoff error rates; we refer to the case when a handoff is required and not done as a missed handoff; we refer to the case when a handoff is not required but done as an unnecessary handoff; handoff is required (resp. is not required) in a small cell when the real residence time in this small cell is bigger or equal (resp. less) to TRT.

Table 2. System-level simulation parameters.

Parameters	Values
Carrier frequency	2.0 GHz
Macrocell ISD	1000 m
Spectrum allocation	10 MHz
Time to trigger	480 ms
Handover preparation/execution time	80 ms
Scanning gap	6 ms

The average number of scanned small cells, denoted by As, and the average residence time in small cells, denoted by At, are computed as follows:

$$As = \frac{Ns}{Nu} \text{ and } At = \frac{Nt}{Nu} \tag{12}$$

where Ns, Nt, and Nu denote the total number of scanned small cells, the total residence time in small cells and the number of UEs respectively. The missed handoff rate, denoted by MH and unnecessary handoff rates, denoted by UH, are computed as follows:

$$MH = \frac{Ne1}{Np} , UH = \frac{Ne2}{Np} \text{ and } Eh = MH + UH \tag{13}$$

where Ne_1, Ne_2 and Np denote the total number of missed handoffs, the total number of unnecessary handoffs and the total number of handoff point crossings respectively. The energy consumption rate, denoted by R, is computed as follows:

$$R = 1 - \frac{Rb}{Bc} \tag{14}$$

where Rb and Bc denote the remaining battery capacity and the battery capacity respectively.

Figure 3 shows the average number of scanned small cells and the average residence time in small cells when varying the threshold residence time (TRT). In Fig. 3a, we observe that for MPSTE (resp. AP1 and AP2), the average number of scanned small cells decreases (resp. remain constant) when TRT increases; this is expected since when TRT increases, the number of estimated residence times bigger than TRT decreases and, thus the average number of scanned small cells decreases. AP1 (resp. AP2) is not impacted due to the fact that its handoff decision is based on a threshold speed (resp. average speed of UEs in the small cell). In Fig. 3a, we observe that when TRT is smaller than 4 min (resp. 7 min), the average number of scanned small cells of AP1 and AP2 is smaller than the average number of scanned small cells of MPSTE. In conclusion, AP1 and AP2 provide a smaller average number of scanning compared to MPSTE; we believe that it is not necessary to perform scanning for small cells for smaller residence time in these cells (e.g., smaller than 5 min in Fig. 3a). Figure 3b shows that, for the three schemes, the average residence time in small cells increases with TRT; this is expected since when TRT increases, UEs spend more time in small cells and, thus the average residence time in small cells increases. We also observe that MPSTE outperforms AP1 and AP2; thus, MPSTE allows to perform better traffic offloading (from macro cells) compared to AP1 and AP2.

Figure 4 shows missed handoff rate (MH) variation with threshold residence time (TRT). We observe that for MPSTE (resp. AP1 and AP2), MH rate increases (resp. remain constant) with TRT. This is expected since when TRT increases, the number of handoffs in small cells increases and, thus the number of handoff that are required and not done increases. We also observe that MPSTE outperforms AP1 and AP2; indeed, MPSTE provides an average MH rate (defined as $\dfrac{\sum_{i=1}^{10}\left(\dfrac{MHi}{i}\right)}{10}$ where i denotes the value of TRT) of 0.02 per minute while AP1 (more efficient than AP2 in this scenario) provides an average MH rate of 0.3 per minute.

The average relative improvement of MPSTE compared to AP1 is about 0.28 (28 %) per minute. We conclude that AP1 and AP2 miss opportunities, compared to MPTSE, to use small cell bandwidth and thus to offload traffic from macro cells; indeed, the goal of using small cells is offloading macro cells.

Figure 5 shows unnecessary handoff rate (UH) when varying threshold residence time (TRT). We observe that for MPSTE (resp. AP1 and AP2), UH increases (resp. remain constant) with TRT.

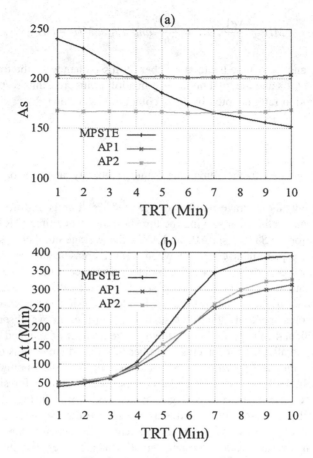

Fig. 3. As and At versus TRT

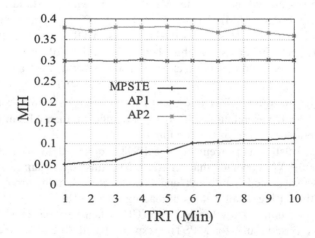

Fig. 4. MH versus TRT

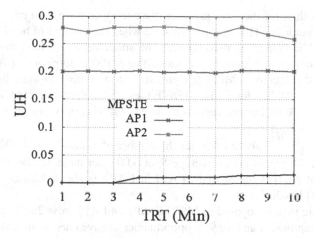

Fig. 5. UH vs versus TRT

This is expected since when TRT increases, the number of handoffs in small cells increases and, thus the number of handoffs, that are not required and done, increases. We also observe that MPSTE outperforms AP1 and AP2; indeed, MPSTE provides an average UH of 0.001 per minute while AP1 (more efficient than AP2 in this scenario) provides an average UH rate (defined as $\dfrac{\sum_{i=1}^{10}\left(UHi_{/i}\right)}{10}$ where i denotes the value of TRT) of 0.2 per minute; the average relative improvement of MPSTE compared to AP1 is about 0.199 (19.9 %) per minutes. We conclude that AP1 and AP2 miss opportunities to save energy by using small cell for short time periods.

Figure 6 shows energy consumption (R) rate when varying the number of handoffs. In order to provide the same number of handoffs of small cells for the three schemes in this scenario, UEs speed is fixed to 1 m/s; in this case, TRT is fixed to 5 min for MPSTE, the speed threshold of AP1 is fixed to 1.1 m/s, and average speed of AP2 is fixed to 1.1 m/s.

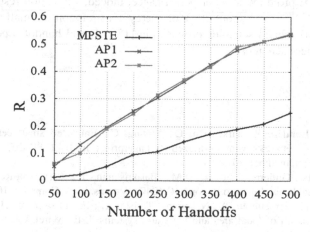

Fig. 6. R versus number of handoffs

We observe that, for the three schemes, the energy consumption rate increases with the number of handoffs. This is expected since when the number of handoffs increases, the energy of UE battery is used to scan the small cells and thus the energy consumption rate increases. Figure 6 also shows that MPSTE outperforms AP1 and AP2; this is due to the fact that MPSTE estimates the best time to start the small cells scanning process. At 500 handoffs, MPSTE uses 25 % of battery capacity while AP1 and AP2 use 53 % of battery capacity; indeed, MPSTE provides an average R rate (defined as $\frac{\sum_{i=50}^{500}\left(Ri_{/i}\right)}{10}$ where i denotes the number of handoffs) of 0.0004 per handoff while AP1 and AP2 provide an average R of 0.001 per handoff; the average relative improvement of MPSTE compared to AP1 and AP2 is about 0.0006 % per handoff (6 % per 10000 handoffs).

We conclude that, compared to MPSTE, AP1 and AP2 miss 28 % opportunities to use small cell bandwidth and 19.9 % opportunities to save energy. In addition, MPSTE provides a considerable reduction of 20 % (resp. 30 %) of energy consumption when considering 250 (resp. 500) handoffs process, compared to AP1 and AP2; definitively MPSTE saves energy and allows better offloading compared to AP1 and AP2.

5 Concluding Remarks

In this paper, we introduced a predictive scanning time based small cell discovery strategy for HetNets. Based on the acquired UE mobility information, the residence time inside a next small cell is computed, which is used for the decision of small cell scamming and handoff. The more recent UEs' movement characteristics located in the next small cell are used to compute the probability distribution function when making the estimation of residence time. When the estimated residence time inside small cell is great than a threshold, we compute the best time to trigger small cell scanning. We evaluated, via simulations, MPSTE and compared it against two related schemes proposed in [13, 14]. Regardless of the given threshold value to decide handoff performing, MPSTE provided a better performance; indeed, simulation results show that our scheme MPSTE saves about 20 % of energy consumption of small cell discovery compared to two related schemes proposed in [13, 14] and handoff operations compared to the others schemes.

References

1. Prasad, A., Lunden, P., Tirkkonen, O., Wijting, C.: Enhanced small cell discovery in heterogeneous networks using optimized RF fingerprints. In: IEEE PIMRC, pp. 2973–2977 London, UK, September 2013
2. Akyildiz, I.F., Gutierrez-Estevez, D.M., Balakrishnan, R., Chavarria-Reyes, E.: LTE-Advanced and the evolution to Beyond 4G (B4G) systems. Phys. Commun. 10, 31–60 (2013)
3. Kishiyama, Y., Benjebbour, A., Nakamura, T., Ishii, H.: Future steps of LTE-A: evolution toward integration of local area and wide area systems. IEEE Wirel. Commun. 20, 12–18 (2013)

4. Damnjanovic, A., Montojo, J., Joonyoung, C., Hyoungju, J., Jin, Y., Pingping, Z.: UE's role in LTE advanced heterogeneous networks. IEEE Commun. Mag. **50**, 164–176 (2012)
5. Prasad, A., Lunden, P., Tirkkonen, O., Wijting, C.: Energy efficient small-cell discovery using received signal strength based radio maps. In: IEEE VTC, pp. 1–5, Dresden, Germany, June 2013
6. Ghosh, A., Ratasuk, R., Mondal, B., Mangalvedhe, N., Thomas, T.: LTE-advanced: next-generation wireless broadband technology [Invited Paper]. IEEE Wirel. Commun. **17**, 10–22 (2010)
7. Nadembega, A., Hafid, A., Taleb, T.: A destination prediction model based on historical data, contextual knowledge and spatial conceptual maps. In: Proceedings of the IEEE ICC, Ottawa, Ontario, Canada, June 2012
8. Nadembega, A., Hafid, A., Taleb, T.: A path prediction model to support mobile multimedia streaming. In: Proceedings of the IEEE ICC, Ottawa, Ontario, Canada, June 2012
9. Barbera, S., Pedersen, K., Michaelsen, P.H., Rosa, C.: Mobility analysis for inter-site carrier aggregation in LTE heterogeneous networks. In: IEEE VTC, pp. 1–5, Las Vegas, NV, USA, September 2013
10. Pedersen, K.I., Michaelsen, P.H., Rosa, C., Barbera, S.: Mobility enhancements for LTE-advanced multilayer networks with inter-site carrier aggregation. IEEE Commun. Mag. **51**, 64–71 (2013)
11. Lopez-Perez, D., Guvenc, I., Xiaoli, C.: Mobility management challenges in 3GPP heterogeneous networks. IEEE Commun. Mag. **50**, 70–78 (2012)
12. Prasad, A., Tirkkonen, O., Lunden, P., Yilmaz, O.N.C., Dalsgaard, L., Wijting, C.: Energy-efficient inter-frequency small cell discovery techniques for LTE-advanced heterogeneous network deployments. IEEE Commun. Mag. **51**, 72–81 (2013)
13. Prasad, A., Lunden, P., Tirkkonen, O., Wijting, C.: Energy-efficient flexible inter-frequency scanning mechanism for enhanced small cell discovery. In: IEEE VTC, pp. 1–5. Dresden, Germany, June 2013
14. Yang, W.-H., Wang, Y.-C., Tseng, Y.-C., Lin, B.-S.P.: Energy-efficient network selection with mobility pattern awareness in an integrated WiMAX and WiFi network. Int'l. J. Commun. Sys **23**, 213–230 (2010)
15. Nadembega, A., Hafid, A., Taleb, T.: Handoff time estimation model for vehicular communications. In: Proceedings of the IEEE ICC, Budapest, Hungary, June 2013
16. Nadembega, A., Hafid, A., Taleb, T.: A framework for mobility prediction and high bandwidth utilization to support mobile multimedia streaming. In: Proceedings of the IEEE ANTS, Chennai, India, December 2013
17. 3GPP TR 36.839, E-UTRA; Mobility Enhancements in Heterogeneous Networks, v. 11.0.0, September 2012
18. 3GPP TS 36.331, E-UTRA: RRC Protocol Specification, v. 10.5.0, March 2012
19. Nadembega, A., Hafid, A., Taleb, T.: An integrated predictive mobile-oriented bandwidth-reservation framework to support mobile multimedia. In: IEEE TWC, vol. 13(12) December 2014
20. Nadembega, A., Hafid, A., Taleb, T.: Mobility prediction-aware bandwidth reservation scheme for mobile networks. In: IEEE TVT, vol. PP(99), August 2014
21. Nadembega, A., Hafid, A., Taleb, T.: DAMP: a destination and mobility path prediction scheme. In: IEEE TVT, vol. PP(99), August 2014
22. Xenakis, D., Passas, N., Merakos, L., Verikoukis, C.: Mobility management for femtocells in LTE-advanced: key aspects and survey of handover decision algorithms. In: IEEE Surveys and Tutorials, vol. 16(1), pp. 64–91, (2014, first quarter)

Emerging Communications, Networking and Computing Technologies for VANETs 2.0

Safety in Vehicular Networks—on the Inevitability of Short-Range Directional Communications

Gérard Le Lann[✉]

INRIA, Paris Rocquencourt, France
gerard.le_lann@inria.fr

Abstract. Safety implies high dependability and strict timeliness under worst-case conditions. These requirements are not met with existing standards aimed at inter-vehicular communications (V2V) in vehicular networks. On-going research targets medium-range omnidirectional V2V communications and short-range directional communications, which we refer to as neighbor-to-neighbor (N2N) communications. Focusing on the latter, we investigate the time-bounded message dissemination (TBMD) problem as it arises in platoons and ad hoc vehicle strings, referred to as cohorts. Informal specifications of TBMD, of a solution, are given. We show how to guarantee cohort-wide dissemination of any N2N message generated by a cohort member, either spontaneously or upon receipt of a V2V message. Dissemination time bounds are given for worst-case conditions regarding N2N channel contention and N2N message losses. These results add to previously demonstrated merits of short-range directional communications as regards safety in vehicular networks.

Keywords: Safety · Vehicular ad hoc networks · Platoons · VANETs · Automated vehicles · Time-bounded and reliable wireless mobile communications · Directional antennas · String control

1 Introduction

Networks of automated (autonomous and communicating) vehicles, such as platoons and VANETs [1], are complex safety-critical cyber-physical systems-of-systems. As of now, numerous safety issues remain open. A few solutions are emerging. However, they fail to meet numerous requirements related to the handling of safety-critical (SC) scenarios. Autonomy essentially rests on perception solutions, namely robotics (radars, lidars, cameras, etc.), which work exclusively in line-of-sight (LOS) conditions, and achieve passive safety only. Active safety in non-LOS (NLOS) conditions is mandatory. With active safety solutions, a vehicle can "tell" other vehicles what it intends to do, rather than letting them guess via perception capabilities (which might be hazardous). Moreover, active safety serves to "enforce" specific behaviors on other vehicles, so that collective behaviors are risk-free. Active safety in NLOS conditions implies highly dependable and time-bounded communications. Existing standards, e.g., IEEE 1609, 802.11p or ETSI ES 202 663, specify omnidirectional radio communications [2]. In this paper, we consider vehicular networks on major roads and

© Springer International Publishing Switzerland 2015
S. Papavassiliou and S. Ruehrup (Eds.): ADHOC-NOW 2015, LNCS 9143, pp. 347–360, 2015.
DOI: 10.1007/978-3-319-19662-6_24

highways. Given that SC scenarios may develop far away from road-side units, only vehicle-to-vehicle (V2V) communications need be examined. In the sequel, "V2V communications" stands for medium-range ($\approx$200 m) omnidirectional communications.

Limitations of passive safety, of V2V communications, relative to dependability and time-bounded terminations, are reviewed in detail in Sect. 2. In Sect. 3, we concentrate on string formations, either planned (platoons) or ad hoc, which happen to instantiate most frequent driving regimes, and we show how to circumvent weaknesses of V2V communications. The companion concepts of short-range ($\approx$30 m) directional neighbor-to-neighbor (N2N) communications and cohorts are presented. In Sect. 4, we introduce and specify the time-bounded message dissemination (TBMD) problem as well as Π, an algorithmic solution. An analytical expression of worst-case cohort-wide dissemination termination bounds is given and numerically illustrated. This paper is a contribution to a series of previous publications which demonstrate the merits of short-range, radio or optical, directional communications.

2   Limitations of Existing and Emerging Solutions

Behaviors of vehicles are under the control of on-board (OB) systems, which comprise perception and radio communications equipment, as well as space-time localization devices (GNSS receivers). Lane changes, on-ramp merging, emergency breaking/stopping, are examples of SC scenarios. For demonstrating safety, one must show that dependability and time-bounded termination requirements are met under worst-case conditions, namely high V2V message loss ratios, high channel contention, high traffic density, and inaccurate space-time data. It is known that GNSS devices may deliver inaccurate data, due to adverse conditions (bad weather, obstructions). Since losses of satellite signals may last several seconds, worst-case geo-positioning inaccuracy γ may be in the order of a few dozens of meters, and worst-case global time (UTC) inaccuracy τ in the order of a few milliseconds.

2.1   Limitations of Passive Safety

Existing autonomous (AU) vehicles can avoid hitting static or moving "obstacles", surrounding vehicles in particular. An AU vehicle adapts its behavior according to its perception of how other vehicles move, and vice versa. Consider AU vehicle V that wants to move to lane j. V's indicators may or may not be "obeyed" by others. Unless vehicles on lane j are "cooperative" (they create a convenient "slot" for V's insertion), V's lane change maneuver is unfeasible. Consider now three adjacent lanes $j-1, j$, and $j+1$, AU vehicle X in lane $j-1$, AU vehicle Y in lane $j+1$, and truck H in lane j. X and Y want to overtake H. They do not "see" each other due to H. Simultaneously, X and Y start a lane change maneuver so as to occupy the same slot ahead of H. At high velocities, this hazardous situation may not be detected in time by perception devices, leading to an accident. In case this hazard is detected in time, both maneuvers are aborted. Tie-breaking (which vehicle goes first, if ever) is conditioned upon the perception of unpredictable and fortuitous motions. In both examples, either the dependability or the time-bounded termination requirement is not met. To computer scientists, this is the well-known "starvation" problem that arises with cyber processes that share resources in a mutually exclusive manner. In vehicular networks, rather than data or processes, asphalt slots are the resources shared by moving processes. It is well known that algorithmic solutions based on passive waiting (perception here) may never terminate. The only algorithmic solutions that achieve time-bounded termination are based on reliable inter-process communications and agreements ("semaphores", "locks", etc.). Similar solutions are needed for achieving active safety in vehicular networks.

2.2 Limitations of V2V Communications

Mobile radio communications are inherently unreliable. Moreover, with the CSMA-CA MAC protocol at the core of existing standardized V2V communications, there are no worst-case upper bounds for channel access delays, thus no such bounds for message delivery delays.

Unreliability. At times, V2V message loss ratios may be too high, even in LOS conditions [3]. With very few exceptions, solutions proposed for Xcast operations (e.g., Multicast, Broadcast, Geocast) rest on a "no acknowledgment" policy, or on positive-acknowledgment-and-retransmission (PAR) protocols [4]. Without acknowledgments, failed message deliveries go unnoticed, which is not acceptable. PAR protocols cannot be used when a sender does not know a priori which vehicles shall receive its message and return acknowledgments. Moreover, these protocols are inadequate for the handling of most SC scenarios. Consider the set of V2V messages and acks exchanged for coordinating a safe lane change. The time budget for delivering V2V messages and acks successfully is in the order of 400 ms (a vehicle moving at 108 km/h would travel 12 m, enough for creating collisions). Given that omnidirectional radio links may be garbled longer than 400 ms, resending messages and/or acknowledgments is useless. Novel solutions specifically designed for achieving reliable time-bounded V2V message deliveries are needed.

No Timeliness. The problem of how to guarantee bounded channel access delays (BCAD) with omnidirectional antennas remains unsolved, when considering realistic worst-case assumptions, i.e. hundreds of contenders, variable number of lanes, space-time inaccuracies γ and τ. We mean strict, non-stochastic, time bounds. CSMA protocols only provide a "best effort" service, which is unsatisfactory. Moreover, unfairness may be experienced by contenders [5, 6], due to numerous causes. For example, since propagation of omnidirectional signals (messages and collisions) is anisotropic, a collision may not be detected by all silent processes within interference range of one of the senders; as a result, computations of backoff periods are incorrect, which leads to amplified unfairness. Reservation-based protocols, a.k.a. scheduling-based protocols, do not solve BCAD [7, 8]. First, access to reservation slots is not collision-free. Second, due to unreliable radio links, not all intended recipients are made aware of reservations heard by others, which results in contention and message collisions. Published variations of TDMA (time-division-multi-access) protocols do not solve BCAD either.

Crux of the problem is how to assign a unique slot to every contender without resorting to collision-prone communications, under realistic assumptions. STDMA [9, 10] which is based on observing that no two contenders may reside at the same space coordinates at the same time, cannot be considered either since inaccuracies γ and τ need be assumed arbitrarily small for STDMA to be correct (or simply efficient). Ditto for SDMA [11] and LCA [12]. An additional unpleasant feature of existing V2V MAC protocols is the high variability of access delays, very small under modest traffic and contention, very large (close to unbounded) under high traffic and contention.

Limited Scope and Relaying. V2V messaging is commonly used in SC scenarios. For example, vehicles involved in an accident do Xcast (M), announcing "accident at location Z, immediate stop or change lane". Given that the distance between senders of M and vehicle strings moving towards Z is much larger than 1-hop range in general, multi-hop relaying is resorted to. Most of the work conducted on geocasting focuses on minimizing the number of relaying vehicles as well as copies of M, ignoring message losses. What if vehicles in charge of relaying do not receive M? Are copies of M delivered to all string members within stipulated time bounds? In cooperative adaptive cruise control (CACC) approaches, where string stability is a major issue [13], consequences of V2V message losses and time-unbounded message deliveries regarding safety are ignored, with few exceptions [14–16]. Furthermore, data collected in platooning experiments show that it is inappropriate to rely on V2V broadcast from the lead vehicle [17].Given that existing V2V communications are not satisfactory, ongoing research investigates novel "deterministic" MAC protocols aimed at V2V communications and transport protocols which "guarantee" message deliveries. Another stream of research focuses on solutions that depart from V2V communications, although based on mastered radio technology. Such an approach is presented in Sect. 3.

3 Cohorts and N2N Communications

3.1 Principles

A cohort is a string of fully automated vehicles circulating in the same lane, bound to meet rigorously defined specifications [18, 19]—see Fig. 1. Cohorts may be preplanned (platoons, with or without a human driver in a lead vehicle) or ad hoc open linear VANETs, possibly short-lived, vehicles joining in or leaving anytime. To the exception of head CH and tail CT, every cohort member has two neighbors. Cohort members share some common knowledge, such as n and n^*, the current number of members and the highest possible number of members, respectively, velocity dependent safe spacing intervals, denoted s_{xy}, highest deceleration rate (dc) and highest acceleration rate (ac). Values for bounds n^*, dc and ac will be set by standardization bodies. Variable n is updated via the TBMD algorithm (see Sect. 4). Below dc or ac, cohort membership does not change. Membership changes occur in case vehicles want to leave or to join, maneuvers that rest on inter-vehicular agreements. Members are assigned consecutive ranks, 1 for CH, and $n \leq n^*$ for CT. Safe inter-cohort spacing intervals (which depend on velocities), denoted $S_{ct/ch}$, are also specified. Due to these intervals, cohorts do not "interfere" with each other as long as rates dc and ac are not exceeded. In the opposite case, e.g., occurrence of a "brick wall" phenomenon within some cohort, other cohorts are immune to rear-end collisions.

A member which violates dc or ac Xcasts a V2V message, signaling the instantiation of emergency conditions, possibly leading to a cohort split (see further). Cohortwide coordination can be accomplished via algorithms resting on N2N directional communications. There is no need to "pollute" the ether with 360° antennas over, e.g., 200 m, for coordinating consecutive members of a given cohort. Very short-range (e.g., 30 m), small beamwidth (e.g., 30°) front-looking and rear-looking directional radio antennas suffice [20], possibly steerable and power controlled. Observe that they would complement radars and lidars regarding spacing control, as well as V2V communications, thereby instantiating diversified redundancy (V2V and N2N communications are allocated different radio spectra), which is essential regarding safety. Most notably, communications among cohort members are not disrupted when V2V channels experience physical jamming [21]. Neighbors exchange N2N messages and beacons, upstream and downstream.

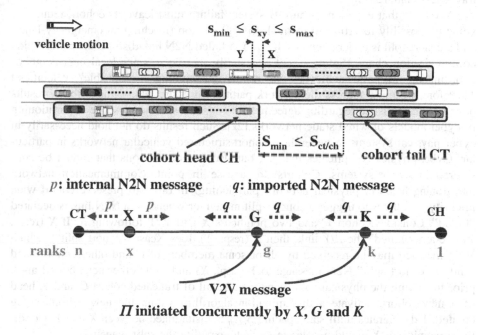

Fig. 1. Cohorts and principles of algorithm Π illustrated

Hop-by-hop N2N message passing also serves to perform a cohort-wide dissemination of any given V2V message, which may not be received by all targeted members. This is one of the motivations for exploring the TBMD problem. Due to the linear structure of cohorts and directional N2N communications, radio interferences may occur only among a limited set of vehicles proximate to each other, which permits to solve BCAD. Thus, the highest delay incurred with transmitting a N2N message or a beacon between two neighbors, in the absence of failures, is known (see further).

3.2 Dependability Within and Across Cohorts

By definition, like any safety-critical cyber-physical system, cohorts are timed systems. Being open systems, some notion of global time (UTC) is needed. Given the limitations of GNSS devices, they must be backed up by OB clocks/timers (see Subsect. 4.2). Since V2V communications are prone to frequent transient failures, it is quite difficult to specify worst-case conditions not leading to impossibility results. Fortunately, given the peculiarities of cohort constructs, it is possible to devise simple solutions to dependability problems arising in cohorts. Let η stand for some finite time interval. Every η, a failure detection algorithm is run by a vehicle OB system, in charge of checking V2V/N2N antennas as well as every other equipment (processors, robotics devices, etc.). Every cohort member must "prove" it is operational via the periodic transmission of a N2N beacon to its neighbor(s). A member or a N2N link is declared "failed" by member V in case no beacon or message has been received by V on this link for η time units. Picking up any value for η smaller than 500 ms would be safe, since behaviors of vehicles are also under the control of robotics (at 90 km/h, vehicles would travel less than 12.5 m).

A vehicle that experiences an OB system failure must leave its cohort, reduce its velocity (possibly reverting to manual driving) or stop (reaching an emergency lane), and a cohort split is undertaken (see below). A failed N2N link disrupts the cohort-wide communication chain. (Safety is not necessarily sacrificed, since local inter-neighbor coordination may be achieved via robotics, assuming "reasonable" velocities.) Therefore, we must address the network partitioning issue. Many impossibility results have been established regarding agreement or consensus in the presence of partitioning in cyber models of wired static networks [22]. Such results do not hold necessarily in cyber-physical systems in general, in cohort-structured vehicular networks in particular. Cyber-physics complicates matters, but also favors solutions that cannot be considered in cyber systems. Cohorts are a case in point: Communication network partitioning leads to physical network partitioning. In order to preserve cohort-wide integrity, it suffices to trigger a cohort split maneuver whenever a N2N link is declared "failed". Consider cohort C and two neighbors X and Y, Y following X. If X (resp., Y) declares "failed" the X/Y link, then X (resp., Y) does Xcast ("cohort split"), which V2V message may be received by all or some members of C and other cohorts, and sends a "cohort split" N2N message to Y (resp., X) and to its other neighbor (if any), prior to starting the physical split maneuver. X, tail of truncated cohort C, and Y, head of a new cohort, activate a dissemination algorithm within the new cohorts being created. Y decelerates until safe spacing $S_{ct/ch}$ is instantiated between X and Y. Conditions permitting, X would accelerate so as to expedite the split maneuver.

The need for N2N communications has been illustrated in [18], where a solution is given to the problem of safe inter-vehicular spacing in the presence of failures of robotics equipment. N2N communications serve to achieve dependability within strings, as well as across strings by disseminating received V2V messages. In [19], we show how to back up V2V communications in lane change scenarios with N2N communications among a limited subset of cohort members. In Sect. 4, with the TBMD problem, we address issues of concern to all members of a cohort.

4 Time-Bounded Message Dissemination in Cohorts

The need for investigating distributed dependable coordination problems (agreement, dissemination) is stressed in [23]. Agreement amounts to deciding on a unique choice among multiple differing proposals, whereas dissemination amounts to delivering a message to every intended vehicle, both in the presence of failures. An informal specification of the TBMD problem is given in Table 1, of a solution (algorithm Π) in Fig. 2. Solving TBMD is interesting from theoretical and practical viewpoints. For example, instabilities/shock waves in CACC regimes are eliminated by resorting to algorithm Π (dissemination of a N2N message that tells "new velocity is ...").

4.1 The Cohort-Wide Time-Bounded Message Dissemination Problem

Processes that instantiate Π are run by OB systems. Consider a cohort Γ and an unknown constantly changing set W of vehicles within Γ's radio range. Imported (imp) and internal (int) are two types of N2N messages subject to dissemination throughout a cohort. At unknown times, some vehicles members of set W Xcast V2V messages. When received by a cohort member, V2V message M is wrapped in a N2N message typed imp used for disseminating M. Due to V2V losses, not all cohort members may receive M. Our goal is to show that whenever a V2V message is heard of by at least 1 member, that message is delivered to all members, in the absence of cohort split. A V2V message carries a unique identity ID, which serves to uniquely identify its copies as an imported N2N message. At unknown times, cohort members may create internal N2N messages spontaneously, which messages are exchanged among neighbors in order to be disseminated. Every internal N2N message is assigned a unique identity by its creator. Some V2V messages ought to be acked/nacked, which entails repetitions. Thus, multiple copies of a given V2V message may be generated and received at different times by various cohort members. Achieving best performance figures implies detecting and suppressing such copies. (Regarding type int N2N messages, see Subsect. 4.2).

For any fault-tolerant distributed algorithm that rests on message-passing, one must specify integer f, $f > 0$, the highest number of losses (messages or acks/nacks) that may be experienced in the course of execution. Worst-case f might be specified in future safety standards for every safety-critical algorithm used by OB systems. Any reasonable valuation of f may be violated. Consequently, it is also mandatory to specify which actions ought to be performed whenever the "up to f" assumption happens to be violated. Failures of N2N links that would violate the "up to f" assumption lead to a cohort split. If algorithm Π is running, Π is aborted, and restarted within newly formed cohorts. Therefore, neither OB system failures nor N2N link failures need be taken into account in the design of Π. Conversely, Π shall withstand N2N link failure patterns that do not violate the "up to f" assumption. In the sequel, m stands for a message meant to be disseminated throughout cohort Γ, regardless of its type (imp or int). The activation of primitive send(m) results in the sending of m to a 1-hop neighbor. The set of neighbors denoted N, for a member of rank r, $r \neq 1$, $r \neq n$, comprises members of ranks $r-1$ and $r + 1$. For CH (resp., CT), N comprises member of rank 2 (resp., of rank $n-1$).

Assumptions   **Table 1.** The cohort-wide TBMD problem

-Timed system model

-Set Γ of n vehicles, forming a cohort. Cohort members create N2N messages at will, which messages are relayed throughout Γ

-Unknown set W of vehicles within Γ's radio range. These vehicles create and Xcast V2V messages at will. Every Xcast is heard of by at least 1 cohort member

-A V2V message received by a cohort member is relayed throughout Γ as a N2N message

-OB system failures

-Unreliable V2V communications: some V2V messages are acked/nacked and may be repeated

-Unreliable N2N communications: up to $f > 0$ losses, not leading to a cohort split

Properties

-Validity: Every N2N message is successfully delivered to every member

-Time-bounded Termination: Cohort-wide dissemination triggered by member X terminates in at most $\Delta_x(f)$ time units

-Termination Awareness: For every N2N message m, every member X is knowledgeable of a dissemination termination time $T_x(m)$

The dissemination of message m terminates successfully when m has been delivered to every cohort member (Validity). An analytical expression of $\Delta_x(f)$ must be given (Time-bounded Termination). Knowing that dissemination terminates does not suffice. Some cyber-physical action $\Psi(m)$ is stated in m, such as, e.g., "merge with left lane", "reduce velocity to 50 km/h". Cohort members need to know when to perform an action, thus the Termination Awareness property. Not an integral part of TBMD, the Synchronicity property (the termination time is the same for all members) is commonly stated in many distributed agreement problems arising in timed systems. As shown further, Synchronicity can always be achieved in the case of internal N2N messages, under certain conditions in the case of imported N2N messages.

4.2   Solving TBMD: Algorithm Π

Algorithm Π may be event-triggered, or time-triggered, e.g., periodically, by any member. Disseminations may overlap and the dissemination of any given message usually entails bi-directional relaying. Solving TBMD implies solving two problems altogether (often addressed separately). One is the MAC-level BCAD problem, the other one is the link-level dissemination problem per se. Thus, one is led to contemplate a cross-layer solution. Timeliness figures depend on how efficient cross-layering is. In the presence of failures, message dissemination does not boil down to mere hop-by-hop relaying. Vehicles keep moving while a message is disseminated. Consequently, the challenge is to achieve termination times $\Delta_x(f)$ that are tight and that do not depend on whether dissemination proceeds upstream or downstream. (Contrary to commonly assumed, followers may need to forward messages to predecessors.)

Intra-Cohort Hop-by-Hop Message Passing. N2N message passing is performed via a point-to-point PAR protocol. Observations made in Subsect. 2.2 do not apply here,

case c_int(m):
begin create m; compute unique identity id(m); call P; record [type(m) = int, t_x(m), T_x(m)] in m; Send (m) to neighbor(s) listed in N(X) end

case c_imp(m):
begin if ID found as id(m) in L_x then discard M else
begin create m copy of M; id(m):= ID; call P; record [type(m) = imp, t_x(m), T_x(m)] in m; Send (m) to neighbor(s) listed in N(X) end end

case r(m):
% from neighbor Y %
begin read m if id(m) in L_x then
begin read A_x(m) if [type(m) = imp and T_y(m) < T_x(m)] then
begin timer_x(m):= current timer_x(m) − (T_x(m) − T_y(m)); T_x(m):= T_y(m); % note 1 %
update A_x(m) end
discard m
end else   % note 2 %
begin
% first receipt of m %
store id(m) in L_x; copy m; timer_x(m):= T_x(m) − (current time − t_x(m)); create A_x(m);
Send (m) to opposite neighbor (if any)
end
end

case aw_x(m):
begin suppress id(m) in L_x; suppress A_x(m) end

Procedure P
begin store id(m) in L_x; t_x(m):= local time value; compute Δ_x(f);
timer_x(m):= Δ_x(f); T_x(m):= t_x(m) + Δ_x(f); create A_x(m) end

Fig. 2. Informal specification of Π for member X

since receivers and senders know each other (they are neighbors). Moreover, fading, interferences and collisions which contribute to garbling V2V communications are less of a problem with short-range directional communications. Therefore, repeating N2N messages or acknowledgments (acks/nacks) is founded. It follows that a N2N message may have multiple copies, which copies must be eliminated in order to avoid wasting communication bandwidth. Let λ stand for the highest N2N hop delay, in the absence of losses. Delay λ includes (1) the delay incurred with transmitting a N2N message, (2) a receiver's OB processing time (e.g., CRC checking). Delay λ is derived from the N2N link bandwidth and the highest length of a N2N message (acks/nacks are shorter). Operation send(m) may fail. Let Send(m) stand for the primitive that serves to perform a successful transmission of m on a 1-hop N2N link (m possibly repeated). A cohort member knows about success when an ack is received. Whenever $f \neq 0$, $Send(m)$ involves a number of sends and acks/nacks, which contribute to augmenting N2N channel contention. Let Λ stand for the duration of $Send(m)$ for a no-loss transmission, and Λ' stand for the duration of $Send(m)$ in the presence of 1 loss (m or ack/nack).

Algorithm Π. Solution Π rests on "sufficiently good" timers and clocks. UTC is provided to OB processes by space-time localization devices (e.g., GNSS) and OB clocks updated often enough. With GNSS outages lasting in the order of 20 s, cheap clocks of intrinsic drift in the order of 0.5×10^{-4} would achieve inaccuracy τ of UTC readings in the order of 1 ms, i.e. a discrepancy in the order of 2 ms for any two vehicles, which suffices for "synchronizing" the physical maneuvers of vehicles. Consider member X, of rank k. Upon arrival at (resp., creation by) X, an imported (resp., internal) N2N message m is tagged with dissemination termination time $T_x(m)$. Let $t_x(m)$ stand for m's arrival/creation time read on X's local clock. Thanks to the cohort construct, X can compute dissemination duration $\Delta_x(f) = h\Lambda + f\Lambda'$, where h stands for the highest number of N2N hops that separate X from CT or CH. Trivially, $h = \max \{k-1, n-k\}$, and $T_x(m) = t_x(m) + \Delta_x(f)$. Message identities that are still current are stored in a list in local OB memory, list denoted L_x for member X. A table denoted $A_x(m)$ is created at X for message m. Identity $id(m)$ in L_x points at $A_x(m)$. $A_x(m)$ and $id(m)$ are saved for $\Delta_x(f)$ time units, discarded afterwards (erased or moved to a data recorder).

Notations and variables (X stands for a cohort member, $n > 1$):
  - $N(X)$: set of X's neighbors   - N2N message types: imp, int
  - $t_x(m)$: m's creation time at X - $timer_x(m)$: timer for m at X
  - $T_x(m)$: m's dissemination termination time for X

Table $A_x(m)$ contains $id(m)$, message m, $t_x(m)$, $timer_x(m)$, and $T_x(m)$.

  - $\Delta_x(f) = h\Lambda + f\Lambda'$: cohort-wide dissemination duration for X, in the presence of up to f faulty N2N transmissions
  - $c\_int(m)$: event that triggers the creation of internal N2N message m
  - $c\_imp(m)$: event that triggers the creation of imported N2N message m (arrival of V2V message M, identity ID)
  - $r(m)$: receipt of m (from a neighbor)
  - $aw_x(m)$: awakening of $timer_x(m)$

5 Discussion

Algorithm Π is relatively simple (correctness proofs are not presented due to lack of space). By definition, being created by a unique member, an internal N2N message carries a unique termination time, which UTC time is copied unchanged by every member. Therefore, Π achieves Synchronicity for such messages. Inaccuracy τ can be made negligible (see above). With τ in the order of 1 ms, distance discrepancies are in the order of 5 cm for any two vehicles moving at 90 km/h. On the contrary, an imported N2N message may be created by more than one member. Consider any two members

P and Q that receive the same V2V message, which is assigned a unique pair $\{m, id(m)\}$. They compute the same arrival times $t_p(m)$ and $t_q(m)$ in case they receive the same copy of that message. Conversely, they compute $t_p(m) \neq t_q(m)$ if they receive different copies. Whatever the case, they are assigned different ranks. It follows that $T_p(m) \neq T_q(m)$. Thus, should the case arise (see note 1), $timer_x(m)$ and $T_x(m)$ are updated to smaller values, in order to maximize the number of members that compute termination times closer to smallest termination time. Nevertheless, Π as given in Fig. 2 does not achieve Synchronicity in general, in the case of imported N2N messages. The discrepancies between computed termination times mostly depend on the wait-for-ack/nack delays and which policy is used to Xcast V2V messages. If Synchronicity is mandatory, one must invoke an agreement algorithm, or a variation of Π (not presented here due to space limitations) whereby copies of an imported N2N message keep being propagated, rather than discarded (see note 2). Synchronicity would then be traded against higher message load, i.e. higher termination times for all members.

Algorithm Π does not impose any particular rule regarding when a member ought to instantiate cyber-physical action $\Psi(m)$ stated in m. Options are (1) upon receipt of m, (2) upon reaching termination time, i.e. when a local timer awakes, (3) some combination of (1) and (2). Similarly, the assignment of a numerical value to f known to all members is "orthogonal" to Π. This mandatory global knowledge may derive from some future standards (static assignment) or may be dynamically updated (while vehicles move) by resorting to a distributed agreement algorithm. The latter approach shall be more trustable given that, very likely, f is an increasing function in n, as well as time and location dependent. Let us observe in passing that this essential question of how to assign "good" values to variables appearing in safety-related performance figures, such as termination bounds, is almost totally ignored in publications devoted to safety issues in vehicular networks. In case a "good" value for f would be exceeded and detected by some member, that member undertakes a cohort split and Π is aborted, to be restarted in cohorts resulting from the split. (Due to space limitations, how to detect a violation of postulated f is not detailed here).

5.1   Time-Bounded Termination

N2N channel contention occurs between members of the same cohort, as well as with members of cohorts circulating in adjacent lanes. Indeed, due to road/highway curvatures, communications triggered by vehicles in adjacent lanes may interfere with each other, despite the small beamwidth and short range of N2N antennas. Thus the need for a MAC protocol which solves the BCAD problem as it arises with cohorts. Previous work on directional MAC protocols is mostly aimed at ad hoc sensor networks (mobility is not considered), with few exceptions [24–26]. Most often, these MAC protocols rest on the RTS/CTS scheme proper to V2V communication standards, which makes them unsuitable.

A MAC protocol designed recently (details are out of the scope of this paper) achieves bounds $\Delta_x(f)$ smaller than bounds given below (tight bound formulae are more involved). Delays Δ and Δ' depend on the MAC protocol, as well as on the number of N2N radio channels available (7 channels with IEEE 802.11p). Let us assume single frequency band N2N radio devices. For creator X of rank k, we have:

$$\Delta_x(f) < 4\lambda[h + 3(f + 2)] \ where \ h = \max\{k - 1, n - k\}.$$

Trivially, when creator X is CT or CH: $\Delta(f) < 4\lambda[n - 1 + 3(f + 2)]$.

These worst-case bounds are close to actual bounds for internal N2N messages. Most often, they are pessimistic for imported messages (imagine that a V2V message is received by 1 every 2 members, which trigger Π quasi-simultaneously).

Assuming a N2N channel bandwidth in the order of 15 Mbits/s, typical values for λ would range between 0.3 ms and 1.2 ms. Let us choose $\lambda = 1$ ms. Consider $n = 20$, and $k = 14$. One finds (bounds in ms):

$$\Delta_x(4) < 124 \quad \Delta_x(0) < 76 \quad \Delta(5) < 160 \quad \Delta(0) < 100$$

Under the Synchronicity option and for internal N2N messages, distances travelled at 90 km/h until cyber-physical actions $\Psi(m)$ start being instantiated would range within bounds smaller than 4 m ($\Delta(5)$) or 1.9 m ($\Delta_x(0)$). Observe that bound $\Delta(0)$ is smaller than the 100 ms figure frequently quoted for delivering a "critical" V2V message, which is a goal set to designers, not a proven bound. In fact, it is easy to demonstrate that the 100 ms requirement cannot be met with CSMA-CA channels, under realistic (between average and worst case) load conditions. A cohort-wide N2N dissemination algorithm such as Π outperforms any CSMA-CA V2V scheme with acks and repetitions whenever $f > 0$. Note that assuming the same f for V2V and N2N communications favors CSMA-CA V2V schemes, since medium-range omnidirectional communications are less reliable than short-range directional communications.

6   Conclusions

In this paper, we provide novel results regarding the merits of short-range neighbor-to-neighbor (N2N) directional communications as regards safety in networks of automated vehicles. Notably, we show how N2N communications can be used for solving problems that have no solutions with medium-range omnidirectional communications. These results add to the existing literature on short-range directional radio and/or optical communications. It may be that time is right for standardizing N2N radio communications. They are based on mastered technologies, similar to standardized V2V radios, thus potentially low-cost. Given the very high level of safety achievable with a combined reliance on cohorts, N2N communications, and dissemination algorithms—in addition to robotics, it may be conjectured that safety authorities and insurance companies would be motivated for supporting the adoption of solutions similar to those presented in this paper.

References

1. Karagiannis, G., et al.: Vehicular networking: a survey and tutorial on requirements, architectures, challenges, standards and solutions. IEEE Comm. Surv. Tutor. **13**(4), 584–616 (2012)

2. Sichitiu, M.L., Kihl, M.: Inter-vehicle communication systems: a survey. IEEE Comm. Surv. Tutor. **10**(2), 88–105 (2008)
3. Karlsson, K., et al.: Field measurements of IEEE 802.11p communication in NLOS environments for a platooning application. In: IEEE VTC Fall, 5 p. (2012)
4. Ros, F.J., et al.: Acknowledgment-based broadcast protocol for reliable and efficient data dissemination in vehicular ad hoc networks. IEEE Trans. Mob. Comput. **11**(1), 33–46 (2012)
5. Kyasanur, P., Vaidya, N.: Selfish MAC layer misbehavior in wireless networks. In: IEEE International Conference on Dependable Systems and Networks (DSN 2003), pp. 173–182 (2003)
6. Tang, J., Cheng, Y., Zhuang, W.: Real-time misbehavior detection in IEEE 802.11-based wireless networks: a analytical approach. IEEE Trans. Mob. Comput. **13**(1), 146–158 (2014)
7. Borgonovo, F., et al.: ADHOC MAC: New MAC architecture for ad hoc networks providing efficient and reliable point-to-point and broadcast services. Wirel. Netw. **10**(4), 359–366 (2004)
8. Bao,L., Garcia-Luna-Aceves, J.: A new approach to channel access scheduling for ad hoc networks. In: 7th ACM International Conference on Mobile Computing and Networking, pp. 210–221 (2001)
9. Grönkvist, J.: Assignment methods for spatial reuse TDMA. In: First IEEE Workshop on Mobile and Ad Hoc Networking and Computing, pp. 119–124 (2000)
10. Amouris, K.: Space-time division multiple access (STDMA) and coordinated, power-aware MACA for mobile ad hoc networks. In: Proceedings of Global Telecommunication Conference vol. 5, pp. 2890–2895 (2001)
11. Bana S.V., Varaiya, P.: Space division multiple access (SDMA) for robust ad hoc vehicle communication networks. In: 4th IEEE International Conference on ITS, pp. 962–967 (2001)
12. Katragadda, S. et al.: A decentralized location-based channel access protocol for inter-vehicle communication. In: IEEE VTC Spring, pp. 1831–1835 (2003)
13. Swaroop, D., Hedrick, J.K.: String stability of interconnected systems. IEEE Trans. Autom. Control **41**(3), 349–357 (1996)
14. Lei, C., et al.: Impact of packet loss on CACC string stability performance. In: 11th International Conference on ITS Telecommunications (ITST 2011), pp. 381–386 (2011)
15. Sepulcre, M. et al., Cooperative vehicle-to-vehicle active safety testing under challenging conditions, Transportation Research, Part C-26, pp. 233–255. Elsevier (2013)
16. Öncü, S., et al.: Cooperative adaptive cruise control: Network-aware analysis of string stability. IEEE Trans. Intell. Transp. Syst. **15**(4), 1527–1537 (2014)
17. Bergenhem, C., et al.: V2V communication quality: measurements in a cooperative automotive platooning application. SAE Int. J. Passeng. Cars – Electron. Electr. Syst. **7**(2), 79 (2014)
18. Le Lann, G.: Cohorts and groups for safe and efficient autonomous driving on highways. In: IEEE Vehicular Networking Conference (VNC), pp. 1–8 (2011)
19. Le Lann, G.: On the power of cohorts - multipoint protocols for fast and reliable safety-critical communications in intelligent vehicular networks. In: ACM/IEEE/IFAC/TRB ICCVE-2012, pp. 35–42 (2012)
20. Ramanathan, R., et al.: Ad hoc networking with directional antennas: A complete system solution. IEEE J. Sel. Areas Commun. **23**(3), 496–506 (2005)
21. Lyamin, N., et al.: Real-time detection of denial-of-service attacks in IEEE 802.11p vehicular networks. IEEE Commun. Lett. **18**(1), 110–113 (2014)
22. Lynch, N.A.: Distributed Algorithms, 872 p. Morgan Kaufmann, San Mateo (1996)

23. Dressler, S.F., et al.: Inter-vehicle communication: Quo vadis. IEEE Commun. Mag. 52(6), 170–177 (2014)
24. Yadumurthy, R.M., et al.: Reliable MAC broadcast protocol in directional and omnidirectional transmissions for vehicular ad hoc networks. In: ACM VANET 2005, pp. 10–19 (2005)
25. Bazan, O., Jaseemuddin, M.: A survey of MAC protocols for wireless ad hoc networks with beamforming antennas. IEEE Comm. Surv. Tutorials 14(2), 216–239 (2012)
26. Little, T.D.C., et al.: Directional communication system for short-range vehicular communications. In: IEEE Vehicular Networking Conference (VNC), pp. 231–238 (2010)

Secure Incentive-Based Architecture for Vehicular Cloud

Kiho Lim, Ismail M. Abumuhfouz, and D. Manivannan[✉]

Department of Computer Science, University of Kentucky,
Lexington, KY, USA
{kiho.lim,imab222}@uky.edu, mani@cs.uky.edu

Abstract. Cloud computing has emerged as a viable technology for supporting utility computing. In the future, vehicles are likely to be equipped with devices that have large computation and communication power as well as large storage. Such computation power and storage are often underutilized. People in several areas such as traffic management, parking management, etc. can benefit from utilizing the unused computation, communication and storage capabilities of the vehicles on the road as well as from the traffic information collected by the vehicles to provide services. In this paper, we propose a secure architecture for the vehicular cloud to support the above-mentioned services. The architecture encourages vehicles to contribute their underutilized resources to the cloud by issuing tokens which can be used by the vehicles to get services from the cloud.

Keywords: Vehicular cloud · VANET · Security in vehicular cloud · Incentive system

1 Introduction

In the last few years, automobile manufacturers have been incorporating technologies in their vehicles that provide safety and entertainment services to their drivers. Such technologies include on board units (OBUs), global positioning systems (GPSs), computational devices, storage devices, communications devices etc. [14]. These devices allow the vehicles to collect information about various phenomena such as, road condition, traffic congestion, delays due to accidents, etc. and share them with other drivers. Moreover, they allow the vehicles to perform complex operations like context management, data filtering and even some cryptographic operations [9,16]. Vehicular Ad hoc Networks (VANETs) help in collecting such information and exchanging them between drivers. For example, accident avoidance warnings could quickly notify drivers of conditions that could cause a collision. Also, in the case of an accident, the scene can be easily constructed by law-enforcement agencies using the velocity information exchanged between vehicles [3].

In the last decade, cloud computing emerged as an economical solution for customers to rent IT infrastructures, platforms or software, instead of investing

© Springer International Publishing Switzerland 2015
S. Papavassiliou and S. Ruehrup (Eds.): ADHOC-NOW 2015, LNCS 9143, pp. 361–374, 2015.
DOI:10.1007/978-3-319-19662-6_25

money to own and maintain such services. The service providers lend such elastic services to customers exactly when they need them, and then they charge them based on their usage. Services provided by the cloud can be broadly divided into three categories: IasS (Infrastructure as a Service), PasS (Platform as a Service), and SaaS (Software as a Service). Although the primary features of the cloud are cost saving, on-demand service, resource pooling, scalability and ease of resources accessing, security and privacy concerns are the major barriers for customers to use the cloud [2,7].

In order to fully utilize the resources of vehicles in VANETs, Olariu et al. [19] proposed the concept of a vehicular cloud, which combines VANET and cloud. A vehicular cloud is a group of largely autonomous vehicles in VANET that contribute their computing, sensing, communication, and physical resources to the cloud. Vehicles' resources and the information exchanged from the vehicles with the cloud can be used in decision making. Vehicular cloud can facilitate in providing services such as parking management, traffic congestion, avoiding accidents, reducing environmental pollution etc. to customers in real time at low cost [10].

By utilizing cloud computing in VANETs, a vehicular cloud could help reduce propagation of redundant data in VANETs and use its resources more efficiently. In current studies on VANETs, multiple vehicles which observe the same phenomena propagate it to other vehicles which can result in redundant propagation of data. This results in vehicles wasting their resources analyzing the redundant data to find relevant information. A vehicular cloud allows vehicles to exchange their collected data with the cloud, where it can be analyzed, verified, organized, stored, and discarded if it is redundant or irrelevant. Cloud then can send needed information to the drivers upon request. Various other applications can benefit from using a vehicular cloud. Following a list of sample applications for vehicular cloud:

- **Intelligent Parking Management:** Millions of vehicles are parked in garages for hours every day. While they are parked, the underutilized resources of the vehicles such as computational and storage facilities could be used to perform tasks coming from the parking management server. Parking management could encourage the drivers of the vehicles to rent the resources of their vehicles and compensate them for that. Compensation could take the shape of free parking, shopping coupons or virtual credits that could be used somewhere else to get a service. Parking lots found in airports, malls, and large companies are examples of where we can use this application [1]. A vehicular cloud could also be used to help finding available parking spots. Drivers and parking management could cooperate to exchange information about empty parking spots and update it to vehicular cloud, so drivers looking for an empty spot could reserve the space through the vehicular cloud [26].
- **Road safety and traffic Management:** Vehicles could collect data about road conditions, traffic, weather etc. and store this data in the cloud. Then, vehicles can periodically receive information from the cloud regarding road

conditions such as, ice on the road, accidents, construction, etc. [11]. Based on this information, drivers may alter their routes.

- **Intelligent Transportation System using Traffic Monitoring:** Current transportation systems use traffic monitoring devices, such as inductive loop detectors (ILDs), videos cameras, radar sensors and others, to measure and monitor the road traffic. A costly ILD (worth around $8,200) is embedded under the road to measure the road traffic by recording a signal every time a vehicle passes over it [19]. The failure rate of these ILDs are very high and the maintenance costs are continuously growing. Hence, a vehicular cloud can be an alternative economical solution to the transportation department for monitoring the traffic.
- **Planned Evacuation:** Vehicular cloud can also help expedite evacuation during disasters like earthquakes, hurricanes and others. Data about disaster could be collected using vehicles inside that disaster area and transferred to the cloud, where it could be analyzed, organized and sent back as useful information to the refugees and the evacuation organizations. Examples of such information are locations of open grocery stores, gas stations, shelters and medical centers [10].

Drivers contribute their vehicles' resources to the vehicular cloud, and in return, they are compensated in the form of incentives for participation. It is expected that vehicles cooperate with the cloud by contributing their vehicles' resources. However, some vehicles may choose not to contribute to the cloud. Vehicles can be enticed to contribute their resources to the cloud by rewarding them with incentives. Several incentive schemes have been proposed in the context of Mobile Ad hoc Networks (MANETs) and VANETs, but none in the context of vehicular cloud. In this paper, we proposed an architecture of a vehicular cloud that uses incentive based scheme to encourage vehicles to contribute their resources, as well as, the information they collect to the cloud. The vehicles in return, receive incentives (tokens) for their contribution. The architecture not only helps in benefiting the drivers of the vehicles, but also the decision makers who need this information. Our solution ensures the security of the entities, operations and messages required to maintain the rewarded incentives. Figure 1 illustrates the architecture proposed in this paper for vehicular cloud. It mainly consists of two integrated parts: A VANET part where a group of stationary or moving vehicles form a cloud and communicate with the conventional cloud through the VANET road side units (RSUs), the other part is the conventional cloud, which contains many entities that are needed in this architecture. These entities include the service provider manager, Token Reward System, Trust Authority, and Revocation Authority. The detailed functionalities of these entities are explained in the system model in Sect. 3.

The rest of the paper is organized as follows. Section 2 surveys the related works. Section 3 introduces the system model, assumptions, problem statement and solution objectives. Section 4 presents our proposed architecture with the reward token system in detail. In Sect. 5, we present an analysis of our scheme. Finally, we conclude in Sect. 6.

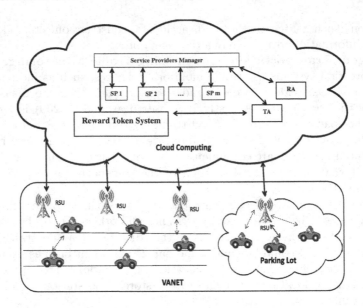

Fig. 1. Incentive-based architecture for vehicular cloud

2   Related Works

Vehicular cloud has received a lot of attention in the last few years [10,19,21]. Vehicles could be organized into a cloud where they utilize their unused resources on-demand to perform tasks. Drivers contribute the information collected from their vehicles and their resources to the cloud, and they receive incentives in return. Some vehicles may be non-cooperative by utilizing their resources only for their own purposes. These are called selfish vehicles (nodes).

Selfish nodes problem has been studied extensively in the context of routing in MANETs and VANETs. Various incentive mechanisms were proposed in [4–6,8,12,13,15,17,18,20,27] to stimulate contribution of nodes and make them more cooperative with the others. Butty et al. [5,6] introduced incentive schemes based on using a tamper proof hardware inside each node. This hardware is responsible to maintain and secure virtual credit gained by the node when it participated in forwarding messages. The sender estimates the rewards based on the number of intermediate nodes.

The authors of [4,17,18] proposed reputation based schemes where nodes observe their neighbors' traffic, record their contribution to the network and isolate nodes which have poor reputation. The high mobility of nodes makes it unfeasible to observe traffic of neighboring vehicles or even build a distributed reputation system based on this. Li et al. [13] presented a receipt counting reward scheme focused on incentive distribution. Their infrastructure requires every source node to obtain permission from an authority for every message they forward before they reward the intermediate nodes. Authors in [8,27] presented a game theory based scheme where they manipulate some parameters such as the amount earned, designation of charging subject etc. to analyze and build the incentive scheme. While all of these schemes seem beneficial for MANETs and VANETs, they are not suitable for vehicular cloud because they were designed to force nodes to participate in routing messages.

Cloud computing promises to provide reliable and elastic services to customers. In order to sustain such objectives, it needs a pool of resources and underutilized resources of vehicles in vehicular cloud can be a pool of resources. Hence, this paper proposes an incentive-based architecture to encourage the vehicles to contribute their resources to the cloud.

3   System Model

In this section, we introduce the system model, assumptions, problem statement and design objectives. The proposed architecture of the vehicular cloud is shown in Fig. 1.

3.1   System Model

The vehicular cloud architecture proposed in this paper consists of the following entities.

- **Service Provider Manager (SPM):** The SPM manages all service providers in the cloud and serves as the representative of service providers so it is responsible for advertising services, making contracts for services, validating proofs of works done by vehicles, and issuing tokens as reward to them as incentive for their participation in the cloud.
- **Reward Token System (RTS):** The RTS serves as the token bank in the cloud. It creates an account for each vehicle when the vehicle is registered and the account is tied to vehicle's pseudo ID. It assigns tokens to the vehicles when they contribute their resources.
- **Revocation Authority (RA):** The RA maintains the revocation list for the misbehaving vehicles. It also Keeps the records of vehicles whose contribution was poor.
- **Trusted Authority (TA):** The TA is in the cloud and is able to communicate with RA, RTS and SPM securely. When a vehicle is registered or renewed, it issues a certificate for the vehicle. And it also manages all private information about vehicles including the certificate, which ties to the vehicle's public key, and (public, private) key pairs. It also helps the SPM in verifying vehicles when needed. The real identity is not given to the SPM, however, when investigations are required by legal authorities, it can reveal the real identity of vehicles to the authorities.
- **Road Side Units (RSUs):** The RSUs are located along the roads and connected by a network so they serve as gateway to the cloud from the VANET.
- **On Board Units (OBUs):** An OBU is a tamper proof device installed on the vehicles. And it has computation, communication capabilities and storage. Also, it is able to check the token balance with the RTS in the cloud.

3.2   Assumptions

We assume that the RA maintains the certificate revocation list of misbehaving vehicles and the certificate of misbehaving vehicles are revoked using revocation protocols for the vehicular cloud network [22–24].

We also make the following assumptions.

1. The RTS, TA, RA are totally trusted and are assumed to be not compromised.
2. When a vehicle is registered, its public/private key pair is assigned and the public keys of the RTS and the SPM are stored in the OBU installed in the vehicles.
3. Vehicles can communicate with the cloud through the RSUs and any vehicle within the transmission range of an RSU can send/receive messages to the RSU.
4. OBUs on vehicles can check their token balance through the RTS.

3.3   Problem Statement and Solution Objectives

Vehicles are equipped with computational resources and storage facilities, but they are often underutilized. There are many people and application managers interested in renting such resources as well as obtaining the information collected on the roads. Since the contribution of the vehicles are optional, some drivers may choose not to do it. Drivers can be enticed to contribute the resources of their vehicles by offering incentives. Several incentive schemes have been proposed to entice selfish nodes in MANETs and VANETs, but they are not suitable for vehicular cloud because they were designed only for handling selfish nodes in routing. Hence, there is a need of an incentive-based architecture to reward the drivers of the vehicles for sharing their resources as well as to help the people who are interested in resources of the vehicles and information collected by the vehicles on the roads.

The proposed scheme should address the following issues: First, the management issue where the rules of every entity involved in the system is determined and the flow of the incentive process is designed, maintained and audited carefully. Second, the schemes should be flexible to deal the dynamicity of vehicles joining or leaving the network, which also include the ability to handle the heterogeneity of the entities and the networks involved, in addition to handling the unpredictable demands of customers and vehicles. Third, security is one of the most important issues that needs to be addressed. This requires to watch, guard, audit and maintain the incentives operations performed inside the scheme, validates the integrity and authenticity of the messages exchanged between the entities, and preserves the privacy of the entities participated in the scheme.

In this paper, we introduce an incentive-based architecture for vehicular cloud and propose the secure token reward system as an incentive scheme to address the issues above. Our scheme is to achieve the following objectives. First, Token transaction should be secure and robust against attacks. Second, Integrity and authenticity of the messages exchanged between entities should be ensured. Third, privacy of vehicles should be protected while contributing their resources for services in cloud and obtaining/using tokens. Fourth, The awarded token can be used for the service in the future.

4   Secure Token Reward System

In this section, we first present the basic idea behind our scheme in the Sect. 4.1 and then describe our secure token reward system for vehicular clouds in detail in the rest of the Sect. 4.

4.1   Basic Idea Behind Our Scheme

The proposed scheme has the following phases:

- **Phase 1: Searching Resources:** When a cloud service provider looks for vehicles for resources, the cloud service provider manager (SPM) broadcasts a message through the cloud on behalf of the service provider. When an interested vehicle receives the message and decides to contribute its own resource to the cloud for the service, it sends a request message for the work to the SPM in the cloud through the road side units (RSUs). Then, the SPM verifies the vehicle (or driver) with the help of the trusted authority (TA) and checks the previous records stored by the revocation authority (RA) (if there's any). If there are more vehicles interested in contributing their resources than what the service provider needs, the SPM picks vehicles based on their previous record. Once the vehicle is verified, the SPM signs a contract for the work between the service provider and the vehicle and sends it to the vehicle, so the vehicle can start the work based on the contract. Here, the vehicles use their pseudo ID in all communications to protect their privacy.
- **Phase 2: Requesting Reward Tokens:** After completing the assigned work, a vehicle sends a message with the proof of the work done to the SPM. This proof message helps the SPM in verifying the completion of the work. Note that all cloud service providers are connected to the SPM and the SPM can handle all messages on behalf of the service providers. After the completion of the work is verified, the SPM sends a reward token request to the reward token system (RTS) so it can send tokens to the vehicle as a compensation for the work. When the reward token request is processed by the RTS, a transaction number is generated and sent to the SPM and the vehicle as a confirmation.
- **Phase 3: Using Tokens for Cloud Service:** In the vehicular cloud, there are various types of services available through cloud service providers. The cloud services generally can be purchased with pay-as-you-go, but the reward token earned by contributing resources into the cloud can also be used as a method of payment for the could services. Since that vehicles are able to check their token balance with the on board units (OBUs), they can simply use the cloud services with tokens if they have enough balance for using the services.

Next, we describe our scheme in detail. The notations used in this scheme are listed in Table 1.

Table 1. Notations

Notation	Description
SPM	Service Provider Manager
RTS	Reward Token System
R	an RSU
V	a vehicle
M	a message
ts	timestamp
Sq	sequence number of a message
C	certificate
sv	cloud service information
$sv\#$	cloud service number
$tr\#$	transaction number for token reward
$proof$	proof of work done
ID_A	identity of entity A
PID_A	pseudo identity of entity A
SK_A	private key of entity A
PK_A	public key of entity A
$H()$	cryptographic one-way hash function
$SIG_A(M)$	signature of message M signed by A's private key
$E(M, K)$	encrypting message M with key K

4.2 Searching Resources

In our scheme, we use hash-based digital signature along with public key encryption to ensure the integrity and the authenticity of messages. When a message is sent, the sender attaches the digital signature to the message. The digital signature is made by encrypting the hash of the message using sender's private key. Since only the sender can generate the digital signature, the authenticity of the message is guaranteed. For integrity, the hash in the digital signature should match with the hash of the message calculated by the receiver. If they match, the receiver is able to verify the authenticity and integrity of the message.

When a cloud service provider needs to use the resources of vehicles, the SPM broadcasts an advertising message M_1 (Fig. 2) through the cloud on behalf of the service provider, where

$$M_1 = ID_{SPM}, sv\#, sv, ts, SIG_{SPM}(m_1)$$

$$where\ SIG_{SPM}(m_1)=E(H(sv\#,sv,ts),SK_{SPM}) \quad (1)$$

The advertising message M_1 includes the service number $sv\#$ and cloud service information sv that contains the ID of the cloud service provider, work type, work requirement, and reward amount. Also, the SPM attaches a digital

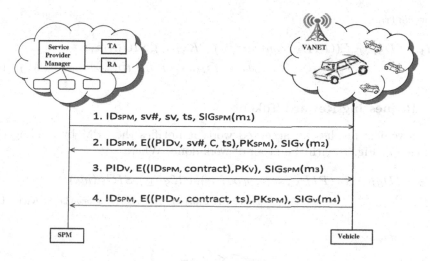

Fig. 2. Contract establishment process

signature that is obtained by computing the hash of the message and encrypting it with its private key. When an interested vehicle V receives the message M_1 and wants to contribute its resource to the cloud, it computes a work request message M_2, where M_2 is given below, by encrypting the service number, certificate and timestamp with the public key of the SPM. Then it sends the message to the SPM in the cloud through a nearby RSU after attaching its digital signature. Here, the contract is an agreement between a vehicle and the service provider for the service and it includes all the details about the work such as work requirement and payment.

$$M_2 = ID_{SPM}, E((PID_V, sv\#, C, ts), PK_{SPM}), SIG_V(m_2)$$

$$\text{where } SIG_V(m_2) = E(H(PID_V, sv\#, C, ts), SK_V) \quad (2)$$

Upon receiving M_2, the SPM is able to verify the vehicle (or driver) by checking its certificate C and pseudo ID PID_V with the help of the TA and decrypt the message. Also, the SPM checks if the vehicle has been blacklisted. Here, a pseudo ID is used for the vehicle to protect its privacy and the TA does not reveal the original identity of the vehicle to the SPM, however, it can be revealed when necessary such as a dispute for a transaction is open.

After the vehicle is authenticated, the SPM generates message M_3 by attaching the contract for the work and its signature obtained by computing the hash of the message as follows.

$$M_3 = PID_V, E((ID_{SPM}, contract, ts), PK_V), SIG_{SPM}(m_3)$$

$$\text{where } SIG_{SPM}(m_3) = E(H(ID_{SPM}, contract, ts), SK_{SPM}) \quad (3)$$

Once the message M_3 is delivered to the vehicle, it first sends an acknowledge M_4 to the SPM and then can start the work for the service based on the requirement in the contract.

$$M_4 = ID_{SPM}, E((PID_V, contract, ts), PK_{SPM}), SIG_V(m_4)$$

$$where\ SIG_V(m_4)=E(H(PID_V,contract,ts),SK_V) \quad (4)$$

4.3 Requesting Reward Tokens

After a vehicle finishes an accepted work, it notifies the SPM by sending the message M_5(Fig. 3) with the proof of work done. Where,

$$M_5 = ID_{SPM}, E((PID_V, sv\#, proof, ts), PK_{SPM}), SIG_V(m_5)$$

$$where\ SIG_V(m)=E(H(PID_V,sv\#,proof,ts),SK_V) \quad (5)$$

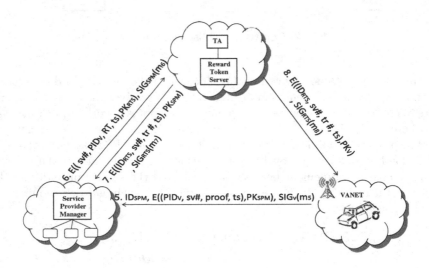

Fig. 3. Token reward process

Here, the *proof* of the work done varies depending on the type of work done and it must be verifiable by the SPM. The SPM is able to verify the completion of the work by checking the service number and the *proof*. Once the work done is verified, the SPM generates a reward token request message M_6, attached with the digital signature to it, and sends it to the RTS, so the reward tokens can be sent to the vehicle as a compensation for the work done, where

$$M_6 = ID_{RTS}, E((sv\#, PID_V, RT, ts), PK_{RTS}), SIG_{SPM}(m_6)$$

$$where\ SIG_{SPM}(m_6)=E(H(sv\#,PID_V,RT,ts),SK_V) \quad (6)$$

After the RTS processes the reward token request, it computes the confirmation message M_7 and M_8 with a transaction number for the SPM and the vehicle respectively as follows

$$M_7 = ID_{SPM}, E((ID_{RTS}, sv\#, tr\#, ts), PK_{SPM}), SIG_{RTS}(m_7)$$

$$where\ SIG_{RTS}(m_7)=E(H(ID_{RTS},sv\#,tr\#,ts),SK_{RTS}) \quad (7)$$

$$M_8 = PID_V, E((ID_{RTS}, sv\#, tr\#, ts), PK_V), SIG_{RTS}(m_8)$$

$$where\ SIG_{RTS}(m_8)=E(H(ID_{RTS},sv\#,tr\#,ts),SK_{RTS}) \quad (8)$$

and sends them to the SPM and the vehicle respectively. Note that the service providers are connected to the SPM so all messages related to a particular service provider are eventually delivered to it accordingly.

4.4 Using Tokens for Cloud Service

In vehicular cloud, several types of services can be provided by service providers. For example, Parking lot data managers can provide to vehicles information about available parking spaces, Transportation system can use the cloud collect traffic information and divert traffic based on the traffic information as well as optimize traffic signals, dynamic traffic signals, etc. [25]. Vehicular cloud can help make services readily available for vehicles on roads and at the parking lots and can be purchased with pay-as-go plan. With our token reward system, the tokens obtained in exchange for sharing their own resources also can be used for the cloud services as a method of the payment. Since a feature of the balance check is available on the OBU, a vehicle simply can use the could services with the tokens by sending a message M_9 to the SPM as follows.

$$M_9 = ID_{SPM}, E((PID_V, sv\#, RT, ts), PK_{SPM}), SIG_V(m_9)$$

$$where\ SIG_V(m_9)=E(H(PID_V,sv\#,RT,ts),SK_V) \quad (9)$$

Once the SPM receives the message from the vehicle, it authenticates the message and sends a token deduction request to the RTS in the same way as requesting for rewarding tokens. If the authenticity of the vehicle is verified and a confirmation for the token deduction is given to the SPM, then it notifies the service provider that now the vehicle can use the cloud service.

5 Security Analysis

In this section, we evaluate the performance of our proposed scheme in terms of security and usability and analyze the token reward system.

To entice the vehicles to contribute their utilized resources, an incentive-based scheme is necessary. Also, dealing with the selfish nodes (i.e., nodes that do not want to contribute their resources but would want to access the services provided by the cloud) is a challenge. The issues related to selfish nodes have not been studied in the context of vehicular cloud. Hence we proposed a secure incentive-based architecture for vehicular cloud to address this issue. In our scheme, we use a reward token system to entice vehicles to participate in sharing their unused resources and discourage selfish nodes. If a driver agrees to share its resource for a cloud service, then reward tokens are given in return after the completion of the work based on the contract. Note that public key cryptography is used for encrypting messages in our scheme and it is not overhead since the size of the messages encrypted are small.

5.1   Message Integrity

Messages are sent with a digital signature generated by the sender using cryptographic one-way hash function. Since the message is discarded upon arrival if the hash of the message does not match, the message integrity is guaranteed. Hence, integrity of token information in the message is guaranteed as well.

5.2   Source Authentication

The SPM is connected to the TA in the cloud. When a request for authentication is received, the TA helps the SPM in authenticating the sender.

5.3   Privacy Preservation

Every vehicle is assigned a unique pseudo ID by the TA. With the pseudo IDs, all private information and real identities of vehicles are protected. However, when a malicious node is detected, the real ID of the malicious vehicle is revealed by the TA to the authorities for legal investigation.

5.4   Usability

Vehicles can earn reward tokens after sharing their resources for cloud services. They can check the token balance with the OBU and use the earned tokens for a cloud service later.

6   Conclusion

In this paper, we proposed an architecture for vehicular cloud and presented an incentive based solution called secure token reward system to entice vehicular nodes for participating in the network. Our scheme is based on the idea that tokens are given as an incentive to the vehicles who contribute their resources for cloud services. The token reward system located in the cloud ensures the integrity of token transaction and efficient management of tokens. Also, the service provider manager is responsible for advertising services, making contracts, validating proofs of work done, and issuing reward tokens on behalf of service providers. Therefore, integrity and authenticity of incentive-related messages are guaranteed and the privacy of vehicles are protected. In addition, tokens received for sharing their resources can be used for obtaining services from the vehicular cloud in the future.

References

1. Arif, S., Olariu, S., Wang, J., Yan, G., Yang, W., Khalil, I.: Datacenter at the airport: reasoning about time-dependent parking lot occupancy. IEEE Trans. Parallel Distrib. Syst. **23**(11), 2067–2080 (2012)
2. Armbrust, M., Fox, A., Griffith, R., Joseph, A.D., Katz, R., Konwinski, A., Lee, G., Patterson, D., Rabkin, A., Stoica, I., et al.: A view of cloud computing. Commun. ACM **53**(4), 50–58 (2010)
3. Bernsen, J., Manivannan, D.: River: a reliable inter-vehicular routing protocol for vehicular ad hoc networks. Comput. Netw. **56**(17), 3795–3807 (2012)
4. Buchegger, S., Le Boudec, J.Y.: Nodes bearing grudges: towards routing security, fairness, and robustness in mobile ad hoc networks. In: Proceedings of 10th Euromicro Workshop on Parallel, Distributed and Network-based Processing, pp. 403–410. IEEE (2002)
5. Butty, L., Hubaux, J.P.: Enforcing service availability in mobile ad-hoc wans. In: Proceedings of the 1st ACM International Symposium on Mobile Ad Hoc Networking and Computing, pp. 87–96. IEEE Press (2000)
6. Buttyán, L., Hubaux, J.P.: Stimulating cooperation in self-organizing mobile ad hoc networks. Mob. Netw. Appl. **8**(5), 579–592 (2003)
7. Buyya, R., Yeo, C.S., Venugopal, S., Broberg, J., Brandic, I.: Cloud computing and emerging it platforms: vision, hype, and reality for delivering computing as the 5th utility. Future Gener. comput. syst. **25**(6), 599–616 (2009)
8. Chen, T., Zhu, L., Wu, F., Zhong, S.: Stimulating cooperation in vehicular ad hoc networks: a coalitional game theoretic approach. IEEE Trans. Veh. Technol. **60**(2), 566–579 (2011)
9. Costa, P., Gavidia, D., Koldehofe, B., Miranda, H., Musolesi, M., Riva, O.: When cars start gossiping. In: Proceedings of the 6th Workshop on Middleware for Network Eccentric and Mobile Applications, pp. 1–4. ACM (2008)
10. Eltoweissy, M., Olariu, S., Younis, M.: Towards autonomous vehicular clouds. In: Zheng, J., Simplot-Ryl, D., Leung, V.C.M. (eds.) ADHOCNETS 2010. LNICST, vol. 49, pp. 1–16. Springer, Heidelberg (2010)
11. Gerla, M.: Vehicular cloud computing. In: Proceedings of the 2012 11th Annual Mediterranean Ad Hoc Networking Workshop (Med-Hoc-Net), pp. 152–155. IEEE (2012)
12. Lee, S.B., Park, J.S., Gerla, M., Lu, S.: Secure incentives for commercial ad dissemination in vehicular networks. IEEE Trans. Veh. Technol. **61**(6), 2715–2728 (2012)
13. Li, F., Wu, J.: Frame: an innovative incentive scheme in vehicular networks. In: Proceedings of IEEE International Conference on Communications, ICC 2009, pp. 1–6. IEEE (2009)
14. Li, L., Song, J., Wang, F.Y., Zheng, N.: Ivs 05: new developments and research trends for intelligent vehicles. IEEE Intell. Syst. **20**(4), 10–14 (2005). http://doi.ieeecomputersociety.org/10.1109/MIS.2005.73
15. Lim, K., Abumuhfouz, I.M.: Stors: secure token reward system for vehicular clouds. In: Proceedings of the IEEE Southeastcon. IEEE (2015)
16. Lim, K., Manivannan, D.: An efficient scheme for authenticated and secure message delivery in vehicular ad hoc networks. In: Proceedings of the 12th IEEE Consumer Communications and Networking Conference (CCNC). IEEE (2015)
17. Liu, Y., Yang, Y.R.: Reputation propagation and agreement in mobile ad-hoc networks. In: 2003 IEEE Wireless Communications and Networking, WCNC 2003, vol. 3, pp. 1510–1515. IEEE (2003)

18. Marti, S., Giuli, T.J., Lai, K., Baker, M.: Mitigating routing misbehavior in mobile ad hoc networks. In: Proceedings of the 6th Annual International Conference on Mobile Computing and Networking, pp. 255–265. ACM (2000)

19. Olariu, S., Khalil, I., Abuelela, M.: Taking vanet to the clouds. Int. J. Pervasive Comput. Commun. **7**(1), 7–21 (2011)

20. Sardis, F., Mapp, G.E., Loo, J., Aiash, M., Vinel, A.: On the investigation of cloud-based mobile media environments with service-populating and qos-aware mechanisms. IEEE Trans. Multimedia **15**, 769–777 (2013)

21. Son, J., Eun, H., Oh, H., Kim, S., Hussain, R.: Rethinking vehicular communications: merging vanet with cloud computing. In: Proceedings of the 2012 IEEE 4th International Conference on Cloud Computing Technology and Science (Cloud-Com), pp. 606–609. IEEE Computer Society (2012)

22. Studer, A., Shi, E., Bai, F., Perrig, A.: Tacking together efficient authentication, revocation, and privacy in vanets. In: Proceedings of 6th Annual IEEE Communications Society Conference on Sensor, Mesh and Ad Hoc Communications and Networks, SECON 2009, pp. 1–9. IEEE (2009)

23. Wang, B., Li, B., Li, H.: Public auditing for shared data with efficient user revocation in the cloud. In: Proceedings of IEEE INFOCOM, pp. 2904–2912. IEEE (2013)

24. Wang, G., Liu, Q., Wu, J., Guo, M.: Hierarchical attribute-based encryption and scalable user revocation for sharing data in cloud servers. Comput. Secur. **30**(5), 320–331 (2011)

25. Whaiduzzamana, M., Sookhaka, M., Gania, A., Buyyab, R.: A survey on vehicular cloud computing. J. Netw. Comput. Appl. **40**, 325–344 (2014)

26. Yan, G., Wen, D., Olariu, S., Weigle, M.C.: Security challenges in vehicular cloud computing. IEEE Trans. Intell. Transp. Syst. **14**(1), 284–294 (2013)

27. Zhong, S., Li, L.E., Liu, Y.G., Yang, Y.R.: On designing incentive-compatible routing and forwarding protocols in wireless ad-hoc networks. Wireless Netw. **13**(6), 799–816 (2007)

EYES: A Novel Overtaking Assistance System for Vehicular Networks

Subhadeep Patra$^{(\boxtimes)}$, Javier H. Arnanz, Carlos T. Calafate,
Juan-Carlos Cano, and Pietro Manzoni

Department of Computer Engineering, Universitat Politècnica de València,
Camino de Vera S/N, 46022 Valencia, Spain
subpat@doctor.upv.es, javiherar@yahoo.com,
{calafate,jucano,pmanzoni}@disca.upv.es

Abstract. Developments in the ITS area are received with great expectation by both consumers and industry. Despite their huge potential benefits, ITS solutions suffer from the slow pace of adoption by manufacturers. In this paper we propose EYES, an ITS system that aims at helping drivers in overtaking. The system autonomously creates a network of the devices running EYES, and provides drivers with a video feed from the vehicle located just ahead, thus presenting a better view of any vehicles coming from the opposite direction and the road ahead. This is specially useful when the front view of the driver is blocked by large vehicles, and thus the decision whether to overtake can be taken based on the visuals provided by the application. We have validated EYES, the proposed overtaking assistance system, in both indoor and realistic scenarios involving vehicular network, and preliminary results allow being optimistic about its effectiveness and applicability.

Keywords: Android application · Real implementation · Video transmission · Live streaming · RTSP · Vehicular network · ITS

1 Introduction

Intelligent Transportation Systems (ITS) are advanced solutions that make use of vehicular and infrastructured networks to provide innovative services related to both traffic and mobility management, and that interface with other models of transport. ITS aims at using the already available transport networks in a smarter manner, resulting in significant coordination and safety improvements. Our goal here is to *integrate smartphones into vehicular networks* to develop ITS applications that can reach out to the masses in a short period of time. The choice of smartphones is not only justified by their wide availability and use, but also because they are evolving towards high performance terminals with multi-core microprocessors packed with sufficiently accurate onboard sensors.

The architecture and application developed has been named EYES. It has been developed for the Android platform, and requires the devices running it to

© Springer International Publishing Switzerland 2015
S. Papavassiliou and S. Ruehrup (Eds.): ADHOC-NOW 2015, LNCS 9143, pp. 375–389, 2015.
DOI:10.1007/978-3-319-19662-6_26

be equipped with at least a GPS and a back camera. The application makes use of the camera to record video and transmit it over the vehicular network, thus providing an enhanced multimedia information aid for overtaking. The location information of the vehicles gathered from the GPS is useful since the transmission of the video feed only occurs between cars travelling in the same direction, and always occurs from the vehicle in front to the vehicle travelling behind. The Android device is to be placed on the vehicle dashboard with the camera facing the windshield, so that a clear view of the road in front and cars coming from the opposite direction can be captured. Once started, the application requires no further user interaction to operate, and it can run in the background. The EYES application can be specially useful in scenarios where the view of the driver is blocked by a larger vehicle, or when a long queue of cars is located ahead and the driver wishes to overtake. In this case, the application will automatically receive the video feed from the vehicle right in front, and play the received feed on screen, thus aiding the driver in deciding the safest moment to overtake. A more detailed explanation of the architecture, design, and implementation issues of the EYES application will be provided in the following sections.

The rest of this paper is organized as follows: In Sect. 2, we survey some works in the literature that are closely related to our own. In Sect. 3, we will present an overview of the developed application. Later, in Sect. 4, we will present a description of EYES, its modules, and some implementation details. The setup used to deploy and validate EYES will be described in detail in Sect. 5. In Sect. 6, we will validate the application in a real testbed. Finally, Sect. 7 concludes this paper summarizing our contributions.

2   State of the Art

Both academia and industry have shown a strong interest in the field of ITS, resulting in the development of many innovative applications. Since our EYES application is targeted for smartphones, thus here in this section we are going to focus the bulk of our attention to some of the most interesting smartphone applications that are related to safe driving.

Most drive safety applications usually aim at warning generation based on onboard location sensors like in the works of Whipple et al. [1], Yang et al. [2], Diewald et al. [3] and Tornell et al. [4]. The application developed by Whipple et al. warns drivers when driving at high speed near schools. Yang et al. were concentrated on finding out the probability of accidents based on the location information. DriveAssist, by Diewald et al., triggers warning messages for certain traffic incidents, while Tornell et al., in their proposed application, display on screen important vehicles like ambulances and police cars on a map view, and they later improved that same solution in [5]. Few other applications used the On Board Diagnostics (OBD-II) [6] interface to detect incidents, like in the work of Zaldivar et al. [7] which aimed at detecting accidents. Wideberg et al. [8] also made use of OBD-II devices to extract safety and environment related information.

Only very few applications concentrated on providing visual aids to the drivers, like in SignalGuru described in [9], which leverages collaborative sensing on windshield-mount smartphones, in order to predict traffic signals' future schedule. The CarSafe App [10] was introduced by You et al., for instance, analyses the images from front and back cameras of smartphones to monitor the driver as well as the road ahead. Another interesting application available for download is iOnRoad [11], which aims at providing driving assistance functions including augmented driving, collision warning and "black-box" like video recording.

Despite the fact that we have found many different drive safety applications for smartphones, only a handful aimed at providing visual aids to the drivers, namely the SignalGuru, CarSafe and iOnRoad. However, none of these smartphone based applications actually provides real-time visual overtaking aid from other cars taking advantage of vehicular networks, even though the idea of video based overtaking assistance systems is not new. Works like the See-Through System [12] which was later improved in [13], although are not targeted for smartphones, are focused on the issue of video based overtaking assistance. Other related works worth mentioning are [14] and [15], which demonstrate the feasibility of such video based assistance systems. Improvement in performance of a video based overtaking assistant on undertaking codec channel adaptation, is shown in [14]. While, [15] is focused on reallocation of wireless channel resources to enhance the visual quality.

Encouraged by the findings from the above mentioned works and in order to fill the necessity of a visual overtaking assistance application that is targeted for the consumer section, which would require no additional hardware, but just simple download and install, we decided to develop EYES. The developed application is targeted for smartphones to achieve rapid acceptance and study the integrability of smartphones to vehicular networks.

3 Overview of the EYES Architecture

The goal of the EYES application is providing assistance during overtaking by streaming real-time video coming from one vehicle to another. The minimum requirements for running EYES is the availability of a device with GPS and back camera, along with a vehicular network for the transmission of video. This section presents an overview of the functionality of the EYES architecture.

The functionality of EYES can be divided in two phases for easy understanding. *Phase one*, i.e. *Server-Client Role Establishment Phase*, involves *electing the sender and the receiver* of the video which is subject to some special tests and validation conditions. And, *phase two*, named as *Video Streaming Phase*, involving the actual *video transmission* between the sender and receiver chosen in phase one.

In the first phase, each device equipped with a back camera and running EYES *broadcasts* an advertisement containing its location and direction, while simultaneously listening for incoming broadcast messages coming from other devices. If a device receives broadcast messages from other devices, it first verifies whether the source of the message is valid. The *validity check* is based on tests which basically involve checking if the source and destination vehicles are traveling one ahead of the other, and in the same direction. For a more detailed description of these validation conditions refer to Sect. 4.1. If several valid sources are found, the device *requests* video from the best source, which is selected based on the distance between sender and receiver devices. The source vehicle, upon receiving the request to send video from the destination vehicle, starts *streaming* the video signal over the vehicular network, in phase two. However, before sending the video, the source double-checks the validation conditions used in phase one. The destination vehicle starts playing the video onscreen as soon as it starts receiving it. The streaming and playback process is stopped only when the vehicle behind successfully overtakes, or when it stops following the vehicle in-front.

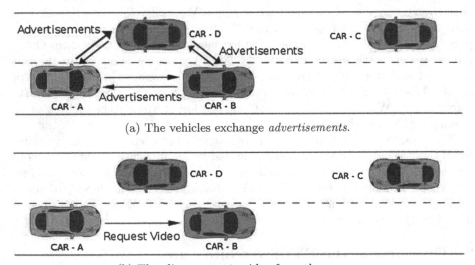

(a) The vehicles exchange *advertisements*.

(b) The client *requests* video from the server.

Fig. 1. EYES: Server-client role establishment phase.

Figure 1 provides more details about phase one. In this example, we have four cars, all of them using EYES. CAR-A and CAR-B are travelling in one direction, while CAR-C and CAR-D travel in the opposite direction. First, the cars broadcast the advertisement to each other as shown in Fig. 1a. Since CAR-C is not within the range of any other car, nobody is able to communicate with it. Each car, upon receiving the advertisement, performs the validity checks to see if the sender of the advertisement is travelling ahead in the same direction and the same lane. In this case, only CAR-A finds the advertisement message from CAR-B to be valid, and thus requests video from it, as depicted in Fig. 1b.

Similarly, Fig. 2 shows that CAR-B, upon receiving the video request from CAR-A, rechecks the validity conditions and starts streaming the video. CAR-A starts receiving the video stream and plays it onscreen immediately for its

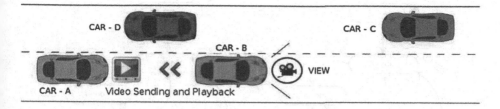

Fig. 2. EYES: Video streaming phase.

driver. It may be noted here that a single device can act both as video source and destination. This is because, while a device is receiving video from another device, it may also be streaming its own video capture to a completely different device.

4   Implementation Details of EYES

From the previous section we already know that EYES has been implemented by splitting its functionality into two distinct phases: (i) The *Server-Client Role Establishment Phase*, in which election of the video source and destination takes place; also, the starting and stopping of the video stream is controlled in this phase; (ii) The *Video Streaming Phase*, on the other hand, encompasses video streaming and playback.

The *client-server role establishment phase* is the most important of the two phases, and is run right from the start of the application until EYES is completely stopped. Even when the *video streaming phase* is transmitting video, the *client-server role establishment phase* keeps working in the background to check if overtaking has occurred and stops the video streaming. Below we describe in detail the two phases of the EYES application.

4.1   Server-Client Role Establishment Phase

This phase, as the name suggests, is in charge of choosing the server or the source of the video, and the client, i.e. the destination, for the *video streaming phase*. The starting and stopping of the video transmission will take place only under some validation conditions, namely the *same direction test* and the *same lane test* for starting the video transmission, while for ending the transmission, the *overtake test* is used. Even though, at the beginning of this phase, the server and client roles are not established, we will use the words server and client to refer to the devices that will be attaining the respective role in the future for the sake of clarity.

Figure 3 shows the different states that a server and client can attain. When the client and the server start, the server is in the *notify* state, as shown in Fig. 3a, and it starts advertising the availability of the video by broadcasting a *hello* message. Besides sending advertisements, the server, while in *notify* state,

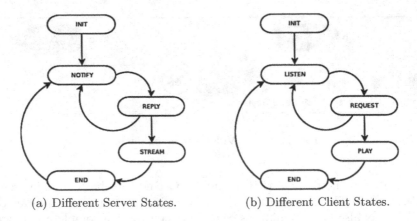

(a) Different Server States.   (b) Different Client States.

Fig. 3. State diagram of the server and client.

also listens for replies to its *hello* message from clients requesting the video feed. A *hello* message contains the location information of the server so that the client, upon receiving it, can determine if the server is ahead of the client and travelling in the same direction. The client remains listening for advertisements from the server for only a certain period of time, while in the *listen* state, as shown in the Fig. 3b.

If the client receives *hello* messages from different servers, the client checks whether the servers are valid, and stores them in a queue of candidate servers. The proposed validity tests include the *same direction test* and the *same lane test* conditions as shown in Fig. 4.

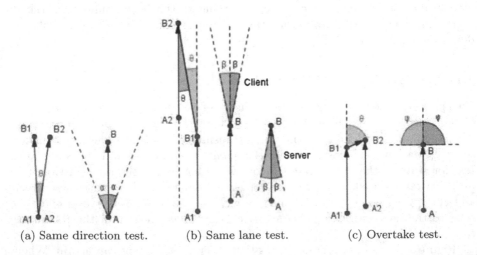

(a) Same direction test.   (b) Same lane test.   (c) Overtake test.

Fig. 4. The different validation conditions in EYES.

The *same direction test* is used to detect whether two vehicles are travelling in the same direction. For understanding the *same direction test*, let us assume we have two cars, one travelling from a point A1 to B1 and the other from A2 to B2 as shown in Fig. 4a. Notice that, even if two cars are travelling in the same direction and speed, its it hard for them to have an overlapping displacement vector, in other words, the angle between the two vectors is not 0. This can happen due to different driving styles and GPS errors. Thus, we measure the angle θ between these two vectors and compare it to a predefined threshold α. If θ is less than α, we can safely assume that the two vehicles are travelling in the same direction. Now, even if two vehicles are travelling in the same direction, it does not necessarily mean that one is ahead of the other, both vehicles may be travelling on different lanes or parallel roads altogether. To check if one is following the other one on the same lane, we perform the *same lane test*, and for this purpose we draw an imaginary line joining the current locations of the two vehicles, as shown in Fig. 4b, where B1 and B2 are the current locations. Then we measure the angle of intersection of this line joining the points B1 and B2 with the displacement vectors of the vehicles. When the measured angle of intersection θ is less than a predefined angle β, then the vehicles are considered to be travelling on the same lane. Being on different lanes will result in a higher value of the measured angle θ, and the *same lane test* will fail. If these two conditions are satisfied, then the two vehicles are assumed to be travelling in the same direction, one following the other.

Table 1. Messages exchanged between the server and client.

Message Type	From → To	Client State	Server State	Message Contents
Hello	S → C	Listen	Notify	Location and Direction
Request	C → S	Request	Notify	Location and Direction
Ready	S → C	Request	Reply	Video sender port
Reject	S → C	Request	Reply	-
Data	S → C	Play	Stream	Location, Direction and Speed
Data-Ack	C → S	Play	Stream	-
End	C → S	Play	Stream	-

The client which was listening for server advertisements, as soon as its listening period is over, chooses the best server from the list of candidate servers based on its distance to the server. The client then tries to connect to the chosen server by sending a *request*, and moves to the *request* state. The server, upon receiving the *request* from the client, also checks its validity by performing the *same direction* and *same lane* tests once again. Before sending the *ready* or *reject* message which denotes whether it is ready to send video being captured by its camera, the server changes its state to *reply*. The server may further choose to change its state back to *notify* or to *stream* modes depending on its own reply. The client, which was previously in the *request* state, only changes its state to *play* if the reply from the server was a *ready* message containing the *video sender port* number, otherwise it may choose to contact some other server. Table 1, details the packet types used during the *server-client role establishment phase*.

In case the server and client are in the *stream* and *play* states respectively, the *video streaming phase* is launched to transmit the video. In fact, the *server-client role establishment phase* remains active even if the *video streaming phase* has been started. The server during this period keeps sending *data* messages containing its location information, direction and speed. This way, its corresponding client can check whether an overtake has occurred, and so the client can request the server to terminate the video stream by sending an *end* message. To find out if an overtake has been successful done, the *overtake test* takes place, as shown in Fig. 4c. This test is similar to the *lane test* condition, the only difference being that the angle θ measured here is the other linear pair of the angle of intersection between the displacement vector and the line formed by joining the current location of the two vehicles. Also, the threshold φ used here is usually a much larger value.

Upon receiving the *data* message from the server, the client, if still has not overtaken as suggested by the *overtake test*, replies the server with *data-ack* to keep the connection alive. Hence, the video streaming is continued. When the video streaming has been stopped, the client switches to the *end* state and later on moves back to the *listen* state once again. The server, on the other hand, can move to the *end* state upon receipt of the *end* message from the client or if the waiting time for a *data-ack* from the client expires. This waiting time is used to detect cases of eventual disconnections.

4.2 Video Streaming Phase

In this phase, the actual video streaming and playback takes place. The *video streaming phase* starts when the client and server of *server-client role establishment phase*, are in the *play* and *stream* states respectively. The sender, which is the source of the video in this phase, is based on the libstreaming [16] API that relies on the RTSP [17] protocol for streaming video over the network. The video is encoded by the sender using the H.264 encoding format before sending. The receiver, on the other hand, is based on libvlc [18] for decoding and displaying the received video. Apart from playing the video on screen, the device at the destination also displays to the driver, the speed information of the vehicle travelling ahead. The speed information of the vehicle ahead is extracted from the *data* message received by the client as part of the *server-client role establishment phase*.

Figure 5 shows that the sender initially awaits for an incoming connection from a receiver on a predefined port, say A. This port A becomes known to the receiver device from the *ready* message send by the server in the *server-client role establishment phase*. The receiver contacts the sender by using a port X, and then the next three steps are followed: step-i is the setup of the video

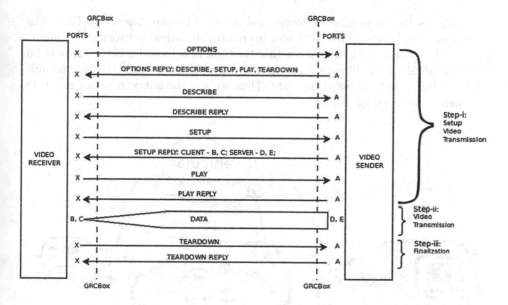

Fig. 5. The video transmission.

transmission; step-ii is when the video data transfer occurs; and step-iii involves the finalization of the video transmission. In step-i, the receiver requests the server for the supported *options*. The sender replies to the receiver with the list of supported options; in our case the sender supports *describe, setup, play and teardown*. The receiver then asks the sender to *describe* the video that it is going to send, and the sender replies. Next, the receiver requests the sender to configure the video streaming, using the *setup* option. As a result, the sender replies to the receiver with the port numbers that will be used for sending and receiving of the audio and video, which in this case are B and C for receiving, and D and E for sending. Then, the receiver tries to open the ports requested by the sender, namely B and C, and then it sends the *play* request. The sender acknowledges the play request from the receiver. In step-ii, *Data* transfer starts between the sender and receiver using the ports chosen in step-i. Notice that UDP is used in this step, whereas in the rest of the steps TCP is used. On overtaking, the receiver requests the sender to stop transmitting the video by sending the *teardown* request in step-iii. The sender acknowledges the teardown request, and the video streaming stops.

5   Deployment

For proper operation, the EYES application assumes the availability of a vehicular network, although the vehicles we use on a daily basis still lack the capability to communicate with one another.

So, for testing our application, we equipped cars with a GRCBox [19] inside them. GRCBox is a low cost connectivity device that allows the integration of smartphones into vehicular networks and is based on the Raspberry Pi. It was developed mainly due to the difficulty in creating an adhoc network using smartphones. Another important feature that the GRCBox provides is the support for V2X communication. The different networks supported by the GRCBox include adhoc, cellular, wifi access points, etc. Thus, we use the adhoc network to create a vehicular network for EYES.

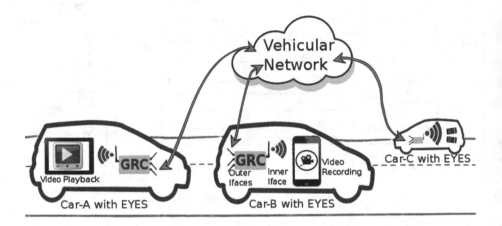

Fig. 6. Example of EYES working together with GRCBox.

Figure 6 shows how EYES works when combined with GRCBox. Each car within the vehicular network has a GRCBox mounted. The smartphones of the passengers within the car are connected to the GRCBox, and the GRCBox which is equipped with Wifi-enabled USB interfaces to communicate in adhoc mode, creates a vehicular network. Even though GRCBox is supposed to be equipped with 802.11p for vehicular communication, we used 802.11a devices instead, as 802.11p-enabled hardware was not at our disposal while setting up the GrcBox to perform the tests. For our future experiments, we intend to get hold of 802.11p compatible hardware to take advantage of the WAVE standard. As shown in the figure, Car-B is ahead of the Car-A, and both of them are travelling in the same direction and running EYES, so the smartphone in Car-B starts recording the video autonomously and sends it to Car-A, by relying on the vehicular network created using the GRCBoxes available within the cars. Concerning the video, it is played onscreen on the device in Car-A as soon as video reception starts.

We have taken advantage of the adhoc communications support of GRCBox to perform our experiments with the EYES application. The Android devices used in combination with the GRCBoxes, are a Nexus 7 and a Samsung Galaxy Note 10.1 (2014 Edition). The Nexus 7 from Google was powered by a quad-core 1.2 GHz processor, ULP GeForce GPU, 1 GB ram and 1.2 MP camera. The Samsung Galaxy Note 10.1, on the other hand, was equipped with a quad-core 1.9 GHz plus quad-core 1.3 GHz processors, 3 GB ram, 8 MP primary camera and 2 MP secondary camera.

6   Validation Test

In the EYES application, the three important conditions evaluated were described in Sect. 4.1, and each of these conditions, namely *same direction test*, *same lane test* and *overtake test*, are dependent on a threshold value. Our aim was to evaluate reasonable values of the threshold angles α, β and φ for two cars of which one follows the other throughout the experiment while travelling along a particular route, so that there is non-stop streaming of video between the cars. The cars were equipped with GRCBox devices, which helps to create a vehicular network, as described in Sect. 5. The Android devices used were the Samsung Galaxy Note 10.1 in the car ahead, and the Nexus 7 in the car that was following it.

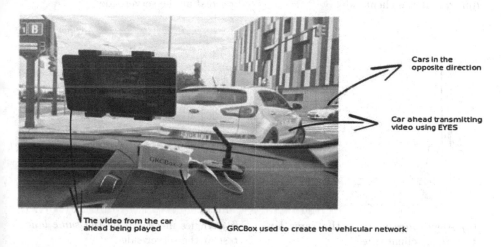

Fig. 7. The experiments with EYES in real scenario.

Figure 7 shows a photo taken during an outdoor test. In this picture, we can see that the front car is trying to take a right turn, and the back car is receiving the video from the car ahead and playing it onscreen. While doing our outdoor tests with EYES, we collected the various angles used in the three different validation tests. Below we can see the graphical representation of the data obtained during the experiment.

Figure 8 shows the results of the test conditions for starting the video transmission between the server and the client. Since these conditions are tested both by the server and the client, we have two sets of results. Notice that the angles shown may not have been measured at the exact same time by the server and client. Figures 8a and 8b show the plot of angles measured by the *same direction test* at the client and the server end, respectively. Most observations for both the client and server lies within 20 degrees, which is satisfactory. It is also noticeable that many peaks occur due to GPS errors, also because the route followed had

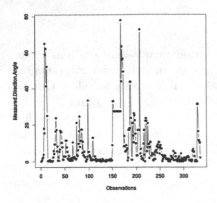

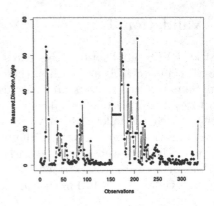

(a) Angles measured by the *same direction test* at the client side.

(b) Angles measured by the *same direction test* at the server side.

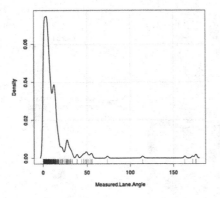

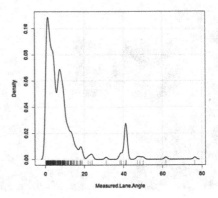

(c) Angles measured by the *same lane test* at the client side.

(d) Angles measured by the *same lane test* at the server side.

Fig. 8. Results of the *same direction* and *same lane* tests.

a lot of turns and curves, and so the two cars were not always on a straight path. Figures 8c and 8d show the density graph for the *same lane test* for the client and the server, respectively. From the two density curves, we can see that most observations for the *same lane test* lie in the range between 20–25 degrees. Notice that this value is too high considering that this test is very sensitive, and used to detect if cars are travelling on the same lane, and so we find that this condition may not be too useful when considering the accuracy of current technology.

Figure 9, shows the density plot for the observations of the *overtake test*. Note that the values used in this graph are 180 - *the observed value*, to make it more simple. Since this test is only performed by the client, we have only its data. We find that the results from this test were pretty much what we expected since all of plotted values are below 90 degrees. Furthermore, it should be noted that,

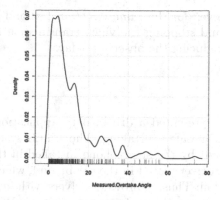

Fig. 9. Results of the *overtake test* from the client.

based on some of our indoor experiments, we have found that the application still has some minor problems related to delay of the order of few seconds between capture and playback.

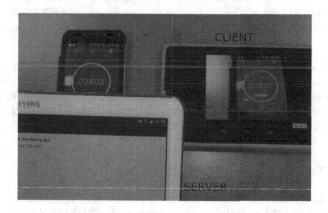

Fig. 10. The delay in one of the indoor experiments.

Figure 10 shows the delay in video recording, transmission and playback during one of our experiments carried out indoors in our laboratory. In this test, the Samsung Note 10.1 was configured to work as the server, recording the video of a stopwatch in front of it. The Nexus 7 tablet was configured to receive the video and display it. When the actual timer hit 24.03 seconds, the video being played at the client side still showed the timer to be at 22.68 seconds. Thus, there was a delay of about 1.35 seconds. Overall, from the experiments carried out, we conclude that the EYES application works correctly even though some delay issues in transmission and playback must be addressed. The validation conditions used in EYES have been validated in a real testbed, and satisfactory results have been achieved for the *same direction test* and *overtake test*, which are used for starting and stopping the video transmission respectively. We are currently working on reducing the observed delay.

7   Conclusions

In this paper, we have presented a drive safety application called EYES that is able to help drivers in safe overtaking. The EYES system provides a real-time video feed captured by the smartphone installed in the vehicle ahead, to the smartphone of the driver seated in the car behind, which displays the video without user intervention. Thus, it provides drivers with important information and helps them to decide whether it is safe to overtake. We have evaluated the different test conditions used in EYES, and found that thresholds in between 20–25 degrees for the *same direction test* and 90 degrees for the *overtake test* are reasonable. Nevertheless, the *same lane test* was found to be useless unless more accurate GPS hardware is made available. Also, the delay of the video transmission and playback is found to be of the order of a few seconds, in some cases. Despite these minor issues, we acknowledge the fact that combining smartphones with vehicular networks indeed opens a new horizon for ITS applications and, in the future, we will focus our attention on improving our EYES application by minimizing the delay and finding an effective substitute for the *same lane* test.

Acknowledgments. This work was partially supported by the *European Commission* under *Svāgata.eu*, the Erasmus Mundus Programme, Action 2 (EMA2) and the *Ministerio de Econom?a y Competitividad, Programa Estatal de Investigaci?n, Desarrollo e Innovaci?n Orientada a los Retos de la Sociedad, Proyectos I+D+I 2014*, Spain, under Grant TEC2014-52690-R.

References

1. Whipple, J., Arensman, W., Boler, M.S.: A public safety application of gps-enabled smartphones and the android operating system. In: 2009 IEEE International Conference on Systems, Man and Cybernetics. SMC 2009, pp. 2059–2061. IEEE (2009)
2. Yang, J., Wang, J., Liu, B.: An intersection collision warning system using wi-fi smartphones in vanet. In: 2011 IEEE Global Telecommunications Conference (GLOBECOM 2011), pp. 1–5. IEEE (2011)
3. Diewald, S., Möller, A., Roalter, L., Kranz, M.: Driveassist-a v2x-based driver assistance system for android. In: Mensch & Computer Workshopband, pp. 373–380 (2012)
4. Tornell, S.M., Calafate, C.T., Cano, J-.C., Manzoni, P., Fogue, M., Martinez, F.J.: Implementing and testing a driving safety application for smartphones based on the emdr protocol. In: Wireless Days (WD), 2012 IFIP, pp. 1–3. IEEE (2012)
5. Patra, S., Tornell, S.M., Calafate, C.T., Cano, J.-C., Manzoni, P.: Messiah: an its drive safety application. In: XXV Jornadas Sarteco, Valladolid, Spain (2014)
6. International Organization for Standardization. Iso 14230–1:1999: Road vehicles, diagnostic systems, keyword protocol 2000 (1999)

7. Zaldivar, J., Calafate, C.T., Cano, J.-C., Manzoni, P.: Providing accident detection in vehicular networks through obd-ii devices and android-based smartphones. In: 2011 IEEE 36th Conference on Local Computer Networks (LCN), pp. 813–819. IEEE (2011)

8. Wideberg, J., Luque, P., Mantaras, D.: A smartphone application to extract safety and environmental related information from the obd-ii interface of a car. Int. J. Veh. Syst. Model. Test. 7(1), 1–11 (2012)

9. Koukoumidis, E., Martonosi, M., Peh, L.-S.: Leveraging smartphone cameras for collaborative road advisories. IEEE Trans. Mob. Comput. 11(5), 707–723 (2012)

10. You, C.-W., Lane, N.D., Chen, F., Wang, R., Chen, Z., Bao, T.J., Montes-de Oca, M., Cheng, Y., Lin, M., Torresani, L., et al.: Carsafe app: alerting drowsy and distracted drivers using dual cameras on smartphones. In: Proceeding of the 11th annual international conference on Mobile systems, applications, and services, pp. 13–26. ACM (2013)

11. ionroad official website. http://www.ionroad.com/. Accessed 8 February 2015

12. Olaverri-Monreal, C., Gomes, P., Fernandes, R., Vieira, F., Ferreira, M: The see-through system: A vanet-enabled assistant for overtaking maneuvers. In Intelligent Vehicles Symposium (IV), 2010 IEEE, pages 123–128. IEEE, 2010

13. Gomes, P., Olaverri-Monreal, C., Ferreira, M.: Making vehicles transparent through v2v video streaming. IEEE Intell. Transp. Syst. 13(2), 930–938 (2012)

14. Vinel, A., Belyaev, E., Egiazarian, K., Koucheryavy, Y.: An overtaking assistance system based on joint beaconing and real-time video transmission. IEEE Trans. Veh. Technol. 61(5), 2319–2329 (2012)

15. Belyaev, E., Mukhanov, P., Vinel, A., Koucheryavy, Y.: The use of automotive radars in video-based overtaking assistance applications. IEEE Intell. Transp. Syst. 14(3), 1035–1042 (2013)

16. The libstreaming android api. https://github.com/fyhertz/libstreaming. Accessed 5 February 2015

17. Schulzrinne, H.: Real time streaming protocol (rtsp) (1998)

18. Libvlc documentation. http://www.videolan.org/vlc/libvlc.html. Accessed 5 February 2015

19. Tornell, S.M., Patra, S., Calafate, C.T., Cano, J.-C., Manzoni, P.: Grcbox: Extending smartphone connectivity in vehicular networks. International Journal of Distributed Sensor Networks (2014)

Study of Probabilistic Worst Case Inter-Beacon Delays Under Realistic Vehicular Mobility Conditions

Alexandre Mouradian[(✉)]

Laboratoire des Signaux Et Systèmes (L2S, UMR8506), Université Paris Sud-CNRS-CentraleSupélec, 91192 Gif-sur-yvette, France
alexandre.mouradian@u-psud.fr

Abstract. Road safety applications are one of the main incentives to deploy vehicular networks. These applications rely on periodic message exchange among vehicles (known as beaconing). The beacon messages contain information about the environment which is used to perceive dangerous situations and alert the drivers. The inter-beacon delay is the time between two consecutive beacons received from a car. It is an essential parameter because, if this delay exceeds the application requirement, the application cannot accurately predict dangerous situations and alert the drivers on time. The worst case inter-beacon delay has thus to be bounded according to the application requirements. Unfortunately, a tight and strict bound is in fact very difficult to obtain for a real network because of the randomness of the collisions among beacons coming from: the unpredictable mobility patterns, random interferences, randomness of the MAC layer backoff, etc.

In this paper, we propose to provide a probabilistic worst-case of the inter-beacon delay under realistic mobility using Extreme Value Theory (EVT). EVT provides statistical tools which allow to make predictions on extreme deviations from the average of a parameter. These statistical predictions can be made based on data gathered from simulation or experimentation. We first introduce the EVT technique. Then we discuss its application to the study of inter-beacon delays. Finally, we apply EVT on the results of extensive vehicular network simulation using a realistic mobility trace: the Cologne trace.

1 Introduction

The road safety applications are seen as an essential motivation for the deployment of Vehicular Ad hoc NETworks (VANETs) [9]. The main goal of these applications is to avoid car crashes and thus reduce the number of road traffic deaths and injuries. We can cite as examples of safety applications [5]:

- traffic signal violation: alert neighbor cars when a user does not stop at a red traffic light;
- electronic brake: alert neighbor vehicles when a car performs an emergency braking;

© Springer International Publishing Switzerland 2015
S. Papavassiliou and S. Ruehrup (Eds.): ADHOC-NOW 2015, LNCS 9143, pp. 390–403, 2015.
DOI:10.1007/978-3-319-19662-6_27

- on-coming traffic warning: notify the driver of on-coming traffic during over-taking maneuvers.

The main building block for safety applications is the exchange of periodical one-hop broadcast messages among neighbor vehicles [9]. These messages are called beacons or heartbeat messages and contain, at least, information on the car positions, directions and speeds. In the rest of this paper, as in [17, 19] we refer to these messages as beacons. Each safety application uses these beacons to build a representation of its environment, analyze the current situation and predict dangerous situations. The period, latency and scope (or range) of the beacons depend on the requirements of the application. For example, according to [9] the electronic brake application needs a beacon frequency of 10 Hz, a maximum latency of 100 ms and a range of 200 m to be effective. In this paper, we focus on the period of these messages: the inter-beacon delay. We are thus interested in the capacity of the MAC layer to broadcast these beacons on a periodical basis which complies with the application requirements. We consider the MAC from the IEEE 802.11p standard, since it is regarded as the standard of choice for VANETs [9].

In order to be effective, critical road safety applications need strong timing guaranties on the delivery of beacons. It is thus of paramount importance that the MAC layer is able to provide worst case inter-beacon delays which are bounded. But more importantly, the bound has to be known and has to correspond to the target application. Nevertheless, most of the performance evaluation techniques of the literature such as stochastic models, simulation and experimentation focus on parameter averages [10] and give very few insights on extreme cases. This is the case, in particular, concerning many performance studies of beaconing in vehicular networks [9, 17, 19]. On the other hand, a strict bound on the inter-beacon delay is difficult, if not impossible, to obtain because of the random nature of the beacon collisions which comes from the mobility patterns, interferences, pathloss, randomness of the MAC layer backoff, etc.

In this paper, we propose to estimate a probabilistic bound on the inter-beacon delay using statistical tools. The probabilistic worst case delay can be viewed as the probability that the maximum delay is less than a given value. This quantity is very useful for the system designer which must ensure that the system can handle the timing requirements of the application with a high probability. An advantage of the statistical approach is that it allows to evaluate directly the real studied system (or a highly detailed simulation model) instead of working on an abstract theoretical model.

The statistical tools we use come from Extreme Value Theory (EVT) which is presented in Sect. 3. The method can be summed up in the three following steps: (1) produce/Gather data (by simulation or experimentation); (2) extract extreme values from the data; (3) fit extracted values to a probability distribution predicted by the theory.

The main contribution of this paper is to use EVT to characterize the distribution of high inter-beacon delays when using the IEEE 802.11p standard under realistic mobility conditions. We first discuss EVT applicability to study inter-beacon delays and then use it on simulation results. The simulations are performed in *ns2* [1] using a highly realistic mobility trace developed by Uppor and Fiore [21]. Up to our knowledge, this study represents the first application of EVT to the study of extreme delays in VANETs and is the first characterization of extreme inter-beacon delays under realistic mobility conditions using the IEEE 802.11p standard.

The remainder of the paper is organized as follows. In Sect. 2 the related work about safety beaconing performance and EVT is presented and commented. In Sect. 3, we introduce EVT main theorems and discuss the application of EVT to the study of large inter-beacon delays. In Sect. 4, we present the simulation setup and the produced data sets. In Sect. 5 we describe the results of the application of EVT to inter-beacon delay simulation data and provide discussion about the method and the obtained probabilistic worst-case inter-beacon delays. Section 6 gives the conclusion remarks and lists future works.

2   Related Work

In this section we present the related work regarding the study of beaconing in VANETs with 802.11p, and on the application of EVT in general and more specifically in computer science.

Many studies concerning the performance of beaconing using 802.11p exist in the literature [3,9,11,16,17,19]. In [9], the authors study the performance of beaconing in VANETs through simulation. They use the *ns2* simulator. In their scenarios, the vehicles are positioned on parallel lines (straight highway scenario) and are not moving. The authors provide the probability of a successful beacon reception for different beaconing frequencies. The results show that the probability of reception decreases rapidly after few hundred meters between the emitter and receiver. In [19], Stanica et al. propose an analytical model of the probability of collision and the reception probability of beacons. They highlight the low reliability of beaconing with 802.11p as with their model, the reception probability is never greater than 0.7. They confirm their findings through simulation. The authors also show that the performance depends on the 802.11 contention window and derive its optimal value. In [16], the repartition of the nodes of the VANET is modeled as a Poisson Point Process. The authors show that above a critical density of nodes, the CSMA scheme of 802.11p behaves like an ALOHA protocol and its performance is thus highly deteriorated. The authors study the reception probability for simple highway scenarios under various traffic conditions and transmission power hypotheses. Again the reliability of beaconing is very low. The authors of [17] simulate beaconing using *ns2* and the realistic Cologne mobility trace [21]. They study the probability of beacon delivery in different areas of the city. They confirm the bad reliability of 802.11p in a realistic mobility scenario. Despite the fact that many studies evaluate the reliability of beaconing with 802.11p, none of the presented study provide insight on the worst case inter-beacon delay under realistic mobility conditions (we can note that [11] is concerned about the inter-beacon delay distribution, but not the worst case). Yet, this parameter is of paramount importance to evaluate the achievable timeliness of safety applications.

In the literature, EVT has been extensively used in various contexts: extreme rainfalls [18], forest fires [2], wind speed [4], financial crashes [8] studies. EVT has also been used in the context of computer networks for the estimation of traffic peaks or bursts [6,13,20]. In [20], the author proposes to study the traffic on an Ethernet network in order to predict traffic peaks. The author shows that the gathered extreme deviations fit very well to a Generalized Pareto distribution as predicted by the theory. In another study [13], the authors apply EVT to the study of traffic throughput in wireless networks and show that the generalized EVT distribution is a better match for large deviation prediction than exponential, gamma or log-normal distributions. In [6] the authors fit Ethernet traffic throughput data to a Weibull distribution (also predicted by EVT). Nevertheless, these applications of EVT focus on network traffic throughput. Here we are more interested in the possibility to use such technique to study large delays in computer networks.

In order to find applications of EVT to the study of large delays in computer science, we have to look to the field of worst case execution times. Works such as [7,14] make use of EVT to derive probabilistic worst case execution times. In our work, we explore the case of inter-beacon delays instead of task executions. We comment, in Sects. 3 and 5, the issues which may arise when applying EVT to this specific case.

The novelty of our work is twofold: we are interested in characterizing large inter-beacon delays behavior instead of their probability of reception in realistic mobility conditions, and we use the EVT method which up to our knowledge have never been apply to the study of delays in networks.

3   Extreme Value Theory Applied to Delays in Networks

3.1   A Brief Introduction to EVT

Extreme Value Theory has been developed during the 20th century and is now a well established tool to study extreme deviations from the average of a measured phenomenon [4]. EVT is built around two main theorems: the Fisher-Tippett-Gnedenko theorem and the Pickands-Balkema-de Haan theorem. As we will detail below, the former is interested in the maximum value of a sequence of variables, whereas the latter focuses on the values of a sequence which are above a given threshold.

The Fisher-Tippett-Gnedenko theorem states that given $\{X_1, \ldots, X_n\}$ a sequence of independent and identically distributed (i.i.d.) variables, the distribution of $M_n = \max\{X_1, \ldots, X_n\}$ the variable representing the maximum value of the sequence converges (for large n) toward one of these three distribution families characterized by their CDF (Cumulative Distribution Function):

– Fréchet:

$$G(x) = e^{-(\frac{x-m}{s})^{-\alpha}}, \text{ for } x > m \tag{1}$$

if the distribution has a heavy tail;

- Gumbel:

$$G(x) = e^{-e^{-\frac{x-m}{s}}} \qquad (2)$$

if the distribution has an exponential tail;
- Weibull

$$G(x) = e^{-(-(\frac{x-m}{s}))^{\alpha}}, \text{ for } x < m \qquad (3)$$

if the distribution has a finite maximum;

with m, s and α the distribution parameters and $\alpha > 0$ in all cases.

The second theorem of EVT is the Pickands-Balkema-de Haan theorem. It states that given $\{X_1, \ldots, X_n\}$ a sequence of i.i.d. variables, their conditional distribution $F_u(y) = P(X - u < y | X > u)$ converges toward a generalized Pareto distribution for large u:

$$\begin{cases} G(y) = 1 - (1 + \frac{(y-m)\gamma}{s})^{-\frac{1}{\gamma}}, \text{ if } \gamma \neq 0 \\ G(y) = 1 - e^{-\frac{y-m}{s}}, \text{ if } \gamma = 0 \end{cases} \qquad (4)$$

To each theorem corresponds a method which can be applied to characterize the distributions of extreme variation of a phenomenon. For the first theorem the method is the Block Maxima (BM) method in which the sequence of measured data is divided into blocks and the maximum of each block is computed. The maxima are then fitted to one of the three previously mentioned distributions. The second theorem of EVT corresponds to the Peak Over Threshold (POT) method. In this method, a threshold value is chosen and the data points which are above the threshold are collected and fitted to a generalized Pareto distribution.

3.2 Application to the Study of Inter-beacon Delays

According to Fisher-Tippett-Gnedenko and the Pickands-Balkema-de Haan theorems, to apply EVT, the sequence of variables must be independent and identically distributed. The question then is: can it be the case for inter-beacon delays in VANETs ?

Let's first consider two nodes A and B. Can we assume that their inter-beacon delays are identically distributed ? We would tend to answer yes if the network environment and measure conditions are the same for both A and B: same channel conditions, same node density, fair access to the medium, etc. Because in this case the packet collisions and packet loss inducing inter-beacon delays would be identically distributed. This leads us to believe that to apply EVT on inter-beacon delays we have to be careful not to have great disparities in the studied network like very dense and very sparse areas in the same network. This statement will be verified in Sect. 5.

Concerning the independence hypothesis, we can argue that it will not always be true in the case of inter-beacon delays. If we consider the successive inter-beacon delays between two nodes with bad channel conditions, they will tend to be correlated (if the channel conditions are stable between two beacons). Nevertheless, in the literature, we find many EVT application cases on data sequences which seem correlated in time. For instance, in [4] it is mentioned that it is possible to fit very accurately maximum wind speeds to an extreme value distribution. Nevertheless, wind speed measures may appear to be temporally correlated as studied in [12]. This shows that EVT may accurately model the large values of a phenomenon even if the measures are partially correlated.

It is not clear from the literature, what are the cases where EVT can or cannot be applied. In Sect. 5, we will consider the i.i.d hypothesis true, try to apply EVT and evaluate how the data fits the model (the fit can be assessed by a statistical test as explained in Sect. 3.4). We will then discuss in which cases EVT is meaningful.

Due to the lack of space, in this paper, we consider only the application of the Fisher-Tippett-Gnedenko theorem. The second theorem of EVT is left as a future work.

3.3 Gathering and Arranging Data for EVT

The first element needed in order to apply EVT is a set of data which is a realization of the sequence of random variables $\{X_1, \ldots, X_n\}$ mentioned in the previous section. In our case it will be measures of inter-beacon delays obtained from simulation. In this paper, we consider technique associated with the Fisher-Tippett-Gnedenko theorem to process the data. This technique is known as Block Maxima (BM). The principle of the BM technique is to divide the data sequence into blocks and to take the maximum of each block. In the literature, a block is often defined as a time interval [4]. According to the Fisher-Tippett-Gnedenko theorem the sequence of block maxima must converge to an extreme value distribution. The observed sequence of block maxima can thus be fitted to one of the Gumbel, Weibull or Fréchet distributions. The type of distribution the data will converge to is difficult to predict from the raw data. In Sect. 5 we thus try to fit the data to the three considered distribution and discuss which is the most accurate model for the worst-case inter-beacon delay.

3.4 Fitting Technique and Statistical Test

In the previous subsection, we have described how to retrieve extreme value data from the original data set. We then have to fit this extreme value data to one of the extreme value distributions predicted by EVT and assess that the fitted distribution is an acceptable representation of the data thanks to a goodness-of-fit statistical test. In this paper we use the Maximum Likelihood Estimation (MLE) technique to estimate the parameters of the distribution, and the Pearson's chi-squared test to assess the goodness-of-fit.

The MLE technique is based on the likelihood function, which is defined as follows:

$$L(\mathbf{x}, \theta) = \prod_{i=1}^{n} f(x_i, \theta) \tag{5}$$

with **x** a vector of observed values, f the pdf we want to fit to the values and θ the vector of parameters for the pdf.

The MLE method consists in finding a vector of parameters θ such that $L(\mathbf{x}, \theta)$ is maximized. In practice, for the problem to be tractable, it is actually the logarithm of the likelihood which is maximized (the logarithm function preserves the optimum). In this paper, in order to solve this optimization problem we use the Nelder-Mead method [15].

Once the optimal distribution parameters are obtained, we have to verify that the fitted distribution is actually a convincing representation for the data. For that purpose, we use two tools: the Pearson chi-squared test and quantile-quantile plots [10]. The former is a statistical test which assesses if there is a statistical difference between an observed data frequency distribution and a theoretical distribution. We can note that this test statistic offers a quality indicator for the fit which can be used in the case several distributions pass the chi-squared test as in [18].

The quantile-quantile plot [10] (or Q-Q plot) is a tool which allows to graphically compare two distributions. In our case, it consists in plotting the quantiles of the collected data against the quantiles of the fitted distribution. If the two distributions match, we should obtain the relation $x = y$. As described in [4,10], the plot consists in a set of points (x, y) where $x \in x_1, \ldots, x_n$ with $x_1 \ldots x_n$ the ordered set of data points (in increasing order) and $y = F^{-1}(\frac{i}{n+1})$ with n the number of data points, $i = 1, \ldots, n$ and F^{-1} the inverse CDF of the fitted distribution (the inverse CDF corresponds to the quantile function).

4   Simulation

In this section we present the simulation parameters and the obtained data sets on which we apply EVT.

4.1   Simulation Setup

We use the discrete-event simulator *ns2* [1] to perform the simulations, the main parameters are described in Table 1. The simulation setup is actually quite classic, we thus focus on the description of the realistic mobility trace and how we use it in the simulator.

The mobility trace we use in the simulations is a realistic micro-mobility trace of the city of Cologne generated by Uppoor and Fiore [21]. The trace covers a $400\,\text{km}^2$ area and a period of 24 h and contains about 700000 vehicle travels. In this paper we use the part of the trace available online which covers the 6am-8am period. In Fig. 1, we plot the instantaneous positions of the nodes at 8am. Each point corresponds to a vehicle position in the coordinate system provided by the trace (the coordinates are expressed in meters). The trace from 6am to 8am contains more than 300000 different vehicles, it is thus not possible to simulate the whole scenario in *ns2*. As we are interested in local communications (one-hop broadcasts), we decide to divide the network and restrict the simulations to

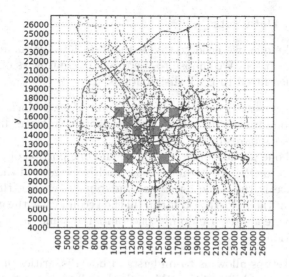

Fig. 1. Snapshot of the Cologne data set

1000×1000 m squares. The considered squares are highlighted in Fig. 1. They contain different types of road traffic (fluid traffic, traffic jams, etc.) and different vehicle densities. Even with this space division of the trace, the number of cars in one square can reach several hundreds and, in some cases, the simulations are either very long (dozens of hours) or not possible (the memory of the machine used for the simulation is not sufficient). We thus divide again the trace, but this time into time intervals. Instead of having 2 h of simulated time we produce subset traces of 200 s. Each of the time blocks also contains different traffic conditions, since the traffic changes over time in the trace.

In each simulation, we monitor the inter-beacon delay as well as the emitter-receiver distance. A simulation typically provides around 2.5 millions measurements of inter-beacon delays (it varies depending on the number of cars present during the simulation).

Table 1. Simulation parameters.

Parameter	Value
Bitrate	6 Mbps
Transmission power	10 dBm
Simulation area	1000×1000 m
Beacon size	400 bytes
MAC and Phy	802.11p
Propagation model	Nakagami m = 1
Beacon frequency	10 Hz

4.2 Obtained Data Sets

For the BM technique, we need to divide the data into blocks and retrieve the maximum of each block. We choose to use three different ways of producing the data which is then fitted to EVT distributions:

1. The first one consists in taking the maximum of each block of 200 s (corresponding to one run of the simulator) for the whole trace duration and all the highlighted squares in Fig. 1.
2. In the second one we take only one square and one portion of time and re-run the simulation for that particular square and block several times.
3. The third case is the same as the second but we change the order of the beacon start of the nodes at each run (in the second case, the order of beacon start dates is generated randomly once and the same order is repeated in every simulation).

These different setups allow us to understand how disparities in the simulation data affects the applicability of the BM method. Indeed, in the first case, the data collected comes from various situations in terms of car traffic amount, network density, etc. Whereas in the two last cases the blocks are more similar to one another (in the second case there are more correlations between the runs because the beacon start dates are the same).

For setup 1, we run simulations for 13 successive blocks of 200 s for each of the 12 considered squares. We take the maximum inter-beacon delay for each run and thus obtain 156 values. For setup 2 and 3 we run 300 simulation of the considered block, so we obtain 300 maximum values. In the results presented in the following sections, we first consider the receivers in a 500 m range from the sender, and we then observe the impact of the range on the maximum delay distribution.

5 Results and Discussion

In this section we present the results of the application of the EVT method presented in Sect. 3 to the data sets presented in Sect. 4.2.

Figure 2a is a histogram representation of the maximum delays for the first data set. First, we have to note that the measured maximum inter-beacon delays are very large compared to the inter-beacon emission period (0.1 s). This is due to two main reasons: the mobility and the broadcast scheme used. When a node broadcasts its beacon using 802.11p, it cannot detect collisions. Indeed, collisions happen at the level of the receivers and the sender does not know which of its neighbors will actually receive the packet and does not wait for acknowledgments in the case of broadcasting. Moreover, broadcast messages are subject to the hidden terminal problem since the RTS/CTS messages are not used for broadcast. These problems have been highlighted several times in the literature [9,16,19]. In order to better comprehend the observed long delays, let's consider the following scenario: first, a node A periodically receives a beacon from

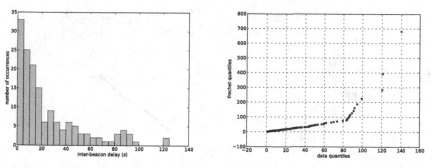

(a) Histogram of the maximum inter-beacon delays

(b) QQ-plot: data against Fréchet distribution

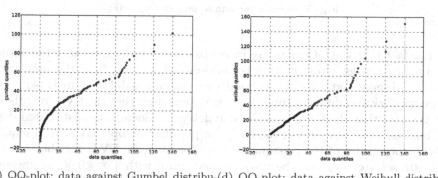

(c) QQ-plot: data against Gumbel distribution

(d) QQ-plot: data against Weibull distribution

Fig. 2. Results for all the squares and blocks and 500 m range

one of its neighbors B, then another node C moves in the neighborhood of A and the beacons from C collide with those of B (in the case of the hidden terminal problem the beacons may constantly collide), then C moves out of range of A again. In this scenario, when C at last get out of the range of A, it puts an end to the collisions with beacons from B. When A receives the first beacon from B after C went out, the inter-beacon delay is approximately equal to the duration of the presence of C in the range of A. In the cologne trace case, this delay can be of the order of tens of seconds as we observe in the simulations results.

Let's now focus on the fitting of the data of the first data set to the different extreme value distribution families. Figures 2b, c and d respectively represent the QQ-plots of the fitted Fréchet, Gumbel and Weibull distributions (fitting has been realized using the MLE method described in Sect. 3). First we observe that none of the distribution fits well to the data. Indeed, none of the graphs show a $x = y$ curve. Nevertheless, we remark that in the Fréchet case the curve is piecewise linear which seems to indicate that there are linear relations between the data quantiles and the fitted Fréchet distribution quantiles. In the case of the fitted Weibull distribution, we observe that for x and y lesser than 60, the points are approximately on the $x = y$ curve.

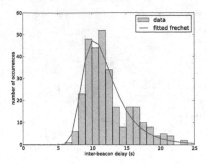 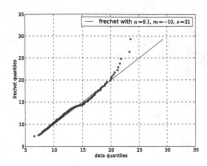

(a) Histogram of the maximum inter-beacon delays

(b) QQ-plot: data against Fréchet distribution

Fig. 3. Results for one square during 200 s

The fact that the QQ-plots are piecewise, seems to indicate that the actual maximum inter-delay distribution is multi-modal. In fact, from the simulation data, we observe that the different modes correspond to different areas (squares) of the network and different time periods. We conclude that the EVT hypothesis which states that the set of inter-beacon delays are identically distributed does not hold for this data set and thus it is not possible to apply EVT. As a matter of fact, the results of the chi-squared tests for all three EVT distributions for this data set are negative (the tests are performed with a $p = 0.05$ significance level).

Figures 3a and b respectively depict the frequency plot of the inter-beacon delays with the expected frequencies from the fitted Fréchet distribution and the QQ-plot of the data against the fitted distribution for the second data set. As stated in Sect. 4.2, this set consists of one square and one portion of 200 s run 300 times. The chosen square is defined as $x \in [11000, 12000]$ and $y \in [11000, 12000]$ (cf. Figure 1) and the considered time block is from 1200 to 1400 s of the original trace. In this case, only the fitted Fréchet distribution successfully passes the Pearson chi-squared test. The QQ-plot (Fig. 3b) shows that the fit is good for the lowest values and of a lower quality for the highest values. We can also still discern at least two modes in the frequency distribution in Fig. 3a (also visible in Fig. 3b, because the points lie over the $x = y$ curve for $x < 15$ and then under until approximately $x = 18$).

Figures 4a and b present the results for the third simulation setup (the third data set): it is the same as the previous (one square for one 200 s interval run multiple times) but the nodes are starting their beacon emission at different dates in each simulation. The starting dates are in fact uniformly distributed in the first second of the simulation. For this data set, all three fitted EVT distributions pass the chi-squared test. Figure 4a depicts the frequency plot of the inter-beacon delays with the expected frequencies from the three EVT distribution. In Fig. 4b, we present only the QQ-plot for the fitted Fréchet distribution because, even if all the distributions pass the chi-squared test, the Fréchet distribution is the best

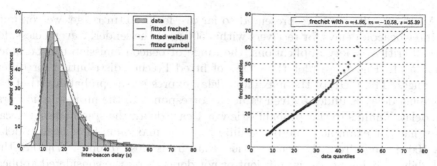

(a) Histogram of the maximum inter-beacon(b) QQ-plot: data against Fréchet distribu-
delays   tion

Fig. 4. Results for one square during 200 s with random beacon start at each run

fit (the one with the lowest test statistic). In this QQ-plot, we observe, as in the
last, that the fit is good for the lower values and less good for higher values.
Nevertheless, for this data set, we do not observe multimodal tendency which
seems to indicate that the *identically distributed* hypothesis holds. We can add
that for larger blocks (more than 200 s) the fit is even better. Unfortunately we
cannot present the results here due to the lack of space. The discussion on the
choice of the size of the block for the BM method will thus be presented in future
works. We can also note that the provided probabilistic bound (corresponding
to the fitted Fréchet distribution) is more pessimistic in third data set case than
in the second as can be seen by comparing the distributions of Figs. 3a and 4a.
In the last case, larger inter-beacon delays are more probable.

These results show that the Fréchet distribution is an adequate model for
the distribution of large inter-beacon delays in VANETs with realistic mobility.
Nevertheless, the distribution models accurately model the data only if the network conditions are coherent in every locations of the network where measures
are taken (as shown by the failure of EVT for data set 1). It means for example,
that for different network densities different fitted Fréchet distributions apply.

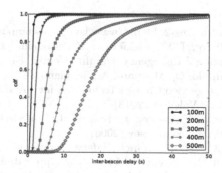

Fig. 5. CDF for different beacon ranges

All the results we have presented so far are for a 500 m range: we compute the inter-beacon delays for receivers within 500 m of the sender. Nevertheless, for most of critical safety applications, the range of beacon emission can be much lower [5]. In Fig. 5, we plot the CDFs of fitted Fréchet distribution for different ranges. The CDF of the maximum delay expresses the probability that the maximum delay is under a given value. It corresponds to the probabilistic worst case delay. Thanks to this probabilistic worst case delay, the system designer can state that, for example, "the probability that the maximum inter-beacon delay in a 100 m range from the sender is less than 5 s is close to one". Whether this probabilistic delay bound is sufficient or not depends on the considered application. In Fig. 5, we can notice that the CDF for higher ranges is lower. This means that the probability that the maximum delay is over a given value is higher for higher ranges. This can be explained by two phenomena: the higher probability of packet loss at longer distances and the hidden terminal problem as explained above in this section.

6   Conclusion and Future Works

In this paper, we study large inter-beacon delays in VANETs under realistic mobility with IEEE 802.11p. Understanding such delays is useful to assert if safety application requirements will be met. We show that large inter-beacon delays are Fréchet distributed. This result can be used in order to evaluate the performance of vehicular safety applications. We also show that the EVT method used to reach that conclusion is applicable for the study of delays in large scale wireless networks such as VANETs. Finally we confirm, in realistic mobility conditions, the results [16,19] which predict the bad performance of IEEE 802.11 broadcast. In future works we plan to: apply the second theorem of EVT to the study of delays in VANETs, evaluate the impact of the choice of the block size (in the BM method) on the quality of the obtained distribution (the goodness of fit), and compare the distribution obtained from various data sets (mobility traces), to evaluate how general the EVT results are.

References

1. The Network Simulator - ns-2. http://www.isi.edu/nsnam/ns/
2. Alvarado, E., Sandberg, D.V., Pickford, S.G.: Modeling Large Forest Fires as Extreme Events. National Emergency Training Center, Emmitsburg (1998)
3. Autolitano, A., Campolo, C., Molinaro, A., Scopigno, R.M., Vesco, A.: An Insight into Decentralized Congestion Control Techniques for VANETs from ETSI TS 102 687 V1. 1.1, IFIP WD, Valencia (2013)
4. Beirlant, J., Goegebeur, Y., Segers, J., Teugels, J.: Statistics of Extremes: Theory And Applications. Wiley, NewJersey (2006)
5. Consortium, C.V.S.C., et al.: Vehicle Safety Communications Project: Task 3 Final Report: Identify Intelligent Vehicle Safety Applications Enabled by DSRC. National Highway Traffic Safety Administration, US Department of Transportation, Washington, DC (2005)

6. Dahab, A.Y., Said, A.M., Hasbullah, H.: Application of extreme value theory to bursts prediction. SPIJ **3**(4), 55 (2009)
7. Edgar, S., Burns, A.: Statistical Analysis of WCET for Scheduling. IEEE RTSS, London (2001)
8. Embrechts, P., Resnick, S.I., Samorodnitsky, G.: Extreme value theory as a risk management tool. North Am. Actuarial J. **3**(2), 30–41 (1999)
9. Hartenstein, H., Laberteaux, K.: VANET: vehicular applications and inter-networking technologies. Wiley Online Library, Chichester (2010)
10. Jain, R.: The Art of Computer Systems Performance Analysis. Wiley, New York (2008)
11. Kloiber, B., Garcia, C., Härri, J., Strang, T.: Update delay: a new information-centric metric for a combined communication and application level reliability evaluation of cam based safety applications. In: ITS World Congress (2012)
12. Koçak, K.: Practical ways of evaluating wind speed persistence. Energy **33**(1), 65–70 (2008)
13. Liu, C., Shu, Y., Liu, J., Yang, O.: Application of Extreme Value Theory to the Analysis of Wireless Network Traffic. IEEE ICC, Glasgow (2007)
14. Lu, Y., Nolte, T., Bate, I., Cucu-Grosjean, L.: A Statistical Response-Time Analysis of Real-Time Embedded Systems, pp. 351–362. IEEE RTSS, Puerto Rico (2012)
15. Nelder, J.A., Mead, R.: A simplex method for function minimization. Comput. J. **7**(4), 308–313 (1965)
16. Nguyen, T.V., Baccelli, F., Zhu, K., Subramanian, S., Wu, X.: A Performance Analysis of CSMA Based Broadcast Protocol in VANETs. IEEE INFOCOM, Turin (2013)
17. Noori, H., Olyaei, B.: A Novel Study on Beaconing for VANET-Based Vehicle to Vehicle Communication: Probability of Beacon Delivery in Realistic Large-scale Urban Area using 802.11p. IEEE SaCoNeT, Paris (2013)
18. Shukla, R.K., Trivedi, M., Kumar, M.: On the proficient use of gev distribution: a case study of subtropical monsoon region in India. Annals. Computer Science Series 8 (2010)
19. Stanica, R., Chaput, E., Beylot, A.L.: Broadcast Communication in Vehicular Ad-hoc Network Safety Applications. IEEE CCNC, Lax Vegas (2011)
20. Uchida, M.: Traffic Data Analysis Based on Extreme Value Theory and its Applications. IEEE GLOBECOM, Dallas (2004)
21. Uppoor, S., Fiore, M.: Large-Scale Urban Vehicular Mobility for Networking Research. IEEE VNC, Amsterdam (2011)

xRadio: An Novel Software Defined Radio (SDR) Platform and Its Exemplar Application to Vehicle-to-Vehicle Communications

Weidong Xiang[1], Fotios Sotiropoulos[2]([⊠]), and Sheng Liu[2]

[1] University of Michigan, Dearborn, Dearborn, MI 48128, USA
xwd@umich.edu
[2] Odyssia Global Communications, New York, NY, USA
arxaios@odyssia.com

Abstract. In this presentation, we introduce a novel software defined radio (SDR) universal wireless platform, xRadio, for fast prototyping of various emerging wireless systems featuring with attracting cost performance ratio when compared to current solutions. xRadio realizes its advancement and integrity based on a compact and right-on-target design strategy, through adopting a cost efficient raspberry PI minicomputer and a field programmable gate array (FPGA) chip from Altera as its core processors. Function modules can be easily realized through C/C++ and/or python program in a Linux environment or programmable logical elements (LEs) of FPGA achieving powerful computation. To evaluate its performance, an onboard unit (OBU) for vehicle-to-vehicle communications based on both long term evolution (LTE) and dedicate short range communications (DSRC) systems is being built up. Corresponding systems design and key performances are tested and validated including bit error rate (BER) to signal-to-noise ratio (SNR) and processing latency, which validates the usability of xRadio. The unprecedented price-to-performance ratio of xRadio has potential to be applied in a broad range of applications ranging from engineering, production and educations.

Keyword: SDR

1 Introduction

Wireless technology continues to be among the cutting-edge technologies of our ages and increasingly intensive research and development (R&D) activities will be evoked worldwide. However, at present most of activities are connected to theoretical analysis and simulation, only a small part of them are able to move forward onto the prototyping stage even through it is highly encouraged by the Government research agent, such as national science foundation (NSF), and heavily expected by industry and companies in the recent years. The main obstruct that prevents this lies on the expensive development kits and environments as well as the high requirements on the backgrounds and skills of developers.

S. Papavassiliou and S. Ruehrup (Eds.): ADHOC-NOW 2015, LNCS 9143, pp. 404–415, 2015.
DOI:10.1007/978-3-319-19662-6_28

In general, the high computation load of baseband processing, such as fast Fourier transformation (FFT), Viterbi decoding or turbo decoding, timing synchronization and channel estimation, make wireless communications systems suitable to be produced through using applications specific integrate circuit (ASIC) once market becomes mature. However, the large amount of invest on engineering and the long period in the development of ASIC becomes the main barriers that can not be passed over for many Universities, Institutes and companies. Only a few large companies and startup companies funded by venture capital are able to play with it. Although ASIC is suitable for massive production of mature wireless communications systems, at their early development and rapid prototyping, a proof-of-conception real-time or quasi real-time platform is seriously in need.

For comparison, here we detail on the status of wireless platform and development kits through two examples: Wireless Open-Access Research Platform (WARP) from Rice University and Universal Software Radio peripheral (USRP) from the Ettus company. WARP is a real-time high-end wireless platform based on field programmable gate array (FPGA) chip from the Xilinx company. More than 30 Universities and companies worldwide have adopted WARP in the past few years. WARP offers open sources for some basic functions of wireless communications systems, such orthogonal frequency multiplexing (OFDM) modulation and demodulation. The main obstacle that prevents WARP to be widely adopted lies on its soaring high price, close to $10 K, complicated hardware modules, some of them unnecessary for most applications, cross platform development environments and years long learning curve for beginners, not quite suitable for University research groups whit frequently updated members and most companies to develop product who expect a short time-to-market. On the other hand, GNU radio is a software defined radio (SDR) development environment featuring in its free open sources of many signal processing modules upon minimal or even zero hardware platform [1]. It is mainly used for research, education and concept-of-proof, instead of real product prototyping. GNU radio applications are written by using the Python programming language in general while the high-speed signal processing by the C++ codes in the formation of floating-point over the general central processing unit (CPU). The computation capacity it offers is not sufficient to accommodate with that required by modern wireless communications systems ranging from dedicate short range communications (DSRC) to long term evolution (LTE). USRP is such a hardware platform to support GNU radio development.

There is a gap between the current high-end real-time wireless platforms and GNU platforms leaving a room for an affordable rapid real-time powerful wireless platform to be created and grows. Motivated by this need, we create a xRadio platform based on recent FPGA chips and amazing Raspberry PI minicomputers resulting in unprecedented price-to-performance ratio. xRadio have two versions with/without integrating FPGA offering options for diverse purposes. The rest of the draft are arranged as below. In Sect. 2, we detail the specifications of xRadio and some test results. Section 3 reports the design and experiment of an onboard unit implementing long term evolution (LTE) and dedicate short range communications (DSRC) based on xRadio. Conclusions are withdrawn finally.

2 Paper Preparation

xRadio targets on the rapid real-time/quasi real-time wireless platform for both prototyping and production featuring with attracting cost-performance ratio. Table 1 compares the main specifications among the WARP [2], USRP [3], and xRadio. Seeing from Table 1, we conclude that xRadio outperforms WARP in having 14-bit ADC/DAC at comparable sample rates as well as up to 30dBm radiation power at an affordable price of WARP. Although the number of LEs of the FPGA chip integrated is about half of those used in WARP, but sufficient to implement modern wireless communications systems. In case more LEs are needed, two or more xRadio can be interconnected for high-end users. Figures 1 and 2 show the diagrams of the xRadio I and II where xRadio I serves as high-end real-time applications and xRadio II for low cost quasi real-time solutions. Power amplifiers can be added on as options, manually, when large radiation power is expected. Figures 3 and 4 shows the picture of xRadio I and xRadio II, respectively.

Table 1. The specifications of xRadio and its comparison with other two SDR platforms

Platforms	Core Processors	ADC/DAC	RF Front Ends	Environments	Applications	Cost($)
WARP	Virtex 6 LX240T (Xilinx 240K LEs)	1) Dual ADC 12bit 100MS/s 2) Dual DAC 12bit 170MS/s	1) Dual bands: 2/5GHz (Max2829) 2) RF power 20dBm	FPGA	Prototyping	10,000
USRP	1) Spartan 3A-DSP 3400 (Xilinx, 47K LEs) 2) OMAP 3 (TI, ARM Cortex-A8) 3) C64 DSP	1) Dual ADC 12bit 64MS/s 2) Dual DAC 14bit 128MS/s	1) Dual bands: 2/5GHz (Max2829) 2) RF power 20dBm	FPGA, ARM and DSP	Prototyping	2,000
xRadio	Cyclone EP4CE115 (Altera 115K LEs)	1) Dual ADC 14bit 65MS/s 2) Dual DAC 14bit 125MS/s	1) Dual bands: 2/5GHz (Max2829) 2) RF power 30dBm	FPGA, Linux	Prototyping and Product	150-500

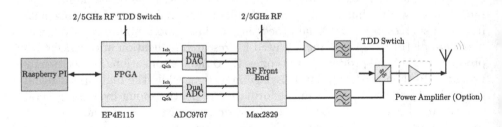

Fig. 1. The system diagram of xRadio I

2.1 SPI Interface

Serial peripheral interface (SPI) is a full duplex synchronous serial data transmission interface developed by Motorola. A SPI interface is equipped by Rasp berry PI minicomputers as its high speed data input and output operating at up to 50 MHz, which is faster than those of general purpose input output interfaces (GPIO) integrated. Devices with SPI interfaces communicate in a master/slave mode where master device initiates transmission and multiple slave devices can be selected through corresponding chip selection (CS) line.

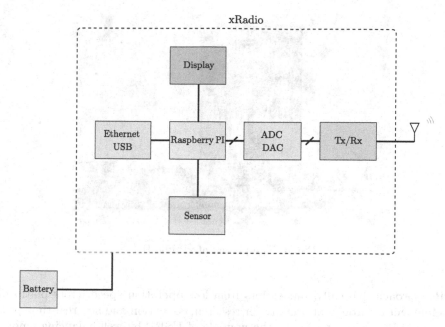

Fig. 2. The system diagram of xRadio II

Fig. 3. The picture of xRadio I

3 Parametrized SDR Implementation Approach

SDR is a design strategy by implementing functions that are modifiable, either adaptively or manually, upon a shared platform. In general, SDR can be implemented by FPGA, digital signal processing (DSP), general purpose processors (GPP), and other programmable system on chip (SoC) and processors. The implementation of functions on DSPs and GPPs are inherently SDR approach but functions operate in a sequence resulting in low speeds based on the current and even near term DSPs and GPPs. The GNU radio based upon USRP is an

Fig. 4. The picture of xRadio II

SDR approach, basically, but suffers from low operation speed, cross platform development environment, mainly for research, education and non real-time prototyping. xRadio is regarded as the upgrade of USRP by well balancing among operation speed, easy to use, cost and flexibility. FPGA can be used for SDR prototyping where real-time and high speed are expected. However, the current methodology for FPGA based SDR is based on either cold codes switch (CCS) or hot codes switch (HCS), which are illustrated in Fig. 5.

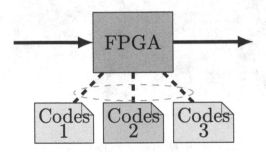

Fig. 5. The diagram of conventional CCS/HCS SDR approach

In CCS mode, functions are defined/redefined by loading corresponding code sets over the same FPGA chip. In such a way, the codes are optimized and function switch is completed online, which is not suitable for the situations requiring fast and frequent function change. In addition, CCS mode requires extra firmware to save the codes and toolkits used to load/unload code sets onto the FPGA.

On the other hand, in HCS mode all of code sets are loaded onto the FPGA and function switch is realized online. Obviously, HCS is an non-optimized design and consumes a great amount of LEs, the decisive factor for the cost of FPGA.

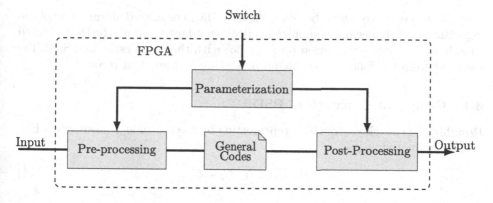

Fig. 6. The diagram of proposed parametrized SDR approach

This is the main reason why there are few SDR products in the market even the conception has been intensively researched for more than decades. In this effort, we present a parametrized SDR (PSDR) approach, which is illustrated in Fig. 6. The basic idea is to merge function codes into a general code set and convert the function switch to processing of parametrization and necessary.

To describe PSDR, we use the following two examples. The first one is the fast Fourier transformation (FFT) processing, which is a basic function for OFDM adopted by both LTE and DSRC and smart utility networks (SUN) based on the IEEE 802.15.4g (draft). However, the block size of the FFT for LTE is 1024; 64 for DSRC and 16 or 8 for SUN. In PSDR, the general code is the 1024 FFT processor where the preprocessing is zero padding forming uniform data and the post-processing the decimation based on the actual FFT size. Another example is quadrature amplitude modulation (QAM) where the modulation level, M, varying from 2, 4, 16 to 64 for DSRC, LTE and SUN. The general code set is to realize 64 QAM mapping where the preprocessing is zero padding to form 6-bit groups and post-processing a simple scaling according to M. Even simple, the two examples clearly illustrate the principle of PSDR.

Furthermore, we consider a complex function of time synchronization for WiFi, DSRC and LTE, which is the first baseband processing used to locate the beginning of the complex samples. Figure 6 shows the diagram of time synchronization based on the PSDR approach. The preprocessing is to upsample the input DSRC complex samples to 20 MSPS through interleaving with a zero sequence in order to allow one general code set to apply to both DSRC and WiFi systems, while bypass WiFi and LTE signals. The function begins with the coarse synchronization which is preformed by detecting the energy jump of receiving signals for both DSRC and WiFi since they are transmitted in a burst mode. However, for LTE the processing is to constantly monitor the energy and compare with a threshold during a predefined period since LTE signal is transmitted continuously. The defined synchronization is thereafter executed to accurately locate the start of an OFDM frame through cross correlation processing. The templates of the short preambles of WiFi and DSRC as well as the cyclic prefix (CP) for LTE are all stored locally and loaded selectively. The synchronization alarm will be released once certain criteria were met. To adjust between the false alarm or missed alarm (none of the algorithms can reduce both, simultaneously), an adaptive filter is further adopted to enhance the synchronization performance with the aid of estimated SNR, the feedback from the following baseband processing and previous record.

3.1   Complexity Analysis of PSDR

Roughly, the complexity of HCS, representing by the size of logic elements (LEs), can be estimated as,

$$C_{HCS} = \sum_{i=1}^{N} C_i + C_s \tag{1}$$

where C_{HCS}, C_i and C_s present the complexity of HCS, the i-th function and switch processing. N is the total number of function modules. We can estimate the complexity for PSDR as below,

$$C_{PSDR} = C_m(1 + \eta) + C_s \tag{2}$$

$$C_m = max\{C_i\} \tag{3}$$

where C_{PDSR}, $C + m$ and C_p express the complexity of PSDR, the most complex function, and the parametrization and preprocessing and post-processing, respectively. η is a percentage factor reflecting the increase in the complexity of the general code set when compared to Cm. The value of η is highly related to the sharable modules ratio (SMR), ρ, which is defined as,

$$\rho = \frac{C_{i,SM}}{C_i} \tag{4}$$

where ρ and $C_{i;SM}$, are the SMR and the sharable complexity of the ith function module. The relation between η and ρ are listed as following,

$$\eta = \frac{\sum_{i=1, i \neq m}^{N} C_i(1 - \eta_i)}{C_m} \tag{5}$$

One can verify that,

$$C_{PSDR} =\rightarrow C_{HSC}, \; if \; \eta_i = 0, \; i = 1, \ldots, N \tag{6}$$

$$C_{PSDR} \approx C_m, \; if \; \eta_i = 1, \; i = 1, \ldots, N \tag{7}$$

where the switch and addition processing are not considered for simplicity.

4   Implementation and Experiments of xRadio

4.1   LTE-R

Long term evolution is one of the two candidates for the fourth generation cellular systems (4G) developed by 3GPP, is probably the most complex commercial radio that has been proposed and being developed in history of information industry featuring in Gigabit data rate, low latency, support for high-mobility and world-wide roaming, high spectrum efficiency, better cover- age at cell edge and many others. The main technical merits of LTE is to adopt two dimensional, in both the frequency and the time domains, parallel transmission technologies, known as (OFDM) and multiple input multiple output (MIMO), respectively. LTE for rail- ways (LTE-R) distinguishes with LTE, obviously, in the applied scenario with relaxed conditions of power consumption, size and cost, but, implying more challenge of high-speed transmission under diverse mobility environments, and low latency requirement for messages associated with train operation in a real-time mode. In follows, we will highlight the perspectives of LTE-R.

1. Mobility: LTE achieves optimum performance when the speed of user utility (UE) is under 15 km/h, high performances within 0–120 km/h and function support within 120–350 km/h. Higher speeds within 350–500 km/h have not be defined yet.
2. Latency: In LTE, the latency of use plane is less than 10 ms.
3. Coverage: LTE offers full function within 5 km, most function within 5–30 km and basic functions within 30 km–100 km.

4.2 DSRC

Vehicular communications and networks are the foundation of nationwide road infrastructure building up vehicle-to-vehicle (V2V) and vehicle-to-infrastructure (V2I) wireless links, robust and timely enough for critical safety message and fast sufficiently for ITS applications. V2V is realized through the connections among the onboard units (OBUs) integrated in vehicles while V2I through OBUs and roadside units (RSUs), serving as access points on the roads. In general, V2V mainly focuses on the safety enhancement while V2I offers drivers and passengers with high-speed wireless access to the Internet, and then the rest of world, for both either work and entertainment, which coins the term of infotainment. DSRC builds upon the IEEE 802.11p protocol, which was formally ratified in 2011. By adopting OFDM modulation scheme, DSRC achieves a data rate of 6–27 Mbit/s and a RSU covers a range of 1000 ft. The Federal Communication Commission (FCC) has assigned the 5.850–5.925 GHz band for the operation of DSRC systems in the United States.

4.3 Doppler Shift

The main challenge is how to deal with Doppler shift when LTE and DSRC is applied to high-speed train applications. The typical railway wireless channel is considered to be consisted of line of sight (LOS) and several non line of sight (NLOS) contributed by the reflectors in the proximity of the base station and the high speed train, which is shown in Fig. 7.

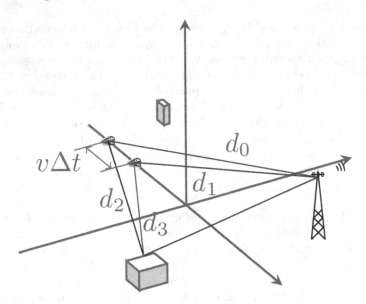

Fig. 7. A typical high speed wireless channel

As shown in the above figure, the locations of the train and base station are at $(a, 0, 0)$ and $(0, b, h)$, respectively. The ith reflector is randomly located at (x_i, y_i, z_i). The Doppler shift can be calculated through,

$$f_{d,LOS} = f_c \frac{v\Delta t + a}{c\sqrt{(v\Delta t + a)^2 + b^2 + h^2}} \tag{8}$$

$$f_{d,NLOS} = -f_c \frac{(x_i - a - v\Delta t)v}{c\sqrt{(x_i - a - v\Delta)^2 + y_i^2 + z_i^2}} \tag{9}$$

where the train moves at the speed of v and c is the light speed.

4.4   Experiments and Results

In this effort, we developed OBU prototypes based on xRadio that is able to operation in both LTE and DSRC using the PSDR approach. The function modules that were developed include main physical layer functions defined by the standard. The MAC functions are implemented by C programs over Raspberry PI. The functions at the transmitter and receiver are illustrated in Figs. 8 and 9. The complexities for the main modules of transmitter and receiver are listed in Tables 2 and 3. A snapshot of one OFDM frame is shown in Fig. 10. The complete physical processing latency including transmission and receiving was measured in Fig. 11, showing as 0.125 ms. To measure the BER, a white Gaussian noises (AWGNs) generator was developed and generated required noise to the baseband OFDM signals within FPGA. The bit errors were calculated corresponding results are shown in Fig. 12 where 4 QAM was adopted and the transmitter and receiver was connected through a cable with sufficient attenuator.

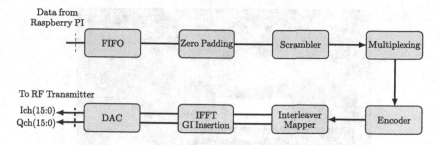

Fig. 8. The diagram of transmitter

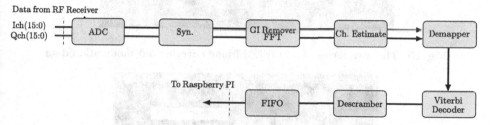

Fig. 9. The diagram of receiver

Table 2. The complexity of the modules of transmitter

Module	Option	Complexity
Ethernet	1 Gbits/s	10429
Scrambler	$x^7 + x^4 + x^1$	9
Encoder		15
Modulation	4/16/46 QAM	24
Interleaver		150
IFFT	8–1024	4182
Guard insertion		66

Table 3. The complexity of the modules of receiver

Module	Option	Complexity
Synchronization		27103
GI remover		57
FFT	8–1024	4092
Channel estimtor		3523
Deinterleaver		320
Demapper	4/16/46 QAM	398
Decoder	Viteber	3082
Descramber	$x^7 + x^4 + 1$	9
Ethernet	1 Gbits/s	10429

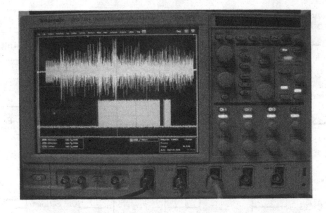

Fig. 10. The waveforms of an OFDM frame carrying out modulation data

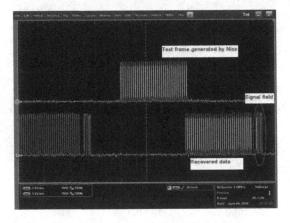

Fig. 11. The modulated and demodulated bit stream at transmitter and receiver

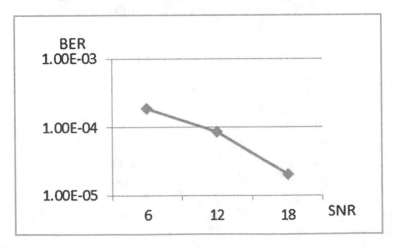

Fig. 12. The measured BER-SNR curve of SDR based on xRadio

5   Conclusion

At an affordable price, xRadio emerges as a rapid easy to use, real-time wireless prototyping approach by integrating recent FPGA, monolithic microwave integrate circuit (MMIC) and minicomputers, which can be applied to a wide range wireless communications systems ranging from LTE, DSRC and SUN with up to Watt radiation powers, achieving attracting performance-cost ratio. Moreover, xRadio has a small factor, making itself suitable for the portable devices for a plenty of indoor and outdoor applications. Meanwhile, PSDR is an advanced SDR approach balancing among operation speed and flexibility. PSDR applies well to homogeneous systems, such as WiFi, DSRC and SUN, since a lot of function blocks are highly correlated. However, how to adopt PSDR to heterogeneous systems, such as DSRC and LTE, which still needs further effort.

Finally, DSRC prototype based on xRadio have been selected to be employed in the Mobility Transformation Center (MTC) Pillar Project, University of Michigan at the implementation scale of up to thousands onboard unit (OBUs) in the city of Ann Arbor, Michigan, USA. More indoor and outdoor experiments will be extensively conducted during the project years from 2014 to 2016.

Acknowledgment. Thanks for the sponsorship of several relevant projects from National Science Foundation (NSF), the CISCO University Grant and the University of Michigan.

References

1. GUN Radio, http://gnuradio.org/redmine/projects/gnuradio/wiki
2. WARP: Wireless Open-Access Research Platform. http://warp.rice.edu/
3. Ettus Research. http://www.ettus.com/

Author Index

Printed in the United States
By Bookmasters